Comprehensive Project Management:

Integrating Optimization Models,

Management Principles, and Computers

Comprehensive Project Management:
Integrating Optimization Models, Management Principles, and Computers

Adedeji B. Badiru
and
P. Simin Pulat

PRENTICE HALL PTR, Englewood Cliffs, New Jersey 07632

Library of Congress Cataloging-in-Publication Data

Badiru, Adedeji Bodunde, 1952–
    Comprehensive project management : integrating optimization
models, management practices, and computers / by Adadeji B. Badiru
and P. Simin Pulat.
      p. cm.
    Includes bibliographical references and index.
    ISBN 0-13-030925-7
    1. Industrial project management. 2. Mathematical optimization.
3. Industrial management—Data processing. I. Pulat, P. Simin.
II. Title.
HD69.P75B327   1994
658.4′04—dc20   94-8035
                                                    CIP

Editorial/production supervision: *Camille Trentacoste*
Cover design: *Design Source*
Manufacturing manager: *Alexis Heydt*
Acquisitions editor: *Mike Hays*
Editorial Assistant: *Kim Intindola*

© 1995 by Prentice Hall PTR
Prentice-Hall, Inc.
A Simon & Schuster Company
Englewood Cliffs, NJ 07632

The publisher offers discounts on this book when ordered in bulk quantities.
For more information, contact:
    Corporate Sales Department
    Prentice Hall PTR
    113 Sylvan Avenue
    Englewood Cliffs, NJ 07632

    Phone: 800-382-3419, Fax: 201-592-2249
    E-mail: dan_rush@prenhall.com

Printed in the United States of America

10 9 8 7 6 5 4 3 2 1

ISBN 0-13-030925-7

PRENTICE-HALL INTERNATIONAL (UK) LIMITED, *London*
PRENTICE-HALL OF AUSTRALIA PTY. LIMITED, *Sydney*
PRENTICE-HALL CANADA INC., *Toronto*
PRENTICE-HALL HISPANOAMERICANA, S.A., *Mexico*
PRENTICE-HALL OF INDIA PRIVATE LIMITED, *New Delhi*
PRENTICE-HALL OF JAPAN, INC., *Tokyo*
SIMON & SCHUSTER ASIA PTE. LTD., *Singapore*
EDITORA PRENTICE-HALL DO BRASIL, LTDA., *Rio de Janeiro*

Contents

SECTION TWO: OPTIMIZATION MODELS

CHAPTER SEVEN: PROJECT MODELING AND OPTIMIZATION   276

CHAPTER EIGHT: FINANCIAL AND ECONOMIC ANALYSES   337

Preface

Project management is the process of managing, allocating, and timing resources to achieve a given goal in an efficient and expedient manner. The objectives that constitute the specified goal may be in terms of time, costs, or technical results. A project can be quite simple or very complex. Project management has now emerged as a separate discipline that is embraced by various other disciplines ranging from business to engineering. Project management techniques are widely used in many areas including construction management, banking, manufacturing, engineering management, marketing, health care delivery systems, transportation, R&D, and public services. Project management represents an excellent basis for integrating various management techniques such as operations research, total quality management, simulation, and so on.

The purpose of this book is to present an integrated approach to project management. The integrated approach covers optimization models, practical management practices, and computer applications. The premise of the book is that both simple and complex projects can be managed better if these three tools are integrated and applied effectively. This book describes how the three categories of tools complement one another to enhance overall project management.

This book is intended to serve as a textbook in colleges and universities for project management and related courses at the senior undergraduate and first-year graduate levels. Specific programs that should be interested in the book include Industrial Engineering, Systems Engineering, Construction Engineering, Operations Research, Operations Management, Production Management, Business Management, and Engineering Management. The book should also appeal to practitioners and consultants because of its practical orientation.

The book is organized into three sections. Section 1 addresses management practices. Section 2 covers optimization models. Section 3 covers computer applications. The chapters are organized hierarchically to illustrate the incremental steps in a project management process. The hierarchy goes from setting project goals to terminating the project.

Chapter 1 presents principles of project management. It covers the basic definition of project management, components of a project, the project management process, selection of a project manager, and an outline of a model for project management functions. Chapter 2 covers project planning in detail. It presents the components of a plan, work breakdown structure, a project feasibility study, communication and coordination strategies, and personnel motivation. Chapter 3 presents project organizational structures including matrix organization, product organization, and functional organization.

Chapter 4 presents project network analysis and basic scheduling techniques including CPM, PERT, and PDM. Basic Gantt charts and their variations are also presented. Chapter 5 discusses resource allocation strategies within the plan-schedule-control cycle. Critical resource diagramming is discussed as well as resource loading and leveling. The use of learning curves in project management is also presented.

Chapter 6 addresses project control. Topics covered include project tracking, reporting, and information transfer approaches. Chapter 7 presents optimization models for project management. Topics include linear programming, network flow problems, the time-cost trade-off model, the assignment problem, the transportation problem, goal programming, and simulated annealing. Chapter 8 covers financial and economic analysis for project management. Topics include basic cash flow analysis, break-even analysis, comparison of project alternatives, cost control criteria, and contract management.

Chapter 9 presents decision analysis for project selection. Topics covered include utility models, the project value model, polar plots, and the analytic hierarchy process. Chapter 10 discusses computer applications for project management. Topics include software selection, simulation of project networks, regression metamodels, simulation approach to AHP, and expert systems. Chapter 11 presents a comprehensive project management case study. Numerous tables and figures are used throughout the book to enhance the effectiveness of the discussions. An extensive bibliography is presented at the end of the book for the reader's further reference. Useful appendixes are also provided.

Adedeji B. Badiru
P. Simin Pulat
Norman, Oklahoma

Acknowledgments

Our sincere appreciation goes to the following individuals for their contributions to our knowledge and interest in project networks: Dr. Salah Elmaghraby and Dr. Gary Whitehouse. The inspiration for the project was provided by our families. We are very grateful for their love, patience, understanding, and confidence in us.

We thank the staff of Prentice Hall PTR for their excellent production work on this manuscript. We thank Mike Hays for his visionary interest right from the beginning. We also thank Camille Trentacoste for her relentless pursuit of higher quality for this book.

Dedication

To our spouses and children:
Iswat, Abi, Ade, and Tunji
Mustafa, Özgur, and Önder

Section 1

Management Practices

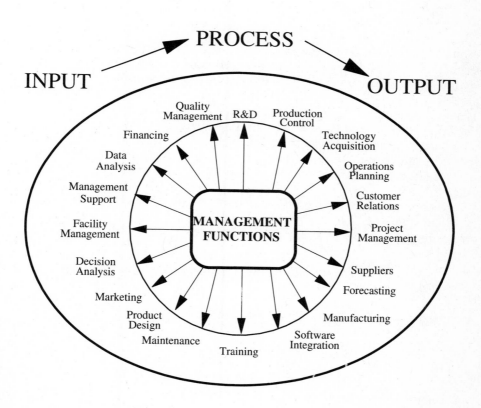

INPUT → PROCESS → OUTPUT

Quality Management · R&D · Production Control · Technology Acquisition · Operations Planning · Customer Relations · Project Management · Suppliers · Forecasting · Manufacturing · Software Integration · Training · Maintenance · Product Design · Marketing · Decision Analysis · Facility Management · Management Support · Data Analysis · Financing

MANAGEMENT FUNCTIONS

1

Principles of Project Management

Project management represents an excellent basis for integrating various management techniques such as operations research, operations management, forecasting, quality control, and simulation. Traditional approaches to project management use these techniques in a disjointed fashion, thus ignoring the potential interplay among the techniques. The need for integrated project management worldwide is evidenced by a 1993 report by the World Bank (World Bank Report 1993). In the report, the bank, which has loaned more than $300 billion to developing countries over the last half century, acknowledges that there has been a dramatic rise in the number of failed projects around the world. Lack of an integrated approach to managing the projects was cited as one of the major causes of failure.

In modern project management, it is essential that related techniques be employed in an integrated fashion so as to maximize the total project output. Badiru (1991a) defines project management as

> The process of managing, allocating, and timing resources to achieve a specific goal in an efficient and expedient manner.

Alternately, we can define project management as

> The systematic integration of technical, human, and financial resources to achieve goals and objectives.

This comprehensive definition requires an integrated approach to project management. This book presents such an integrated approach. To accomplish the goal of project management, an integrated use of managerial, mathematical, and computer tools must be developed. The first step in the project management process is to

SET GOALS

1.1 MANAGEMENT BY PROJECT

Project management continues to grow as an effective means of managing functions in any organization. Project management should be an enterprise-wide endeavor. McFarlane (1993) defines enterprise-wide project management as the application of project management techniques and practices across the full scope of the enterprise. This concept is also referred to as management by project (MBP). Management by project is a recent concept (Sharad 1986) that employs project management techniques in various functions within an organization. MBP recommends pursuing endeavors as project-oriented activities. It is an effective way to conduct any business activity. It represents a disciplined approach that defines any work assignment as a project. Under MBP, every undertaking is viewed as a project that must be managed just like a traditional project. The characteristics required of each project so defined are

1. An identified scope and a goal
2. A desired completion time
3. Availability of resources
4. A defined performance measure
5. A measurement scale for review of work

An MBP approach to operations helps in identifying unique entities within functional requirements. This identification helps determine where functions overlap and how they are interrelated, thus paving the way for better planning, scheduling, and control. Enterprise-wide project management facilitates a unified view of organizational goals and provides a way for project teams to use information generated by other departments to carry out their functions.

The use of project management continues to grow rapidly. The need to develop effective management tools increases with the increasing complexity of new technologies and processes. The life cycle of a new product to be introduced into a competitive market is a good example of a complex process that must be managed with integrative project management approaches. The product will encounter management functions as it goes from one stage to the next. Project management will be needed throughout the design and production stages of the product. Project management will be needed in developing marketing, transportation, and delivery strategies for the product. When the product finally gets to the customer, project management will be needed to integrate its use with those of other products within the customer's organization.

The need for a project management approach is established by the fact that a project will always tend to increase in size even if its scope is narrowing. The following three literary laws are applicable to any project environment:

Parkinson's law.   Work expands to fill the available time or space.

Peter's principle.   People rise to the level of their incompetence.

Murphy's law.   Whatever can go wrong will.

An integrated project management approach can help diminish the impact of these laws through good project planning, organizing, scheduling, and control.

1.2 THE INTEGRATED APPROACH

Project management tools can be classified into three major categories:

1. Qualitative tools. These are the managerial tools that aid in the interpersonal and organizational processes required for project management.
2. Quantitative tools. These are analytical techniques that aid in the computational aspects of project management.
3. Computer tools. These are software and hardware tools that simplify the process of planning, organizing, scheduling, and controlling a project. Software tools can help in both the qualitative and quantitative analyses needed for project management.

Although individual books dealing with management principles, optimization models, and computer tools are available, there are few guidelines for the integration of the three areas for project management purposes. In this book, we integrate these three areas for a comprehensive guide to project management. The book introduces the *Triad Approach* to improve the effectiveness of project management with respect to schedule, cost, and performance constraints. Figure 1–1 illustrates this emphasis. The approach considers not only the management of the project itself but also the management of all the functions that support the project.

It is one thing to have a quantitative model, but it is a different thing to be able to apply the model to real-world problems in a practical form. The systems approach presented in this book illustrates how to make the transition from model to practice.

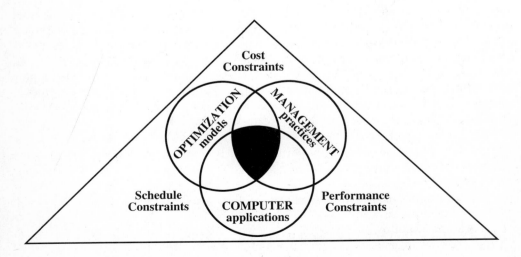

Figure 1-1.   Integration of Project Management Tools

A systems approach helps increase the intersection of the three categories of project management tools and, hence, improve overall management effectiveness. Crisis should not be the instigator for the use of project management techniques. Project management approaches should be used upfront to prevent avoidable problems rather than to fight them when they develop. What is worth doing is worth doing well, right from the beginning.

1.3 PROJECT MANAGEMENT BODY OF KNOWLEDGE

The Project Management Institute (PMI) defines the project management body of knowledge (PMBOK) as those topics, subject areas, and processes that are used in conjunction with sound project management principles to collectively execute a project (PMI 1987). Eight major functional areas are identified in the PMBOK: *scope, quality, time, cost, risk, human resources, contract/procurement,* and *communications.*

Scope management refers to the process of directing and controlling the entire scope of the project with respect to a specific goal. The establishment and clear definition of project goals and objectives form the foundation of scope management. The scope and plans form the baseline against which changes or deviations can be monitored and controlled. A project that is out of scope may be out of luck as far as satisfactory completion is concerned. Topics essential for scope management are covered in Chapters 1, 2, 3, and 9.

Quality management involves ensuring that the performance of a project conforms to specifications with respect to the requirements and expectations of the project stakeholders and participants. The objective of quality management is to minimize deviation from the actual project plans. Quality management must be performed throughout the life cycle of a project, not just by a final inspection of the product. Techniques useful for quality management are covered in Chapters 1, 5, and 6.

Time management involves the effective and efficient use of time to facilitate the execution of a project expeditiously. Time is often the most noticeable aspect of a project. Consequently, time management is of utmost importance in project management. The first step of good time management is to develop a project plan that represents the process and techniques needed to execute the project satisfactorily. The effectiveness of time management is reflected in the schedule performance. Hence, scheduling is a major focus in project management. Chapters 4 and 7 present techniques needed for effective time management.

Cost management is a primary function in project management. Cost is a vital criterion for assessing project performance. Cost management involves having an effective control over project costs through the use of reliable techniques of estimation, forecasting, budgeting, and reporting. Cost estimation requires collecting relevant data needed to estimate elemental costs during the life cycle of a project. Cost planning involves developing an adequate budget for the planned work. Cost control involves the continual process of monitoring, collecting, analyzing, and reporting cost data. Chapter 8 discusses cost management concepts and techniques.

Risk management is the process of identifying, analyzing, and recognizing the various risks and uncertainties that might affect a project. Change can be expected in any project environment. Change portends risk and uncertainty. Risk analysis outlines possible future events and their likelihood of occurrence. With the information from risk analysis, the project team can be better prepared for change with good planning and control actions. By identifying the various project alternatives and their associated risks, the project team can select the most appropriate courses of action. Techniques relevant for risk management are presented in Chapters 4, 5, 6, 8, and 9.

Human resources management recognizes the fact that people make things happen. Even in highly automated environments, human resources are still a key element in accomplishing goals and objectives. Human resources management involves the function of directing human resources throughout a project's life cycle. This requires the art and science of behavioral knowledge to achieve project objectives. Employee involvement and empowerment are crucial elements of achieving the quality objectives of a project. The project manager is the key player in human resources management. Good leadership qualities and interpersonal skills are essential for dealing with both internal and external human resources associated with a project. The legal and safety aspects of employee welfare are important factors in human resources management. Chapters 1, 2, 3, 4, 5, 6, and 10 present topics relevant to human resources management.

Contract/procurement management involves the process of acquiring the necessary equipment, tools, goods, services, and resources needed to successfully accomplish project goals. The buy, lease, or make options available to the project must be evaluated with respect to time, cost, and technical performance requirements. Contractual agreements in written or oral form constitute the legal document that defines the work obligation of each participant in a project. Procurement refers to the actual process of obtaining the needed services and resources. Concepts and techniques useful for contract/procurement management are presented in Chapters 2, 5, and 8.

Communications management refers to the functional interface among individuals and groups within the project environment. This involves proper organization, routing, and control of information needed to facilitate work. Good communication is in effect when there is a common understanding of information between the communicator and the target. Communications management facilitates unity of purpose in the project environment. The success of a project is directly related to the effectiveness of project communication. From the authors' experience, most project problems can be traced to a lack of proper communication. Guidelines for improving project communication are presented in Chapters 1, 2, 3, and 6. Chapter 11 presents a case study that illustrates how the various elements in the project management body of knowledge can be integrated.

1.4 PROJECT MANAGEMENT PROCESS

Organize, prioritize, and optimize the project. The project management process consists of several steps starting from problem definition and going through project termination. Figure 1–2 presents the major steps. A brief overview of the steps is presented in this section. Some of the major steps are discussed in detail in subsequent chapters.

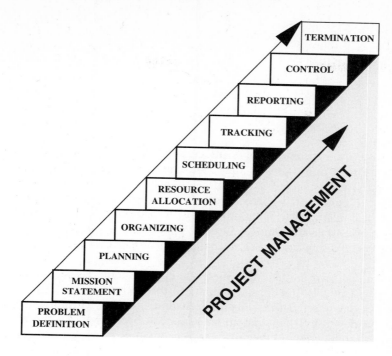

Figure 1-2. Project Management Steps

1.4.1 Problem Identification

Problem identification is the stage where a need for a proposed project is identified, defined, and justified. A project may be concerned with the development of new products, implementation of new processes, or improvement of existing facilities.

1.4.2 Project Definition

Project definition is the phase at which the purpose of the project is clarified. A *mission statement* is the major output of this stage. For example, a prevailing low level of productivity may indicate a need for a new manufacturing technology. In general, the definition should specify how project management may be used to avoid missed deadlines, poor scheduling, inadequate resource allocation, lack of coordination, poor quality, and conflicting priorities.

1.4.3 Project Planning

A plan represents the outline of the series of actions needed to accomplish a goal. Project planning determines how to initiate a project and execute its objectives. It may be a simple statement of a project goal or it may be a detailed account of procedures to be followed during the project. Project planning is discussed in detail in Chapter 2. Planning can be summarized as

- Objectives
- Project definition
- Team organization
- Performance criteria (time, cost, quality)

1.4.4 Project Organizing

Project organization specifies how to integrate the functions of the personnel involved in a project. Organizing is usually done concurrently with project planning. Directing is an important aspect of project organization. Directing involves guiding and supervising the project personnel. It is a crucial aspect of the management function. Directing requires skillful managers who can interact with subordinates effectively through good communication and motivation techniques. A good project manager will facilitate project success by directing his or her staff, through proper task assignments, toward the project goal.

Workers perform better when there are clearly defined expectations. They need to know how their job functions contribute to the overall goals of the project. Workers should be given some flexibility for self-direction in performing their functions. Individual worker needs and limitations should be recognized by the manager when directing project functions. Directing a project requires skills dealing with motivating, supervising, and delegating.

1.4.5 Resource Allocation

Project goals and objectives are accomplished by allocating resources to functional requirements. Resources can consist of money, people, equipment, tools, facilities, information, skills, and so on. These are usually in short supply. The people needed for a particular task may be committed to other ongoing projects. A crucial piece of equipment may be under the control of another team. Chapter 5 addresses resource allocation in detail.

1.4.6 Project Scheduling

Timeliness is the essence of project management. Scheduling is often the major focus in project management. The main purpose of scheduling is to allocate resources so that the overall project objectives are achieved within a reasonable time span. Project objectives are generally conflicting in nature. For example, minimization of the project completion time and minimization of the project cost are conflicting objectives. That is, one objective is improved at the expense of worsening the other objective. Therefore, project scheduling is a multiple-objective decision-making problem.

In general, scheduling involves the assignment of time periods to specific tasks within the work schedule. Resource availability, time limitations, urgency level, required performance level, precedence requirements, work priorities, technical constraints, and other factors complicate the scheduling process. Thus, the assignment of a time slot to a task does not necessarily ensure that the task will be performed satisfactorily in accordance with the schedule. Consequently, careful control must be developed and maintained throughout the project scheduling process. Chapter 4 covers project scheduling in detail. Project scheduling involves

- Resource availability analysis (human, material, money)
- Scheduling techniques (CPM, PERT, Gantt charts)

1.4.7 Project Tracking and Reporting

This phase involves checking whether or not project results conform to project plans and performance specifications. Tracking and reporting are prerequisites for project control. A properly organized report of the project status will help identify any deficiencies in the progress of the project and help pinpoint corrective actions.

1.4.8 Project Control

Project control requires that appropriate actions be taken to correct unacceptable deviations from expected performance. Control is actuated through measurement, evaluation, and corrective action. Measurement is the process of measuring the relationship between planned performance and actual performance with respect to project objectives. The variables to be measured, the measurement scales, and the measuring approaches should be clearly specified during the planning stage. Corrective actions may involve rescheduling, reallocation of resources, or expedition of task performance. Project control is discussed in detail in Chapter 6. Control involves

- Tracking and reporting
- Measurement and evaluation
- Corrective action (plan revision, rescheduling, updating)

1.4.9 Project Termination

Termination is the last stage of a project. The phaseout of a project is as important as its initiation. The termination of a project should be implemented expeditiously. A project should not be allowed to drag on after the expected completion time. A terminal activity should be defined for a project during the planning phase. An example of a terminal activity may be the submission of a final report, the power-on of new equipment, or the signing of a release order. The conclusion of such an activity should be viewed as the completion of the project. Arrangements may be made for follow-up activities that may improve or extend the outcome of the project. These follow-up or spinoff projects should be managed as new projects but with proper input-output relationships within the sequence of projects.

1.5 PROJECT MANAGEMENT OUTLINE

An outline of the functions to be carried out during a project should be made during the planning stage of the project. A model for such an outline is presented below. It may be necessary to rearrange the contents of the outline to fit the specific needs of a project.

1.5.1 1. Planning

I. Specify project background
   A. Define current situation and process
      1. Understand the process
      2. Identify important variables
      3. Quantify variables
   B. Identify areas for improvement
      1. List and discuss the areas
      2. Study potential strategy for solution
II. Define unique terminologies relevant to the project
   A. Industry-specific terminologies
   B. Company-specific terminologies
   C. Project-specific terminologies
III. Define project goal and objectives
   A. Write mission statement
   B. Solicit inputs and ideas from personnel
IV. Establish performance standards
   A. Schedule
   B. Performance
   C. Cost
V. Conduct formal project feasibility study
   A. Determine impact on cost
   B. Determine impact on organization
   C. Determine project deliverables
VI. Secure management support

1.5.2 2. Organizing

I. Identify project management team
   A. Specify project organization structure
      1. Matrix structure
      2. Formal and informal structures
      3. Justify structure
   B. Specify departments involved and key personnel
      1. Purchasing
      2. Materials management
      3. Engineering, design, manufacturing, and so on
   C. Define project management responsibilities
      1. Select project manager
      2. Write project charter
      3. Establish project policies and procedures
II. Implement Triple C Model
   A. Communication
      1. Determine communication interfaces
      2. Develop communication matrix

B. Cooperation
    1. Outline cooperation requirements
C. Coordination
    1. Develop work breakdown structure
    2. Assign task responsibilities
    3. Develop responsibility chart

1.5.3 3. Scheduling and Resource Allocation

I. Develop master schedule
  A. Estimate task duration
  B. Identify task precedence requirements
    1. Technical precedence
    2. Resource-imposed precedence
    3. Procedural precedence
  C. Use analytical models
    1. CPM
    2. PERT
    3. Gantt chart
    4. Optimization models

1.5.4 4. Tracking, Reporting, and Control

I. Establish guidelines for tracking, reporting, and control
  A. Define data requirements
    1. Data categories
    2. Data characterization
    3. Measurement scales
  B. Develop data documentation
    1. Data update requirements
    2. Data quality control
    3. Establish data security measures
II. Categorize control points
  A. Schedule audit
    1. Activity network and Gantt charts
    2. Milestones
    3. Delivery schedule
  B. Performance audit
    1. Employee performance
    2. Product quality
  C. Cost audit
    1. Cost containment measures
    2. Percent completion versus budget depletion
III. Identify implementation process
  A. Comparison with targeted schedules
  B. Corrective course of action

    1. Rescheduling

    2. Reallocation of resources

IV. Terminate the project

    A. Performance review

    B. Strategy for follow-up projects

    C. Personnel retention and releases

V. Document project and submit final report

1.6 SELECTING THE PROJECT MANAGER

The role of a manager is to use available resources (manpower and tools) to accomplish goals and objectives. A project manager has the primary responsibility of ensuring that a project is implemented according to the project plan. The project manager has a wide span of interaction within and outside the project environment. He or she must be versatile, assertive, and effective in handling problems that develop during the execution phase of the project. Selecting a project manager requires careful consideration because the selection of the project manager is one of the most crucial project functions. The project manager should be someone who can: get the job done promptly and satisfactorily; possess both technical and administrative credibility; be perceived as having the technical knowledge to direct the project; be current with the technologies pertinent to the project requirements; and be conversant with the industry's terminologies. The project manager must also be a good record keeper. Since the project manager is the vital link between the project and upper management, he or she must be able to convey information at various levels of detail. The project manager should have good leadership qualities, although leadership is an after-the-fact attribute. Therefore, caution should be exercised in extrapolating prior observations to future performance when evaluating candidates for the post of project manager.

    The selection process should be as formal as a regular recruiting process. A pool of candidates may be developed through nominations, applications, eligibility records, shortlisted group, or administrative appointment. The candidates should be aware of the nature of the project and what they would be expected to do. Formal interviews may be required in some cases, particularly those involving large projects. In a few cases, the selection may have to be made by default if there are no other suitably qualified candidates. Default appointment of a project manager implies that no formal evaluation process has been carried out. Political considerations and quota requirements often lead to default selection of project managers. As soon as a selection is made, an announcement should be made to inform the project team of the selection. The desirable attributes a project manager should possess are:

- Inquisitiveness
- Good labor relations
- Good motivational skills
- Availability and accessibility
- Versatility with company operations

- Good rapport with senior executives
- Good analytical and technical background
- Technical and administrative credibility
- Perseverance toward project goals
- Excellent communication skills
- Receptiveness to suggestions
- Good leadership qualities
- Good diplomatic skills
- Congenial personality

1.7 SELLING THE PROJECT PLAN

The project plan must be sold throughout the organization. Different levels of detail will be needed when presenting the project to various groups in the organization. The higher the level of management, the lower the level of detail. Top management will be more interested in the global aspects of the project. For example, when presenting the project to management, it is necessary to specify how the overall organization will be affected by the project. When presenting the project to the supervisory level staff, the most important aspect of the project may be the operational level of detail. At the worker or operator level, the individual will be more concerned about how he or she fits into the project. The project manager or analyst must be able to accommodate these various levels of detail when presenting the plan to both participants in and customers of the project. Regardless of the group being addressed, the project presentation should cover the following elements with appropriate levels of detail:

- Executive summary
- Introduction
- Project description
  - Goals and objectives
  - Expected outcome
- Performance measures
- Conclusion

The use of charts, figures and tables is necessary for better communication with management. A presentation to middle-level managers may follow a more detailed outline that might include the following:

- Objectives
- Methodologies
- What has been done
- What is currently being done
- What remains to be done

- Problems encountered to date
- Results obtained to date
- Future work plan
- Conclusions and recommendations

1.8 STAFFING THE PROJECT

Once the project manager has been selected and formally installed, one of his first tasks is the selection of the personnel for the project. In some cases, the project manager simply inherits a project team that was formed before he was selected as the project manager. In that case, the project manager's initial responsibility will be to ensure that a good project team has been formed. The project team should be chosen on the basis of skills relevant to the project requirements and team congeniality. The personnel required may be obtained either from within the organization or from outside sources. If outside sources are used, a clear statement should be made about the duration of the project assignment. If opportunities for permanent absorption into the organization exist, the project manager may use that fact as an incentive both in recruiting for the project and in running the project. An incentive for internal personnel may be the opportunity for advancement within the organization.

Job descriptions should be prepared in unambiguous terms. Formal employment announcements may be issued or direct contacts through functional departments may be utilized. The objective is to avoid having a pool of applicants that is either too large or too small. If job descriptions are too broad, many unqualified people will apply. If the descriptions are too restrictive, very few of those qualified will apply. Some skill tolerance or allowance should be established. Since it is nearly impossible to obtain the perfect candidate for each position, some preparation should be made for in-house specialized skill development to satisfy project objectives. Typical job classifications in a project environment include the following:

- Project administrator
- Project director
- Project coordinator
- Program manager
- Project manager
- Project engineer
- Project assistant
- Project specialist
- Task manager
- Project auditor

Staff selection criteria should be based on project requirements and the availability of a staff pool. Factors to consider in staff selection include

- Recommendation letters and references
- Salary requirements
- Geographical preference
- Education and experience
- Past project performance
- Time frame of availability
- Frequency of previous job changes
- Versatility for project requirements
- Completeness and directness of responses
- Special project requirements (quotas, politics, etc.)
- Overqualification (Overqualified workers tend to be unhappy at lower job levels.)
- Organizational skills

An initial screening of the applicants on the basis of the above factors may help reduce the applicant pool to a manageable level. If company policy permits, direct contact over the telephone or in person may then be used to further prune the pool of applicants. A direct conversation usually brings out more accurate information about applicants. In many cases, people fill out applications by writing what they feel the employer wants to read rather than what they want to say. Direct contact can help determine if applicants are really interested in the job, whether they will be available when needed, or whether they possess vital communication skills.

Confidentiality of applications should be maintained, particularly for applicants who do not want a disclosure to their current employers. References should be checked out and the information obtained should be used with utmost discretion. Interviews should then be arranged for the leading candidates. Final selection should be based on the merits of the applicants rather than mere personality appeal. Both the successful and the unsuccessful candidates should be informed of the outcome as soon as administrative policies permit.

In many technical fields, personnel shortage is a serious problem. The problem of recruiting in such circumstances becomes that of expanding the pool of applicants rather than pruning the pool. It is a big battle among employers to entice highly qualified technical personnel from one another. Some recruiters have even been known to resort to unethical means in the attempt to lure prospective employees. Project staffing involving skilled manpower can be enhanced by the following:

- Employee exchange programs
- Transfer from other projects
- In-house training for new employees
- Use of temporary project consultants
- Diversification of in-house job skills

- Cooperative arrangements among employers
- Continuing education for present employees

Committees may be set up to guide the project effort from the recruitment stage to the final implementation stage. Figure 1–3 shows a generic organizational chart for the project office and the role of a project committee. The primary role of a committee should be to provide supporting consultations to the project manager. Such a committee might use the steering committee model which is formed by including representatives from different functional areas. The steering committee should serve as an advisory board for the project. A committee may be set up under one of the following two structures:

1. **Ad hoc committee.** This is set up for a more immediate and specific purpose (e.g., project feasibility study).
2. **Standing committee.** This is set up on a more permanent basis to oversee ongoing project activities.

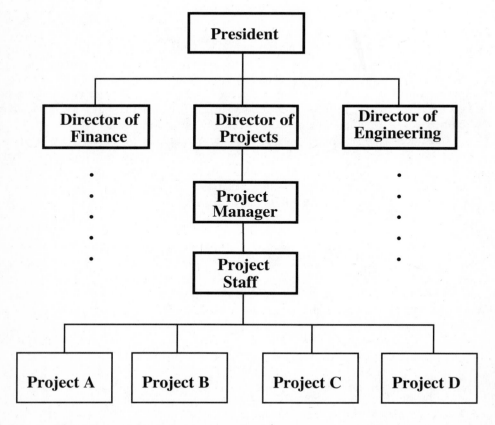

Figure 1-3.   Organization of the Project Office

1.9 PROJECT DECISION ANALYSIS

Badiru (1991a) presents decision steps for project management. The steps facilitate a proper consideration of the essential elements of decisions in a project environment. These essential elements include the problem statement, information, performance measure, decision model, and an implementation of the decision. The steps recommended for project decisions are outlined next.

1.9.1 Step 1. Problem Statement

A problem involves choosing between competing, and probably conflicting, alternatives. The components of problem solving in project management include

- Describing the problem (goals, performance measures)
- Defining a model to represent the problem
- Solving the model
- Testing the solution
- Implementing and maintaining the solution

Problem definition is very crucial. In many cases, *symptoms* of a problem are more readily recognized than its *cause* and *location*. Even after the problem is accurately identified and defined, a benefit/cost analysis may be needed to determine if the cost of solving the problem is justified.

1.9.2 Step 2. Data and Information Requirements

Information is the driving force for the project decision process. Information clarifies the relative states of past, present, and future events. The collection, storage, retrieval, organization, and processing of raw data are important components for generating information. Without data, there can be no information. Without good information, there cannot be a valid decision. The essential requirements for generating information are

- Ensuring that an effective data collection procedure is followed
- Determining the type and the appropriate amount of data to collect
- Evaluating the data collected with respect to information potential
- Evaluating the cost of collecting the required data

For example, suppose a manager is presented with a recorded fact that says, "*Sales for the last quarter are 10,000 units.*" This constitutes ordinary data. There are many ways of using the above data to make a decision depending on the manager's value system. An analyst, however, can ensure the proper use of the data by transforming it into information, such as, "*Sales of 10,000 units for last quarter are within x percent of the targeted value.*" This type of information is more useful to the manager for decision making.

1.9.3 Step 3. Performance Measure

A performance measure for the competing alternatives should be specified. The decision maker assigns a perceived worth or value to the available alternatives. Setting measures of performance is crucial to the process of defining and selecting alternatives. Some performance measures commonly used in project management are project cost, completion time, resource usage, and stability in the workforce.

1.9.4 Step 4. Decision Model

A decision model provides the basis for the analysis and synthesis of information and is the mechanism by which competing alternatives are compared. To be effective, a decision model must be based on a systematic and logical framework for guiding project decisions. A decision model can be a verbal, graphical, or mathematical representation of the ideas in the decision-making process. A project decision model should have the following characteristics:

- Simplified representation of the actual situation
- Explanation and prediction of the actual situation
- Validity and appropriateness
- Applicability to similar problems

The formulation of a decision model involves three essential components:

Abstraction: Determining the relevant factors
Construction: Combining the factors into a logical model
Validation: Assuring that the model adequately represents the problem

The basic types of decision models for project management are described next.

Descriptive models.   These models are directed at describing a decision scenario and identifying the associated problem. For example, a project analyst might use a critical path method (CPM) network model to identify bottleneck tasks in a project.

Prescriptive models.   These models furnish procedural guidelines for implementing actions. The Triple C approach, for example, is a model that prescribes the procedures for achieving communication, cooperation, and coordination in a project environment.

Predictive models.   These models are used to predict future events in a problem environment. They are typically based on historical data about the problem situation. For example, a regression model based on past data may be used to predict future productivity gains associated with expected levels of resource allocation. Simulation models can be used when uncertainties exist in the task durations or resource requirements.

Satisficing models. These are models that provide trade-off strategies for achieving a satisfactory solution to a problem within given constraints. Goal programming and other multicriteria techniques provide good satisficing solutions. For example, these models are helpful in cases where time limitations, resource shortages, and performance requirements constrain the implementation of a project.

Optimization models. These models are designed to find the best available solution to a problem subject to a certain set of constraints. For example, a linear programming model can be used to determine the optimal product mix in a production environment.

In many situations, two or more of the above models may be involved in the solution of a problem. For example, a descriptive model might provide insights into the nature of the problem; an optimization model might provide the optimal set of actions to take in solving the problem; a satisficing model might temper the optimal solution with reality; a prescriptive model might suggest the procedures for implementing the selected solution; and a predictive model might predict the expected outcome of implementing the solution.

1.9.5 Step 5. Making the Decision

Using the available data, information, and the decision model, the decision maker will determine the real-world actions that are needed to solve the stated problem. A sensitivity analysis may be useful for determining what changes in parameter values might cause a change in the decision.

1.9.6 Step 6. Implementing the Decision

A decision represents the selection of an alternative that satisfies the objective stated in the problem statement. A good decision is useless until it is implemented. An important aspect of a decision is to specify how it is to be implemented. Selling the decision and the project to management requires a well-organized persuasive presentation. The way a decision is presented can directly influence whether or not it is adopted. The presentation of a decision should include at least the following: an executive summary, technical aspects of the decision, managerial aspects of the decision, resources required to implement the decision, cost of the decision, the time frame for implementing the decision, and the risks associated with the decision.

1.10 CONDUCTING PROJECT MEETINGS

Meetings are one avenue for information flow in project decision making. Effective management of meetings is an important skill for any managerial staff. Employees often feel that meetings waste time and obstruct productivity. This is because most meetings are poorly organized, improperly managed, called at the wrong time, or even unnecessary. In some organizations, meetings are conducted as routine requirements rather than from necessity. Meetings are essential for communication and decision making. Unfortunately, many meetings accomplish nothing and waste everyone's time. A meeting of 30 people wasting only 30 minutes in effect wastes 15 full hours of employee time. That much time, in a corporate setting, may amount to thousands of dollars in lost time. It does not make sense to use a one-hour meeting to discuss a task that will take only five minutes to perform. That is like hiring someone at a $50,000 annual salary to manage an annual budget of $20,000. Stewart (1993), in a humorous column about management meetings, writes:

> "Management meetings are rapidly becoming this country's biggest growth industry. As nearly as I can determine, the working day of a typical middle manager consists of seven hours of *meetings*, plus lunch. Half a dozen years ago at my newspaper, we hired a new middle management editor with an impressive reputation. Unfortunately, I haven't met her yet. On her first day at work, she went into a *meeting* and has never come out." Stewart concludes his satire with "I'm expected to attend the next meeting. I'm not sure when it's scheduled exactly. I think they're having a meeting this afternoon about that."

In the past, when an employee had a request, he went to the boss, who would say yes or no right away. The whole process might have taken less than one minute of the employee's day. Nowadays several hierarchies of meetings may need to be held to review the request. Thus, we may have a departmental meeting, a middle-management staff meeting, upper-management meeting, executive meeting, steering committee meeting, ad hoc committee meeting, and a meeting with outside consultants all for the purpose of reviewing that simple request. Badiru (1988a) presents the following list of observations about project meetings:

1. Most of the information passed out at meetings can be more effectively disseminated through an ordinary memo. The proliferation of desktop computers and electronic mail should be fully exploited to replace most meetings.
2. The point of diminishing return for any meeting is equal to the number of people who are actually needed for the meeting. The more people at a meeting, the lower the meeting's productivity. The extra attendees serve only to generate unconstructive and conflicting ideas that only impede the meeting.
3. Not being invited to a meeting could be viewed as an indication of the high value placed on an individual's time within the organization.
4. Regularly scheduled meetings with specific time slots often become a forum for social assemblies.
5. The optimal adjournment time of a meeting is equal to the scheduled start time plus five times the number of agenda items minus the start-up time. Mathematically, this is expressed as

$$L = (T + 5N) - S$$

where

L = optimal length in minutes
T = scheduled time
N = number of agenda items
S = meeting start-up time (i.e., time taken to actually call the meeting to order)

Since it is difficult to do away with meetings (the necessary and the unnecessary), we must attempt to maximize their output. Some guidelines for running meetings more effectively are presented next.

1. Do premeeting homework.
   - List topics to be discussed (agenda).
   - Establish the desired outcome for each topic.
   - Determine how the outcome will be verified.
   - Determine who really needs to attend the meeting.
   - Evaluate the suitability of meeting time and venue.
   - Categorize meeting topics (e.g., announcements, important, urgent).
   - Assign a time duration to each topic.
   - Verify that the meeting is really needed.
   - Consider alternatives to the meeting (e.g., memo, telephone, electronic mail).
2. Circulate a written agenda prior to the meeting.
3. Start meeting on time.
4. Review agenda at the beginning.
5. Get everyone involved; if necessary, employ direct questions and eye contact.
6. Keep to the agenda; do not add new items unless absolutely essential.
7. Be a facilitator for meeting discussions.
8. Quickly terminate conflicts that develop from routine discussions.
9. Redirect irrelevant discussions back to the topic of the meeting.
10. Retain leadership and control of the meeting.
11. Recap the accomplishments of each topic before going to the next. Let those who have made commitments (e.g., promise to look into certain issues) know what is expected of them.
12. End meeting on time.
13. Prepare and distribute minutes. Emphasize the outcome and success of the meeting.

Huyler and Crosby (1993) present an analysis of the economic impact of poorly managed meetings. They provided guidelines for project managers to improve meetings. They suggest evaluating meetings by asking the following post-meeting questions:

1. What did we do well in this meeting?
2. What can we improve next time?

Despite the shortcomings of poorly managed meetings, meetings offer a suitable avenue for group decision making.

1.11 GROUP DECISION MAKING

Many decision situations are complex and poorly understood. No one person has all the information to make all decisions accurately. As a result, crucial decisions are made by a group of people. Some organizations use outside consultants with appropriate expertise to make recommendations for important decisions. Other organizations set up their own internal consulting groups without having to go outside the organization. Decisions can be made through linear responsibility, in which case one person makes the final decision based on inputs from other people. Decisions can also be made through shared responsibility, in which case a group of people share the responsibility for making joint decisions. The major advantages of group decision making are

1. **Ability to share experience, knowledge, and resources.** Many heads are better than one. A group will possess greater collective ability to solve a given decision problem.
2. **Increased credibility.** Decisions made by a group of people often carry more weight in an organization.
3. **Improved morale.** Personnel morale can be positively influenced because many people have the opportunity to participate in the decision-making process.
4. **Better rationalization.** The opportunity to observe other people's views can lead to an improvement in an individual's reasoning process.

Some disadvantages of group decision making are

1. **Difficulty in arriving at a decision.** Individuals may have conflicting objectives.
2. **Reluctance of some individuals in implementing the decision.**
3. **Potential for conflicts among members of the decision group.**
4. **Loss of productive employee time.**

1.11.1 Brainstorming

Brainstorming is a way of generating many new ideas. In brainstorming, the decision group comes together to discuss alternate ways of solving a problem. The members of the brainstorming group may be from different departments, may have different backgrounds and training, and may not even know one another. The diversity of the participants helps create a stimulating environment for generating different ideas from different viewpoints. The technique encourages free outward expression of new ideas no matter how far-fetched the ideas might appear. No criticism of any new idea is permitted during the brainstorming session. A major concern in brainstorming is that extroverts may take control of the discussions. For this reason, an experienced and respected individual should manage the brainstorming discussions. The group leader establishes the procedure for proposing ideas, keeps the discussions in line with the group's mission, discourages disruptive statements, and encourages the participation of all members.

After the group runs out of ideas, open discussions are held to weed out the unsuitable ones. It is expected that even the rejected ideas may stimulate the generation of other ideas which may eventually lead to other favored ideas. Guidelines for improving brainstorming sessions are presented as follows:

- Focus on a specific decision problem.
- Keep ideas relevant to the intended decision.
- Be receptive to all new ideas.
- Evaluate the ideas on a relative basis after exhausting new ideas.
- Maintain an atmosphere conducive to cooperative discussions.
- Maintain a record of the ideas generated.

1.11.2 Delphi Method

The traditional approach to group decision making is to obtain the opinion of experienced participants through open discussions. An attempt is made to reach a consensus among the participants. However, open group discussions are often biased because of the influence or subtle intimidation from dominant individuals. Even when the threat of a dominant individual is not present, opinions may still be swayed by group pressure. This is called the "bandwagon effect" of group decision making.

The Delphi method, developed by Gordon and Helmer (1964), attempts to overcome these difficulties by requiring individuals to present their opinions anonymously through an intermediary. The method differs from the other interactive group methods because it eliminates face-to-face confrontations. It was originally developed for forecasting applications, but it has been modified in various ways for application to different types of decision making. The method can be quite useful for project management decisions. It is particularly effective when decisions must be based on a broad set of factors. The Delphi method is normally implemented as follows:

1. **Problem definition.** A decision problem that is considered significant is identified and clearly described.
2. **Group selection.** An appropriate group of experts or experienced individuals is formed to address the particular decision problem. Both internal and external experts may be involved in the Delphi process. A leading individual is appointed to serve as the administrator of the decision process. The group may operate through the mail or gather together in a room. In either case, all opinions are expressed anonymously on paper. If the group meets in the same room, care should be taken to provide enough room so that each member does not have the feeling that someone may accidentally or deliberately observe their responses.
3. **Initial opinion poll.** The technique is initiated by describing the problem to be addressed in unambiguous terms. The group members are requested to submit a list of major areas of concern in their specialty areas as they relate to the decision problem.
4. **Questionnaire design and distribution.** Questionnaires are prepared to address the areas of concern related to the decision problem. The written responses to the questionnaires are collected and organized by the administrator. The administrator aggregates the responses in a statistical format. For example, the average, mode, and median of the responses may by computed. This analysis is distributed to the decision group. Each member can then see how his or her responses compare with the anonymous views of the other members.

5. Iterative balloting. Additional questionnaires based on the previous responses are passed to the members. The members submit their responses again. They may choose to alter or not to alter their previous responses.
6. Silent discussions and consensus. The iterative balloting may involve anonymous written discussions of why some responses are correct or incorrect. The process is continued until a consensus is reached. A consensus may be declared after five or six iterations of the balloting or when a specified percentage (e.g., 80 percent) of the group agrees on the questionnaires. If a consensus cannot be declared on a particular point, it may be displayed to the whole group with a note that it does not represent a consensus.

In addition to its use in technological forecasting, the Delphi method has been widely used in other general decision making. Its major characteristics of anonymity of responses, statistical summary of responses, and controlled procedure make it a reliable mechanism for obtaining numeric data from subjective opinion. The major limitations of the Delphi method are

1. Its effectiveness may be limited in cultures where strict hierarchy, seniority, and age influence decision-making processes.
2. Some experts may not readily accept the contribution of nonexperts to the group decision-making process.
3. Since opinions are expressed anonymously, some members may take the liberty of making ludicrous statements. However, if the group composition is carefully reviewed, this problem may be avoided.

1.11.3 Nominal Group Technique

The nominal group technique is a silent version of brainstorming. It is a method of reaching consensus. Rather than asking people to state their ideas aloud, the team leader asks each member to jot down a minimum number of ideas, for example, five or six. A single list of ideas is then written on a chalkboard for the whole group to see. The group then discusses the ideas and weeds out some iteratively until a final decision is made. The nominal group technique is easier to control. Unlike brainstorming where members may get into shouting matches, the nominal group technique permits members to silently present their views. In addition, it allows introversive members to contribute to the decision without the pressure of having to speak out too often.

In all of the group decision-making techniques, an important aspect that can enhance and expedite the decision-making process is to require that members review all pertinent data before coming to the group meeting. This will ensure that the decision process is not impeded by trivial preliminary discussions. Some disadvantages of group decision making are

1. Peer pressure in a group situation may influence a member's opinion or discussions.
2. In a large group, some members may not get to participate effectively in the discussions.
3. A member's relative reputation in the group may influence how well his or her opinion is rated.

4. A member with a dominant personality may overwhelm the other members in the discussions.
5. The limited time available to the group may create a time pressure that forces some members to present their opinions without fully evaluating the ramifications of the available data.
6. It is often difficult to get all members of a decision group together at the same time.

Despite the noted disadvantages, group decision making definitely has many advantages that may nullify the shortcomings. The advantages as presented earlier will have varying levels of effect from one organization to another. The Triple C principle presented in Chapter 2 may also be used to improve the success of decision teams. Team work can be enhanced in group decision making by adhering to the following guidelines:

1. Get a willing group of people together.
2. Set an achievable goal for the group.
3. Determine the limitations of the group.
4. Develop a set of guiding rules for the group.
5. Create an atmosphere conducive to group synergism.
6. Identify the questions to be addressed in advance.
7. Plan to address only one topic per meeting.

For major decisions and long-term group activities, arrange for team training which allows the group to learn the decision rules and responsibilities together. The steps for the nominal group technique are

1. Silently generate ideas, in writing.
2. Record ideas without discussion.
3. Conduct group discussion for clarification of meaning, not argument.
4. Vote to establish the priority or rank of each item.
5. Discuss vote.
6. Cast final vote.

1.11.4 Interviews, Surveys, and Questionnaires

Interviews, surveys, and questionnaires are important information gathering techniques. They also foster cooperative working relationships. They encourage direct participation and inputs into project decision-making processes. They provide an opportunity for employees at the lower levels of an organization to contribute ideas and inputs for decision making. The greater the number of people involved in the interviews, surveys, and questionnaires, the more valid the final decision. The following guidelines are useful for conducting interviews, surveys, and questionnaires to collect data and information for project decisions:

1. Collect and organize background information and supporting documents on the items to be covered by the interview, survey, or questionnaire.

2. Outline the items to be covered and list the major questions to be asked.
3. Use a suitable medium of interaction and communication: telephone, fax, electronic mail, face-to-face, observation, meeting venue, poster, or memo.
4. Tell the respondent the purpose of the interview, survey, or questionnaire, and indicate how long it will take.
5. Use open-ended questions that stimulate ideas from the respondents.
6. Minimize the use of yes or no type of questions.
7. Encourage expressive statements that indicate the respondent's views.
8. Use the who, what, where, when, why, and how approach to elicit specific information.
9. Thank the respondents for their participation.
10. Let the respondents know the outcome of the exercise.

1.11.5 Multivote

Multivoting is a series of votes used to arrive at a group decision. It can be used to assign priorities to a list of items. It can be used at team meetings after a brainstorming session has generated a long list of items. Multivoting helps reduce such long lists to a few items, usually three to five. The steps for multivoting are

1. Take a first vote. Each person votes as many times as desired, but only once per item.
2. Circle the items receiving a relatively higher number of votes (i.e., majority vote) than the other items.
3. Take second vote. Each person votes for a number of items equal to one-half the total number of items circled in step 2. Only one vote per item is permitted.
4. Repeat steps 2 and 3 until the list is reduced to three to five items depending on the needs of the group. It is not recommended to multivote down to only one item.
5. Perform further analysis of the items selected in step 4, if needed.

1.12 PROJECT LEADERSHIP

Some leaders lead by demonstrating good examples. Others attempt to lead by dictating. People learn and act best when good examples are available to emulate. Examples learned in childhood can last a lifetime. A leader should have a spirit of performance which stimulates his or her subordinates to perform at their own best. Rather than dictating what needs to be done, a good leader would show what needs to be done. Showing, in this case, does not necessarily imply an actual physical demonstration of what is to be done. Rather, it implies projecting a commitment to the function at hand and a readiness to participate as appropriate. Traditional managers manage workers to work. So, there is no point of convergence or active participation. Modern managers team up with workers to get the job done. Figure 1–4 presents a leadership model for project management. The model suggests starting by listening and asking questions,

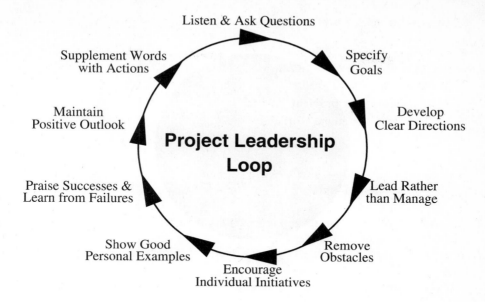

Figure 1-4.   Project Leadership Loop

specifying objectives, developing clear directions, removing obstacles, encouraging individual initiatives, learning from past experiences, and repeating the loop by listening some more and asking more questions.

Good leadership is an essential component of project management. Project leadership involves dealing with managers and supporting personnel across the functional lines of the project. It is a misconception to think that a leader leads only his or her own subordinates. Leadership responsibilities can cover vertically up or down functions. A good project leader can lead not only his or her subordinates but also the entire project organization including the highest superiors. Kostner and Strbiak (1993) suggest a 3-D leadership model that consists of self-leadership, team leadership, and leadership-oriented teamwork. Leadership involves recognizing an opportunity to make improvement in a project and taking the initiative to implement the improvement. In addition to inherent personal qualities, leadership style can be influenced by training, experience, and dedication. Some pitfalls to avoid in project leadership are

POLITICS AND EGOTISM

- Forget personal ego
- Don't glamorize personality
- Focus on the big picture of project goals
- Build up credibility with successful leadership actions
- Cut out the politics and develop a spirit of cooperation

PREACHING VERSUS IMPLEMENTING

- Back up words with action
- Adopt a "do as I do" attitude
- Avoid a "do as I say" attitude
- Participate in joint problem solving
- Develop and implement workable ideas

1.13 PERSONNEL MANAGEMENT

Positive personnel management and interactions are essential for project success. Effective personnel management can enhance team building and coordination. The following guidelines are offered for personnel management in a project environment.

1. Leadership style
   - Lead the team rather than manage the team.
   - Display self-confidence.
   - Establish self-concept of your job functions.
   - Engage in professional networking without being pushy.
   - Be discrete with personal discussions.
   - Perform a self-assessment of professional strengths.
   - Dress professionally without being flashy.
   - Be assertive without being autocratic.
   - Keep up with the developments in the technical field.
   - Work hard without overexerting.
   - Take positive initiative where others procrastinate.
2. Supervision
   - Delegate when appropriate.
   - Motivate subordinates with vigor and an objective approach.
   - Set goals and prioritize them.
   - Develop objective performance appraisal mechanisms.
   - Discipline promptly, as required.
   - Don't overmanage.
   - Don't shy away from mentoring or being mentored.
   - Establish credibility and decisiveness.
   - Don't be intimidated by difficult employees.
   - Use empathy in decision-making processes.
3. Communication
   - Be professional in communication approaches.
   - Do homework about the communication needs.
   - Contribute constructively to meaningful discussions.
   - Exhibit knowledge without being patronizing.
   - Convey ideas effectively to gain respect.

- Cultivate good listening habits.
- Incorporate charisma into communication approaches.
4. Handling conflicts
   - Learn the politics and policies of the organization.
   - Align project goals with organizational goals.
   - Overcome fear of confrontation.
   - Form a mediating liaison among peers, subordinates, and superiors.
   - Control emotions in tense situations.
   - Don't take office conflicts home and don't take home conflicts to work.
   - Avoid a power struggle but claim functional rights.
   - Handle mistakes honestly without being condescending.

1.14 QUALITY MANAGEMENT

Project objectives can best be achieved by establishing and adhering to standards. The desire for better quality worldwide has led to the need for unified international quality standards. The International Organization for Standardization (IOS) has prepared the quality standard known as *ISO 9000*. The IOS is a special international agency for standardization composed of the national standards bodies of several countries. The organization is also known as International Standards Organization (ISO). The term ISO is more prevalent because people tend to confuse ISO with the "ISO" in ISO 9000. ISO 9000 has significant implication on how modern project management is practiced.

1.14.1 What Is ISO 9000?

ISO 9000 is a set of five individual but related international standards on quality management and quality assurance. The standards were developed to help companies effectively document the quality system elements required to maintain an efficient quality system. The standards were originally published in 1987. They are not specific to any particular industry, product, or service.

The short designation for ISO 9000 was borrowed from the Greek word *isos*, which means *equal*. *Isos* is used as the root for many words having to do with equality, thus, *isometric*, meaning of equal measures or dimensions; *isonomy*, meaning equality of people in the eye of the law; *isothermal*, meaning presence of equal temperatures; and *isotropic*, meaning invariant with respect to direction. ISO 9000 is intended to convey the idea of the *invariance* that is possible when a standard is available. When we have a standard for a project, the project is expected to possess uniformity of performance with respect to producing identical or invariant units of a product; that is, *iso-product* or *iso-units*. In an attempt to justify the name ISO 9000, people generally misinterpret the organization's name, IOS, as ISO. There is actually little connection between ISO and ISO 9000. The five individual standards that make up the ISO 9000 series are explained next.

ISO 9000

Title: *Quality management and quality assurance standards: Guidelines for use.* This is the road map that provides guidelines for selecting and using 9001, 9002, 9003, and 9004. A supplementary publication, ISO 8402, provides quality related definitions.

ISO 9001

Title: *Quality Systems: Model for quality assurance in design/development, production, installation and servicing.* This is the most comprehensive standard. It presents a model for quality assurance for design, manufacturing, installation and servicing systems.

ISO 9002

Title: *Quality Systems: Model for quality assurance in production and installation.* This presents a model for quality assurance in production and installation.

ISO 9003

Title: *Quality Systems: Model for quality assurance in final inspection and test.* This presents a model for quality assurance in the final inspection and test.

ISO 9004

Title: *Quality Management and Quality Systems Elements: Guidelines.* This provides guidelines to users in the process of developing in-house quality systems.

1.14.2 Purpose of ISO 9000

The ISO 9000 standards help in determining capable suppliers with effective quality assurance systems. The standards help reduce buyers' quality costs through confidence and assurance in suppliers' quality practices. Compliance with an ISO 9000 standard provides a means for contractual agreement between the buyer and the supplier. Companies that are certified and registered as meeting the ISO standards will be perceived as viable suppliers to their customers. Those that are not will be perceived as providing less desirable products and services.

The standards are designed to address a variety of quality management scenarios. For example, if a supplier has only a manufacturing facility with no design or development function, then ISO 9002 would be used to evaluate the quality system. Each country has its own quality system standards that relate to the ISO 9000 standards. Each individual company is encouraged to formally register for compliance with the standards. In fact, a request for a supplier's ISO 9000 registration number has become an important element when companies make their selection of suppliers.

The ISO 9000 series standards define the minimum requirements a supplier must meet to assure its customers that they are receiving high quality products. This has had a major impact on companies around the world. Through the ISO standards, suppliers can now be evaluated consistently and uniformly. The ISO 9000 series has been adopted in the United States by the American National Standards Institute (ANSI) and the American Society for Quality Control (ASQC) as ANSI/ASQC Q90 standards. The European equivalent of ISO 9000, named the EN 290000 series, is also now having world impact.

1.14.3 ISO 9000 Audit and Registration Process

The implementation of ISO 9000 is carried out through a third-party process. The third party, usually a local standards organization, for example, Underwriters Laboratory (UL), acts as an independent body in evaluating a supplier's quality system. Some of the key elements that the third-party auditors check for in a company are

- Whether the company has a documentation process. Does the documentation provide adequate guidelines for workers?
- Whether everyone in the company is following the documented process. Is everyone aware of updates and changes to the documentation?
- How materials are selected. Are appropriate materials selected for specific processes?
- How suppliers' deliveries are inspected in-house. Is the company getting what it wants from suppliers?
- The calibration and metrology processes. Are calibrations done properly? Are measurements being made accurately?
- The procedure for taking corrective actions. Are avenues available for identifying, reporting, and correcting problems?
- The internal self-auditing process. Are problems overlooked when they are identified? Is there a formal process review policy? Is the company defensive about obvious quality problems?

A successful ISO 9000 audit is a prerequisite for ISO 9000 registration. Thus, registration affirms that a company is meeting acceptable quality standards. Even after registration, the auditors come back periodically to make sure that the standards continue to be met. Good quality documentation helps each employee know exactly what is expected of him or her with respect to the quality of products and services. This awareness can positively affect morale and provide the impetus for further personal commitment to quality in a project environment.

1.15 INTEGRATED SYSTEMS APPROACH

The traditional concepts of systems analysis are applicable to the project process. The definitions of a project system and its components are presented next.

System.   A project system consists of interrelated elements organized for the purpose of achieving a common goal. The elements are organized to work synergistically to generate a unified output that is greater than the sum of the individual outputs of the components.

Program.   A program is a very large and prolonged undertaking. Such endeavors often span several years. Programs are usually associated with particular systems. For example, we may have a space exploration program within a national defense system.

Project.   A project is a time-phased effort of much smaller scope and duration than a program. Programs are sometimes viewed as consisting of a set of projects. Government projects are often called *programs* because of their broad and comprehensive nature. Industry tends to use the term *project* because of the short-term and focused nature of most industrial efforts.

Task.   A task is a functional element of a project. A project is composed of a sequence of tasks that all contribute to the overall project goal.

Activity.   An activity can be defined as a single element of a project. Activities are generally smaller in scope than tasks. In a detailed analysis of a project, an activity may be viewed as the smallest, practically indivisible work element of the project. For example, we can regard a manufacturing plant as a system. A plantwide endeavor to improve productivity can be viewed as a program. The installation of a flexible manufacturing system is a project within the productivity improvement program. The process of identifying and selecting equipment vendors is a task, and the actual process of placing an order with a preferred vendor is an activity. The systems structure of a project is illustrated in Figure 1–5.

The emergence of systems development has had an extensive effect on project management in recent years. A system can be defined as a collection of interrelated elements brought together to achieve a specified objective. In a management context,

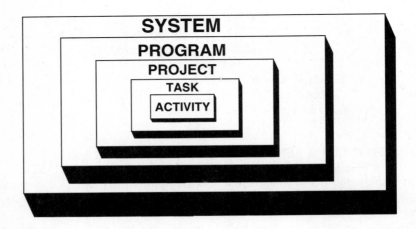

Figure 1-5.   Systems Structure of a Project

the purposes of a system are to develop and manage operational procedures and to facilitate an effective decision-making process. Some of the common characteristics of a system include:

1. Interaction with the environment
2. Objective
3. Self-regulation
4. Self-adjustment

Representative components of a project system are the organizational subsystem, planning subsystem, scheduling subsystem, information management subsystem, control subsystem, and project delivery subsystem. The primary responsibilities of project analysts involve ensuring the proper flow of information throughout the project system. The classical approach to the decision process follows rigid lines of organizational charts. By contrast, the systems approach considers all the interactions necessary among the various elements of an organization in the decision process.

The various elements (or subsystems) of the organization act simultaneously in a separate but interrelated fashion to achieve a common goal. This synergism helps to expedite the decision process and to enhance the effectiveness of decisions. The supporting commitments from other subsystems of the organization serve to counterbalance the weaknesses of a given subsystem. Thus, the overall effectiveness of the system is greater than the sum of the individual results from the subsystems.

The increasing complexity of organizations and projects makes the systems approach essential in today's management environment. As the number of complex projects increase, there will be an increasing need for project management professionals who can function as systems integrators. Project management techniques can be applied to the various stages of implementing a system as shown in the following guidelines:

1. Systems definition. Define the system and associated problems using keywords that signify the importance of the problem to the overall organization. Locate experts in this area who are willing to contribute to the effort. Prepare and announce the development plan.
2. Personnel assignment. The project group and the respective tasks should be announced, a qualified project manager should be appointed, and a solid line of command should be established and enforced.
3. Project initiation. Arrange an organizational meeting during which a general approach to the problem should be discussed. Prepare a specific development plan and arrange for the installation of needed hardware and tools.
4. System prototype. Develop a prototype system, test it, and learn more about the problem from the test results.
5. Full system development. Expand the prototype to a full system, evaluate the user interface structure, and incorporate user training facilities and documentation.
6. System verification. Get experts and potential users involved, ensure that the system performs as designed, and debug the system as needed.
7. System validation. Ensure that the system yields expected outputs. Validate the system by evaluating performance level, such as percentage of success in so many

trials, measuring the level of deviation from expected outputs, and measuring the effectiveness of the system output in solving the problem.

8. **System integration.** Implement the full system as planned, ensure the system can coexist with systems already in operation, and arrange for technology transfer to other projects.

9. **System maintenance.** Arrange for continuing maintenance of the system. Update solution procedures as new pieces of information become available. Retain responsibility for system performance or delegate to well-trained and authorized personnel.

10. **Documentation.** Prepare full documentation of the system, prepare a user's guide, and appoint a user consultant.

Systems integration permits sharing of resources. Physical equipment, concepts, information, and skills may be shared as resources. Systems integration is now a major concern of many organizations. Even some of the organizations that traditionally compete and typically shun cooperative efforts are beginning to appreciate the value of integrating their operations. For these reasons, systems integration has emerged as a major interest in business. Systems integration may involve the physical integration of technical components, objective integration of operations, conceptual integration of management processes, or a combination of any of these.

Systems integration involves the linking of components to form subsystems and the linking of subsystems to form composite systems within a single department and/or across departments. It facilitates the coordination of technical and managerial efforts to enhance organizational functions, reduce cost, save energy, improve productivity, and increase the utilization of resources. Systems integration emphasizes the identification and coordination of the interface requirements among the components in an integrated system. The components and subsystems operate synergistically to optimize the performance of the total system. Systems integration ensures that all performance goals are satisfied with a minimum expenditure of time and resources. Integration can be achieved in several forms including the following:

1. **Dual-use integration:** This involves the use of a single component by separate subsystems to reduce both the initial cost and the operating cost during the project life cycle.

2. **Dynamic resource integration:** This involves integrating the resource flows of two normally separate subsystems so that the resource flow from one to or through the other minimizes the total resource requirements in a project.

3. **Restructuring of functions:** This involves the restructuring of functions and reintegration of subsystems to optimize costs when a new subsystem is introduced into the project environment.

Systems integration is particularly important when introducing new technology into an existing system. It involves coordinating new operations to coexist with existing operations. It may require the adjustment of functions to permit the sharing of resources, development of new policies to accommodate product integration, or realignment of managerial responsibilities. It can affect both hardware and software components of an organization. Presented below are guidelines and important questions relevant for systems integration.

- What are the unique characteristics of each component in the integrated system?
- How do the characteristics complement one another?
- What physical interfaces exist among the components?
- What data/information interfaces exist among the components?
- What ideological differences exist among the components?
- What are the data flow requirements for the components?
- Are there similar integrated systems operating elsewhere?
- What are the reporting requirements in the integrated system?
- Are there any hierarchical restrictions on the operations of the components of the integrated system?
- What internal and external factors are expected to influence the integrated system?
- How can the performance of the integrated system be measured?
- What benefit/cost documentations are required for the integrated system?
- What is the cost of designing and implementing the integrated system?
- What are the relative priorities assigned to each component of the integrated system?
- What are the strengths of the integrated system?
- What are the weaknesses of the integrated system?
- What resources are needed to keep the integrated system operating satisfactorily?
- Which section of the organization will have primary responsibility for the operation of the integrated system?
- What are the quality specifications and requirements for the integrated systems?

The integrated approach to project management recommended in this book is represented by the flowchart in Figure 1–6. The process starts with a managerial analysis of the project effort. Goals and objectives are defined, a mission statement is written, and the statement of work is developed. After these, traditional project management approaches, such as the selection of an organization structure, are employed. Conventional analytical tools including the critical path method (CPM) and the precedence diagramming method (PERT) are then mobilized. The use of optimization models is then called upon as appropriate. Some of the parameters to be optimized are cost, resource allocation, and schedule length. It should be understood that not all project parameters will be amenable to optimization. The use of commercial project management software should start only after the managerial functions have been completed. Some project management software have built-in capabilities for the planning and optimization needs.

A frequent mistake in project management is the rush to use a project management software without first completing the planning and analytical studies required by the project. Project management software should be used as a management tool, the same way a word processor is used as a writing tool. It will not be effective to start using

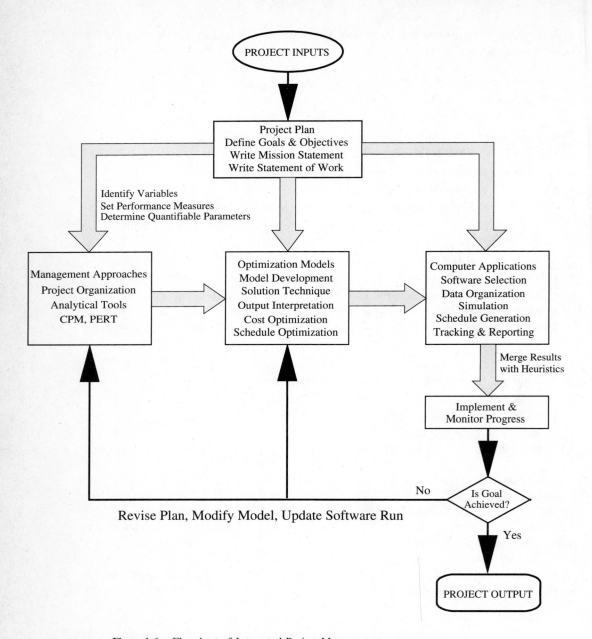

Figure 1-6. Flowchart of Integrated Project Management

the word processor without first organizing the thoughts about what is to be written. Project management is much more than just the project management software. If project management is carried out in accordance with the integration approach presented in Figure 1–6, the odds of success will be increased. Of course, the structure of the flowchart should not be rigid. Flows and interfaces among the blocks in the flowchart may need to be altered or modified depending on specific project needs.

1.16 INTERFACE OF RESEARCH AND PRACTICE

Much archival research is being done in various areas of project management, but few of the research results make the transition from research to practice. One reason for this is that most of these results are published in journals that are rarely read by practitioners. Meanwhile, the practice of project management continues to recycle the same age-old concepts that may no longer be valid in modern project environments. It is important to institute a marriage of the traditional management practices and modern research results. Such coupling will facilitate more practical and timely tools for project management. In Figure 1–7, we present a model for the interface of project management research and practice. The profit-driven objectives of practice should be integrated with the knowledge-driven goals of research.

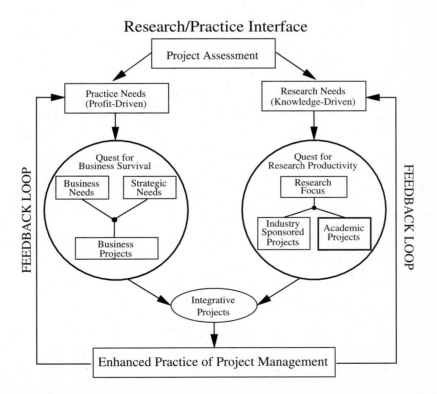

Figure 1-7.   Model for the Interface of Research and Practice

1.17 EXERCISES

1.1. Sketch a diagram to illustrate the relationships among a system, a program, a project, a task, and an activity in a project environment.

1.2. Define the functions of project planning, scheduling, and control.

1.3. What project parameters or variables may be important for defining qualitative responsibility for project management?

1.4. What project parameters or variables may be important for defining quantitative accountability for project management?

1.5. Give examples to illustrate the differences between a *project specialist* and a *technical specialist*.

1.6. How can integrated project management tools be used to develop a project implementation strategy?

1.7. Discuss the differences between a leader and a manager.

1.8. Develop a tabulated comparison of the advantages and disadvantages of group decision making.

1.9. How is project management different from traditional management functions?

1.10. Define a project system.

1.11. Expand the flowchart presented in Figure 1–6 to include other important functions within the project management cycle.

1.12. Select your favorite leader in your organization. Discuss how he or she satisfies the characteristics listed in the project leadership loop.

1.13. Select your least favorite leader in your organization. Identify which characteristics in the leadership loop he or she she does not have. Discuss how you can counsel this individual on being a better leader.

2

Project Planning

This chapter addresses project planning requirements. The required components of a plan are discussed. Motivation is discussed as one of the factors that influence project plans. The important considerations in project feasibility studies are presented. Guidelines are presented for issuing or responding to requests for proposals. Budgeting is discussed as a planning tool for project efforts. Work breakdown structure is discussed as it relates to getting more accurate work data for project plans. The need for legal considerations in project plans is also discussed. The planning stage of project management can be represented at this point as

$$\text{SET GOALS} \rightarrow \text{PLAN}$$

2.1 PROJECT PLANNING OBJECTIVES

The key to a successful project is good planning. Project planning provides the basis for the initiation, implementation, and termination of a project. It sets guidelines for specific project objectives, project structure, tasks, milestones, personnel, cost, equipment, performance, and problem resolutions. An analysis of what is needed and what is available should be conducted in the planning phase of new projects. The availability of technical expertise within the organization and outside the organization should be reviewed. If subcontracting is needed, the nature of the contract should undergo a thorough analysis. The question of whether or not the project is needed at all should be addressed. The "make," "buy," "lease," "sub-contract," or "do nothing" alternatives should be compared as part of the project planning process. Wilson (1993) presents the seven following guidelines for project plans:

1. Use project plans to coordinate rather than to control.
2. Make use of different personalities within the project environment.

39

3. Preschedule frequent revisions to project plans.
4. Empower workers to estimate their own work.
5. Describe value-creating tasks rather than activities.
6. Define specific and tangible milestones.
7. Use checklists, matrices, and other supplements to project plans.

In the initial stage of project planning, the internal and external factors that influence the project should be determined and given priority weights. Examples of internal influences on project plans include the following:

- Infrastructure
- Project scope
- Labor relations
- Project location
- Project leadership
- Organizational goal
- Management approach
- Technical personnel supply
- Resource and capital availability

In addition to internal factors, a project plan can be influenced by external factors. An external factor may be the sole instigator of a project or it may manifest itself in combination with other external and internal factors. Such external factors include

- Public needs
- Market needs
- National goals
- Industry stability
- State of technology
- Industrial competition
- Government regulations

2.2 TIME-COST-PERFORMANCE CRITERIA FOR PLANNING

Project goals determine the nature of project planning. Project goals may be specified in terms of time (schedule), cost (resources), or performance (output). A project can be simple or complex. While simple projects may not require the whole array of project management tools, complex projects may not be successful without all the tools. Project management techniques are applicable to a wide collection of problems ranging from manufacturing to medical services.

The techniques of project management can help achieve goals relating to better product quality, improved resource utilization, better customer relations, higher productivity, and fulfillment of due dates. These can be expressed in terms of the following project constraints:

- *Performance specifications*
- *Schedule requirements*
- *Cost limitations*

Project planning determines the nature of actions and responsibilities needed to achieve the project goal. It entails the development of alternate courses of action and the selection of the best action to achieve the objectives making up the goal. Planning determines what needs to be done, by whom, for whom, and when. Whether it is done for long-range (strategic) purposes or short-range (operational) purposes, planning should be one of the first steps of project management.

Badiru (1991a) outlined the strategic levels of planning. Decisions involving strategic planning lay the foundation for the successful implementation of projects. Planning forms the basis for all actions. Strategic decisions may be divided into three strategy levels: *supralevel planning, macrolevel planning,* and *microlevel planning.*

Supralevel planning.   Planning at the supralevel deals with the big picture of how the project fits the overall and long-range organizational goals. Questions faced at this level concern potential contributions of the project to the welfare of the organization, its effect on the depletion of company resources, required interfaces with other projects within and outside the organization, risk exposure, management support for the project, concurrent projects, company culture, market share, shareholder expectations, and financial stability.

Macrolevel planning.   Planning decisions at the macrolevel address the overall planning within the project boundary. The scope of the project and its operational interfaces should be addressed at this level. Questions faced at the macrolevel include goal definition, project scope, availability of qualified personnel, resource availability, project policies, communication interfaces, budget requirements, goal interactions, deadline, and conflict resolution strategies.

Microlevel planning.   The microlevel deals with detailed operational plans at the task levels of the project. Definite and explicit tactics for accomplishing specific project objectives are developed at the microlevel. The concept of MBO (management by objective) may be particularly effective at this level. MBO permits each project member to plan his or own work at the microlevel. Factors to be considered at the microlevel of project decisions include scheduled time, training requirements, required tools, task procedures, reporting requirements, and quality requirements.

Project decisions at the three levels defined above will involve numerous personnel within the organization with various types and levels of expertise. In addition to the conventional roles of the project manager, specialized roles may be developed within the project scope. Such roles include the following:

1. *Technical specialist*: This person will have responsibility for addressing specific technical requirements of the project. In a large project, there will typically be several technical specialists working together to solve project problems.

2. *Operations integrator*: This person will be responsible for making sure that all operational components of the project interface correctly to satisfy project goals. This person should have good technical awareness and excellent interpersonal skills.

3. *Project specialist*: This person has specific expertise related to the specific goals and requirements of the project. Even though a technical specialist may also serve as a project specialist, the two roles should be distinguished. A general electrical engineer may be a technical specialist on the electronic design components of a project. However, if the specific setting of the electronics project is in the medical field, then an electrical engineer with expertise in medical operations may be needed to serve as the project specialist.

2.3 COMPONENTS OF A PLAN

Planning is an ongoing process that is conducted throughout the project life cycle. Initial planning may relate to overall organizational efforts. This is where specific projects to be undertaken are determined. Subsequent planning may relate to specific objectives of the selected project. In general, a project plan should consist of the following components:

1. Summary of project plan. This is a brief description of what is planned. Project scope and objectives should be enumerated. The critical constraints on the project should be outlined. The types of resources required and available should be specified. The summary should include a statement of how the project complements organizational and national goals, budget size, and milestones.

2. Objectives. The objectives should be very detailed in outlining what the project is expected to achieve and how the expected achievements will contribute to the overall goals of a project. The performance measures for evaluating the achievement of the objectives should be specified.

3. Approach. The managerial and technical methodologies of implementing the project should be specified. The managerial approach may relate to project organization, communication network, approval hierarchy, responsibility, and accountability. The technical approach may relate to company experience on previous projects and currently available technology.

4. Policies and procedures. Development of a project policy involves specifying the general guidelines for carrying out tasks within the project. Project procedure involves specifying the detailed method for implementing a given policy relative to the tasks needed to achieve the project goal.

5. Contractual requirements. This portion of the project plan should outline reporting requirements, communication links, customer specifications, performance specifications, deadlines, review process, project deliverables, delivery schedules, internal and external contacts, data security, policies, and procedures. This section should be as detailed as practically possible. Any item that has the slightest potential of creating problems later should be documented.

6. Project schedule. The project schedule signifies the commitment of resources against time in pursuit of project objectives. A project schedule should specify when

the project will be initiated and when it is expected to be completed. The major phases of the project should be identified. The schedule should include reliable time estimates for project tasks. The estimates may come from knowledgeable personnel, past records, or forecasting. Task milestones should be generated on the basis of objective analysis rather than arbitrary stipulations. The schedule in this planning stage constitutes the master project schedule. Detailed activity schedules should be generated under specific project functions.

7. Resource requirements. Project resources, budget, and costs are to be documented in this section of the project plan. Capital requirements should be specified by tasks. Resources may include personnel, equipment, and information. Special personnel skills, hiring, and training should be explained. Personnel requirements should be aligned with schedule requirements so as to ensure their availability when needed. Budget size and source should be presented. The basis for estimating budget requirements should be justified and the cost allocation and monitoring approach should be shown.

8. Performance measures. Measures of evaluating project progress should be developed. The measures may be based on standard practices or customized needs. The method of monitoring, collecting, and analyzing the measures should also be specified. Corrective actions for specific undesirable events should be outlined.

9. Contingency plans. Courses of actions to be taken in the case of undesirable events should be predetermined. Many projects have failed simply because no plans have been developed for emergency situations. In the excitement of getting a project under way, it is often easy to overlook the need for contingency plans.

10. Tracking, reporting, and auditing. These involve keeping track of the project plans, evaluating tasks, and scrutinizing the records of the project.

Planning for large projects may include a statement about the feasibility of subcontracting part of the project work. Subcontracting may be needed for various reasons including lower cost, higher efficiency, or logistical convenience.

2.4 MOTIVATING THE PERSONNEL

Motivation is an essential component of implementing project plans. National leaders, public employees, management staff, producers, and consumers may all need to be motivated about project plans that affect a wide spectrum of society. Those who will play active direct roles in the project must be motivated to ensure productive participation. Direct beneficiaries of the project must be motivated to make good use of the outputs of the project. Other groups must be motivated to play supporting roles to the project.

Motivation may take several forms. For projects that are of a short-term nature, motivation could be either impaired or enhanced by the strategy employed. Impairment may occur if a participant views the project as a mere disruption of regular activities or as a job without long-term benefits. Long-term projects have the advantage of giving participants enough time to readjust to the project efforts.

2.4.1 Motivation Concepts

Frederick Taylor (1911) stated that "management is knowing exactly what you want men to do, and seeing to it that it is done in the best and cheapest way." Koontz and O'Donnel (1959) defined management as the function of getting things done through people. McGregor (1960) stated that successful management depends significantly on the ability to predict and control human behavior. An effective manager should be interested in both results and the people he or she works with. Whatever definition is used, management ultimately involves some human elements with behavioral and motivational implications. In order to get a worker to work effectively, he or she must be motivated. Some workers are inherently self-motivating. There are other workers for whom motivation is an external force that must be managerially instilled. McGregor (1960) viewed worker performance under two basic concepts of theory X and theory Y.

The axiom of theory X.   Theory X assumes that the worker is essentially uninterested and unmotivated to perform his or her work. Motivation must be instilled into the worker by the adoption of external motivating agents. A theory X worker is inherently indolent and requires constant supervision and prodding to get him or her to perform. To motivate a theory X worker, a mixture of managerial actions may be needed. The actions must be used judiciously, based on the prevailing circumstances. Examples of motivation approaches under theory X are

- Rewards to recognize improved effort
- Strict rules to constrain worker behavior
- Incentives to encourage better performance
- Threats to job security associated with performance failure

The axiom of theory Y.   Theory Y assumes that the worker is naturally interested and motivated to perform his or her job. The worker views the job function positively and uses self-control and self-direction to pursue project goals. Under theory Y, management has the task of taking advantage of the worker's positive intuition so that his or her actions coincide with the objectives of the project. Thus, a theory Y manager attempts to use the worker's self-direction as the principal instrument for accomplishing work. In general, theory Y facilitates the following:

- Worker-designed job methodology
- Worker participation in decision making
- Cordial management-worker relationship
- Worker individualism within acceptable company limits

There are proponents of both theory X and theory Y and managers who operate under each or both can be found in any organization. The important thing to note is that whatever theory one subscribes to, the approach to worker motivation should be conducive to the achievement of the overall goal of the project.

Hierarchy of needs.   The needs of project participants must be taken into consideration in any project planning. Maslow (1954) presented what is usually known as Maslow's *hierarchy of needs*. He stresses that human needs are ordered in an hierarchical fashion consisting of five categories:

1. Physiological needs: The needs for the basic things of life, such as food, water, housing, and clothing. This is the level where access to money is most critical.
2. Safety needs: The needs for security, stability, and freedom from threat of physical harm. The fear of adverse environmental impact may inhibit project efforts.
3. Social needs: The needs for social approval, friends, love, affection, and association. For example, public service projects may bring about a better economic outlook that may enable individuals to be in a better position to meet their social needs.
4. Esteem needs: The needs for accomplishment, respect, recognition, attention, and appreciation. These needs are important not only at the individual level but also at the national level.
5. Self-actualization needs: These are the needs for self-fulfillment and self-improvement. They also involve the availability of opportunity to grow professionally. Work improvement projects may lead to self-actualization opportunities for individuals to assert themselves socially and economically. Job achievement and professional recognition are two of the most important factors that lead to employee satisfaction and better motivation.

Hierarchical motivation implies that the particular motivation technique utilized for a given person should depend on where the person stands in the hierarchy of needs. For example, the need for esteem takes precedence over physiological needs when the latter are relatively well satisfied. Money, for example, cannot be expected to be a very successful motivational factor for an individual who is already on the fourth level of the hierarchy of needs. The hierarchy of needs emphasizes the fact that things that are highly craved in youth tend to assume less importance later in life.

Hygiene factors and motivators.   Herzberg (1968) presents a motivation concept that takes a look at the characteristics of work itself. He claims that there are two motivational factors classified as the *hygiene factors* and *motivators*. He states that the hygiene factors are necessary but not sufficient conditions for a contented worker. The negative aspects of the factors may lead to a disgruntled worker, whereas their positive aspects do not necessarily enhance the satisfaction of the worker. Examples include

1. Administrative policies. Bad policies can lead to the discontent of workers while good policies are viewed as routine with no specific contribution to improving worker satisfaction.
2. Supervision. A bad supervisor can make a worker unhappy and less productive while a good supervisor cannot necessarily improve worker performance.
3. Working conditions. Bad working conditions can enrage workers, but good working conditions do not automatically generate improved productivity.

4. Salary. Low salaries can make a worker unhappy, disruptive, and uncooperative, but a raise will not necessarily provoke him to perform better. While a raise in salary will not necessarily increase professionalism, a reduction in salary will most certainly have an adverse effect on morale.
5. Personal life. Miserable personal life can adversely affect worker performance, but a happy life does not imply that he or she will be a better worker.
6. Interpersonal relationships. Good peer, superior, and subordinate relationships are important to keep a worker happy and productive, but extraordinarily good relations do not guarantee that he or she will be more productive.
7. Social and professional status. Low status can force a worker to perform at *his* or *her* level whereas high status does not imply performance at a higher level.
8. Security. A safe environment may not motivate a worker to perform better, but an unsafe condition will certainly impede productivity.

Motivators are motivating agents that should be inherent in the work itself. If necessary, work should be redesigned to include inherent motivating factors. Some guidelines for incorporating motivators into jobs are as follows:

1. Achievement: The job design should give consideration to opportunities for worker achievement and avenues to set personal goals to excel.
2. Recognition: The mechanism for recognizing superior performance should be incorporated into the job design. Opportunities for recognizing innovation should be built into the job.
3. Work content: The work content should be interesting enough to motivate and stimulate the creativity of the worker. The amount of work and the organization of the work should be designed to fit a worker's needs.
4. Responsibility: The worker should have some measure of responsibility for how his or her job is performed. Personal responsibility leads to accountability which invariably yields better work performance.
5. Professional growth: The work should offer an opportunity for advancement so that the worker can set his or her own achievement level for professional growth within a project plan.

The above examples may be described as job enrichment approaches with the basic philosophy that work can be made more interesting in order to induce an individual to perform better. Normally, work is regarded as an unpleasant necessity (a necessary evil). A proper design of work will encourage workers to become anxious to go to work to perform their jobs.

Management by objective. Management by objective (MBO) is the management concept whereby a worker is allowed to take responsibility for the design and performance of a task under controlled conditions. It gives workers a chance to set their own objectives in achieving project goals. Workers can monitor their own progress and take corrective actions when needed without management intervention. Workers under the concept of theory Y appear to be the best suited for the MBO concept. MBO has some disadvantages which include the possible abuse of the freedom to self-direct and possible disruption of overall project coordination. The advantages of MBO include

1. It encourages workers to find better ways of performing their jobs.
2. It avoids oversupervision of professionals.
3. It helps workers become better aware of what is expected of them.
4. It permits timely feedback on worker performance.

Management by exception.   Management by exception (MBE) is an after-the-fact management approach to control. Contingency plans are not made and there is no rigid monitoring. Deviations from expectations are viewed as exceptions to the normal course of events. When intolerable deviations from plans occur, they are investigated, and then an action is taken. The major advantage of MBE is that it lessens the management workload and reduces the cost of management. However, it is a dangerous concept to follow especially for high-risk technology-based projects. Many of the problems that can develop in complex projects are such that after-the-fact corrections are expensive or even impossible. As a result, MBE should be carefully evaluated before adopting it.

The above motivational concepts can be implemented successfully for specific large projects. They may be used as single approaches or in a combined strategy. The motivation approaches may be directed at individuals or groups of individuals, locally or at the national level.

2.5 PROJECT FEASIBILITY STUDY

The feasibility of a project can be ascertained in terms of technical factors, economic factors, or both. A feasibility study is documented with a report showing all the ramifications of the project.

Technical feasibility.   Technical feasibility refers to the ability of the process to take advantage of the current state of the technology in pursuing further improvement. The technical capability of the personnel as well as the capability of the available technology should be considered.

Managerial feasibility.   Managerial feasibility involves the capability of the infrastructure of a process to achieve and sustain process improvement. Management support, employee involvement, and commitment are key elements required to ascertain managerial feasibility.

Economic feasibility.   This involves the feasibility of the proposed project to generate economic benefits. A benefit-cost analysis and a break-even analysis are important aspects of evaluating the economic feasibility of new industrial projects. The tangible and intangible aspects of a project should be translated into economic terms to facilitate a consistent basis for evaluation.

Financial feasibility.   Financial feasibility should be distinguished from economic feasibility. Financial feasibility involves the capability of the project organization to raise the appropriate funds needed to implement the proposed project. Project financing can be a major obstacle in large multiparty projects because of the level of capital required. Loan availability, credit worthiness, equity, and loan schedule are important aspects of financial feasibility analysis.

Cultural feasibility. Cultural feasibility deals with the compatibility of the proposed project with the cultural setup of the project environment. In labor-intensive projects, planned functions must be integrated with local cultural practices and beliefs. For example, religious beliefs may influence what an individual is willing to do or not do.

Social feasibility. Social feasibility addresses the influences that a proposed project may have on the social system in the project environment. The ambient social structure may be such that certain categories of workers may be in short supply or nonexistent. The effect of the project on the social status of the project participants must be assessed to ensure compatibility. It should be recognized that workers in certain industries may have certain status symbols within the society.

Safety feasibility. Safety feasibility is another important aspect that should be considered in project planning. Safety feasibility refers to an analysis of whether the project is capable of being implemented and operated safely with minimal adverse effects on the environment. Unfortunately, environmental impact assessment is often not adequately addressed in complex projects. As an example, the North America Free Trade Agreement (NAFTA) among the United States, Canada, and Mexico was temporarily suspended early in 1993 because of the legal consideration of the potential environmental impacts of the projects to be undertaken under the agreement. The agreement finally passed later in the year and took effect January 1994.

Political feasibility. A politically feasible project may be referred to as a "politically correct project." Political considerations often dictate the direction for a proposed project. This is particularly true for large projects with national visibility that may have significant government inputs and political implications. For example, political necessity may be a source of support for a project regardless of the project's merits. On the other hand, worthy projects may face insurmountable opposition simply because of political factors. Political feasibility analysis requires an evaluation of the compatibility of project goals with the prevailing goals of the political system.

2.5.1 Scope of Feasibility Analysis

In general terms, the elements of a feasibility analysis for a project should cover the following items:

1. Need analysis. This indicates a recognition of a need for the project. The need may affect the organization itself, another organization, the public, or the government. A preliminary study is then conducted to confirm and evaluate the need. A proposal of how the need may be satisfied is then made. Pertinent questions that should be asked include

   • Is the need significant enough to justify the proposed project?

- Will the need still exist by the time the project is completed?
- What are the alternate means of satisfying the need?
- What are the economic, social, environmental, and political impacts of the need?

2. **Process work.** This is the preliminary analysis done to determine what will be required to satisfy the need. The work may be performed by a consultant who is an expert in the project field. The preliminary study often involves system models or prototypes. For technology-oriented projects, artist's conceptions and scaled-down models may be used for illustrating the general characteristics of a process. A simulation of the proposed system can be carried out to predict the outcome before the actual project starts.

3. **Engineering and design.** This involves a detailed technical study of the proposed project. Written quotations are obtained from suppliers and subcontractors as needed. Technological capabilities are evaluated as needed. Product design, if needed, should be done at this stage.

4. **Cost estimate.** This involves estimating project cost to an acceptable level of accuracy. Levels of around −5 percent to +15 percent are common at this stage of a project plan. Both the initial and operating costs are included in the cost estimation. Estimates of capital investment and recurring and nonrecurring costs should also be contained in the cost estimate document. Sensitivity analysis can be carried out on the estimated cost values to see how sensitive the project plan is to the estimated cost values.

5. **Financial analysis.** This involves an analysis of the cash flow profile of the project. The analysis should consider rates of return, inflation, sources of capital, payback periods, break-even point, residual values, and sensitivity. This is a critical analysis since it determines whether or not and when funds will be available to the project. The project cash flow profile helps support the economic and financial feasibility of the project.

6. **Project impacts.** This portion of the feasibility study provides an assessment of the impact of the proposed project. Environmental, social, cultural, political, and economic impacts may be some of the factors that will determine how a project is perceived by the public. The value-added potential of the project should also be assessed. A value-added tax may be assessed based on the price of a product and the cost of the raw material used in making the product. A tax so collected may be viewed as a contribution to government coffers.

7. **Conclusions and recommendations.** The feasibility study should end with the overall outcome of the project analysis. This may indicate an endorsement or disapproval of the project. Recommendations on what should be done should be included in this section of the feasibility study.

.6 PROJECT PROPOSALS

Once a project is shown to be feasible, the next step is to issue a *request for proposal (RFP)* depending on the funding sources involved. Proposals are classified as either "solicited" or "unsolicited." Solicited proposals are those written in response to a request for a proposal while unsolicited ones are those written without a formal invitation from the funding source. Many companies prepare proposals in response to inquiries received from potential clients. Many proposals are written under competitive bids. If an RFP is issued, it should include statements about project scope, funding level, performance criteria, and deadlines.

The purpose of the RFP is to identify companies that are qualified to successfully conduct the project in a cost-effective manner. Formal RFPs are sometimes issued to only a selected list of bidders who have been preliminarily evaluated as being qualified. These may be referred to as *targeted* RFPs. In some cases, general or open RFPs are issued and whoever is interested may bid for the project. This, however, has been found to be inefficient in many respects. Ambitious, but unqualified, organizations waste valuable time preparing losing proposals. The receiving agency, on the other hand, spends much time reviewing and rejecting worthless proposals. Open proposals do have proponents who praise their "equal opportunity" approach.

In industry, each organization has its own RFP format, content, and procedures. The request is called by different names including PI (procurement invitation), PR (procurement request), RFB (request for bid), or IFB (invitation for bids). In some countries, it is sometimes referred to as request for tender (RFT). Irrespective of the format used, an RFP should request information on bidder's costs, technical capability, management, and other characteristics. It should, in turn, furnish sufficient information on the expected work. A typical detailed RFP should include

1. Project background: Need, scope, preliminary studies and results.
2. Project deliverables and deadlines: What products are expected from the project, when the products are expected, and how the products will be delivered should be contained in this document.
3. Project performance specifications: Sometimes, it may be more advisable to specify system requirements rather than rigid specifications. This gives the system or project analysts the flexibility to utilize the most updated and most cost-effective technology in meeting the requirements. If rigid specifications are given, what is specified is what will be provided regardless of the cost and the level of efficiency.
4. Funding level: This is sometimes not specified because of nondisclosure policies or because of budget uncertainties. However, whenever possible, the funding level should be indicated in the RFP.
5. Reporting requirements: Project reviews, format, number and frequency of written reports, oral communication, financial disclosure, and other requirements should be specified.
6. Contract administration: Guidelines for data management, proprietary work, progress monitoring, proposal evaluation procedure, requirements for inventions, trade secrets, copyrights, and so on should be included in the RFP.
7. Special requirements (as applicable): Facility access restrictions, equal opportunity/affirmative actions, small business support, access facilities for the handicapped, false statement penalties, cost sharing, compliance with government regulations, and so on should be included if applicable.
8. Boilerplates (as applicable): These are special requirements that specify the specific ways certain project items are handled. Boilerplates are usually written based on organizational policy and are not normally subject to conditional changes. For

example, an organization may have a policy that requires that no more than 50 percent of a contract award will be paid prior to the completion of the contract. Boilerplates are quite common in government-related projects. Thus, large projects may need boilerplates dealing with environmental impacts, social contributions, and financial requirements.

2.6.1 Proposal Preparation

Whether responding to an RFP or preparing an unsolicited proposal, a proposing organization must take care to provide enough detail to permit an accurate assessment of a project proposal. The proposing organization will need to find out the following:

- Project time frame
- Level of competition
- Organization's available budget
- Organization of the agency
- Person to contact within the agency
- Previous contracts awarded by the agency
- Exact procedures used in awarding contracts
- Nature of the work done by the funding agency

The proposal should present the detailed plan for executing the proposed project. The proposal may be directed to a management team within the same organization or to an external organization. However, the same level of professional preparation should be practiced for both internal and external proposals. The proposal contents may be written in two parts: technical section and management section.

1. Technical Section of Project Proposal
    (a) Project background
        - Expertise in the project area
        - Project scope
        - Primary objectives
        - Secondary objectives
    (b) Technical approach
        - Required technology
        - Available technology
        - Problems and their resolutions
        - Work breakdown structure
    (c) Work statement
        - Task definitions and list
        - Expectations

      (d) Schedule
- Gantt charts
- Milestones
- Deadlines

      (e) Project deliverables

      (f) Value of the project
- Significance
- Benefit
- Impact

2. Management Section of Project Proposal

      (a) Project staff and experience
- Staff vita

      (b) Organization
- Task assignment
- Project manager, liaison, assistants, consultants, and so on

      (c) Cost analysis
- Personnel cost
- Equipment and materials
- Computing cost
- Travel
- Documentation preparation
- Cost sharing
- Facilities cost

      (d) Delivery dates
- Specified deliverables

      (e) Quality control measures
- Rework policy

      (f) Progress and performance monitoring
- Productivity measurement

      (g) Cost control measures

An executive summary or cover letter may accompany the proposal. The summary should briefly state the capability of the proposing organization in terms of previous experience on similar projects; unique qualification of the project personnel; advantages of the organization over other potential bidders; and reasons why the project should be awarded to the bidder.

2.6.2 Proposal Incentives

In some cases, it may be possible to include an incentive clause in a proposal in an attempt to entice the funding organization. An example is the use of cost sharing arrangements. Other frequently used project proposal incentives include bonus and penalty clauses, employment of minorities, public service, and contribution to charity. If incentives are allowed in project proposals, their nature should be critically reviewed. If not controlled, a project incentive arrangement may turn out to be an opportunity for an organization to buy itself into a project contract.

2.7 BUDGET PLANNING

After the planning for a project has been completed, the next step is the allocation of the resources required to implement the project plan. This is referred to as budgeting or capital rationing. Budgeting is the process of allocating scarce resources to the various endeavors of an organization. It involves the selection of a preferred subset of a set of acceptable projects due to overall budget constraints. Budget constraints may result from restrictions on capital expenditures, shortage of skilled personnel, shortage of materials, or mutually exclusive projects. The budgeting approach can be used to express the overall organizational policy. The budget serves many useful purposes including the following:

- Performance measure
- Incentive for efficiency
- Project selection criterion
- Expression of organizational policy
- Plan of resource expenditure
- Catalyst for productivity improvement
- Control basis for managers and administrators
- Standardization of operations within a given horizon

The preliminary effort in the preparation of a budget is the collection and proper organization of relevant data. The preparation of a budget for a project is more difficult than the preparation of budgets for regular and permanent organizational endeavors. Recurring endeavors usually generate historical data which serve as inputs to subsequent estimating functions. Projects, on the other hand, are often one-time undertakings without the benefits of prior data. The input data for the budgeting process may include inflationary trends, cost of capital, standard cost guides, past records, and forecast projections. Budget data collection may be accomplished by one of several available approaches including top-down budgeting and bottom-up budgeting.

2.7.1 Top-Down Budgeting

Top-down budgeting involves collecting data from upper-level sources such as top and middle managers. The cost estimates supplied by the managers may come from their judgments, past experiences, or past data on similar project activities. The cost estimates are passed to lower-level managers, who then break the estimates down into specific work components within the project. These estimates may, in turn, be given to line managers, supervisors, and so on to continue the process. At the end, individual activity costs are developed. The top management issues the global budget while the line worker generates specific activity budget requirements.

One advantage of the top-down budgeting approach is that individual work elements need not be identified prior to approving the overall project budget. Another advantage of the approach is that the aggregate or overall project budget can be reasonably accurate even though specific activity costs may contain substantial errors. There is, consequently, a keen competition among lower-level managers to get the biggest slice of the budget pie.

2.7.2 Bottom-Up Budgeting

Bottom-up budgeting is the reverse of top-down budgeting. In this method, elemental activities, their schedules, descriptions, and labor skill requirements are used to construct detailed budget requests. The line workers who are actually performing the activities are requested to furnish cost estimates. Estimates are made for each activity in terms of labor time, materials, and machine time. The estimates are then converted to dollar values. The dollar estimates are combined into composite budgets at each successive level up the budgeting hierarchy. If estimate discrepancies develop, they can be resolved through intervention to senior management, junior management, functional managers, project managers, accountants, or financial consultants. Analytical tools such as learning curve analysis, work sampling, and statistical estimation may be used in the budgeting process as appropriate to improve the quality of cost estimates.

All component costs and departmental budgets are combined into an overall budget and sent to top management for approval. A common problem with bottom-up budgeting is that individuals tend to overstate their needs with the notion that top management may cut the budget by some percentage. It should be noted, however, that sending erroneous and misleading estimates will only lead to a loss of credibility. Properly documented and justified budget requests are often spared the budget ax. Honesty and accuracy are invariably the best policies for budgeting.

2.7.3 Zero-Base Budgeting

Zero-base budgeting is a budgeting approach that bases the level of project funding on previous performance. It is normally applicable to recurring programs especially in the public sector. Accomplishments in past funding cycles are weighed against the level of resource expenditure. Programs that are stagnant in terms of their accomplishments relative to budget size do not receive additional budgets. Programs that have suffered decreasing yields are subjected to budget cuts or even elimination. On the other hand, programs that experience increments in accomplishments are rewarded with larger budgets.

A major problem with zero-base budgeting is that it puts participants under tremendous data collection, organization, and program justification pressures. Too much time may be spent documenting program accomplishments to the extent that productivity improvement on current projects may be sacrificed. For this reason, the approach has received only limited use in practice. However, proponents believe it is a good means of making managers and administrators more conscious of their management responsibilities. In a project management context, the zero-base budgeting approach may be used to eliminate specific activities that have not contributed to project goals in the past.

2.8 PROJECT BREAKDOWN STRUCTURE

Project breakdown structure (PBS) refers to the breakdown of a project for planning, scheduling, and control purposes. The breakdown is often referred to as a *work breakdown structure (WBS)*. This represents a family tree hierarchy of project operations required to accomplish project objectives. Tasks that are contained in the WBS collectively describe the overall project. The tasks may involve physical products (e.g., steam generators), services (e.g., testing), and data (e.g., reports, sales data). The WBS serves to describe the link between the end objective and the operations required to reach that objective. It shows work elements in the conceptual framework for planning and controlling. The objective of developing a WBS is to study the elemental components of a project in detail. It permits the implementation of the "divide and conquer" concepts. Overall project planning and control can be improved by using a WBS approach. A large project may be broken down into smaller subprojects which may, in turn, be systematically broken down into task groups.

Individual components in a WBS are referred to as WBS elements and the hierarchy of each is designated by a level identifier. Elements at the same level of subdivision are said to be of the same WBS level. Descending levels provide increasingly detailed definition of project tasks. The complexity of a project and the degree of control desired determine the number of levels in the WBS. An example of a WBS is shown in Figure 2–1.

Each WBS component is successively broken down into smaller details at lower levels. The process may continue until specific project activities are reached. The basic approach for preparing a WBS is as follows:

Level 1:   Level 1 contains only the final project purpose. This item should be identifiable directly as an organizational budget item.
Level 2:   Level 2 contains the major subsections of the project. These subsections are usually identified by their contiguous location or by their related purposes.
Level 3:   Level 3 contains definable components of the level 2 subsections.

Subsequent levels are constructed in more specific detail depending on the level of control desired. If a complete WBS becomes too crowded, separate WBSs may be drawn for the level 2 components. A *specification of work (SOW)* or WBS summary should normally accompany the WBS. A statement of work is a narrative of the work to be done. It should include the objectives of the work, its nature, resource requirements, and tentative schedule. Each WBS element is assigned a code that is used for its identification throughout the project life cycle. Alphanumeric codes may be used to indicate element level as well as component group.

.9 LEGAL CONSIDERATIONS IN PROJECT PLANNING

Today, managing a project is tougher and more complicated than in the past. The workforce is more volatile, technology is more dynamic, and society is less predictable. Governmental changes, institutional changes, and personnel changes are some of the factors that complicate project management with respect to legal requirements. The number of legal issues that arise is increasing at an alarming rate. The job of project management has, consequently, become more strenuous. Any prudent manager of today should give serious considerations to the legal implications of project operations. Many

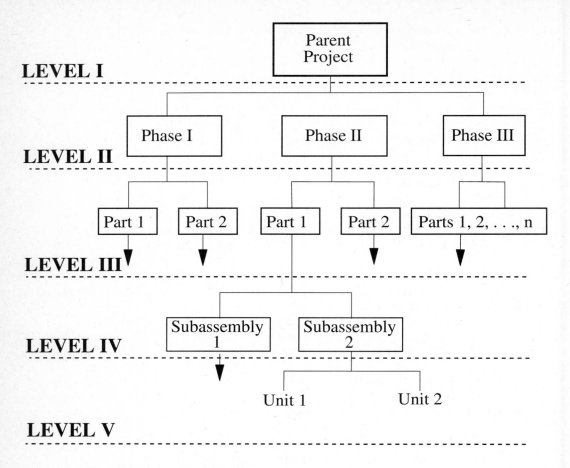

Figure 2-1. Project Breakdown Structure

organizations that have failed to recognize legal consequences have paid dearly for their mistakes.

There are several examples of project errors and legal problems. Some of the families of the astronauts killed in the space shuttle challenger sued NASA for millions of dollars. Some of the managerial staff involved in the Chernobyl accident in the Soviet Union lost their jobs and were legally convicted for dereliction of duty and criminal negligence, and several were sentenced to stiff jail terms. The Love Canal incident near Niagara Falls is still haunting residents and those responsible. The U.S. Justice Department filed a law suit against Hooker Chemicals & Plastics Company, the company responsible for dumping the Love Canal industrial waste. In 1980, New York State sued the same company for $635 million for negligence. The Three Mile Island accident of 28 March, 1979 in Pennsylvania caused the loss of $2 billion and an erosion of public confidence in the nuclear industry. The Bhopal disaster in India on 3 December 1984 left over 3,000 dead, some 250,000 disabled, and is still costing the Union Carbide Company a great number of legal problems. The oil spill in Alaska will have legal repercussions for many years to come.

With the emergence of new technology and complex systems (e.g., genetic engineering) it is only prudent to anticipate dangerous and unmanageable events. The key to preventing disasters is thoughtful planning and cautious preparation. Industrial projects are particularly prone to legal problems, the most pronounced of which are related to environmental damage. Industrial project planning should include a comprehensive evaluation of the potential legal aspects of the project.

2.10 INFORMATION FLOW FOR PROJECT PLANNING

Information flow is very crucial in project planning. Information is the driving force for project decisions. The value of information is measured in terms of the quality of the decisions that can be generated from the information. What appears to be valuable information to one user may be useless to another. Similarly, the timing of information can significantly affect its decision-making value. The same information that is useful in one instance may be useless in another. Some of the crucial factors affecting the value of information include accuracy, timeliness, relevance, reliability, validity, completeness, clearness, and comprehensibility. Proper information flow in project management ensures that tasks are accomplished when, where, and how they are needed. Figure 2–2 illustrates the flow of information for decision making in project management.

Information starts with raw data (e.g., numbers, facts, specifications). The data may pertain to raw material, technical skills, or other factors relevant to the project goal. The data is processed to generate information in the desired form. The information feedback model acts as a management control process that monitors project status and generates appropriate control actions. The contribution of the information to the project goal is fed back through an information feedback loop. The feedback information is used to enhance future processing of input data to generate additional information. The final information output provides the basis for improved management decisions. The key questions to ask when requesting, generating, or evaluating information for project management are

Figure 2-2.   Information Flow for Project Decision Making

- What data is needed to generate the information?
- Where is the data going to come from?
- When will the data be available?
- Is the data source reliable?
- Is there enough data?
- Who needs the information?
- When is the information needed?
- In what form is the information needed?
- Is the information relevant to project goals?
- Is the information accurate, clear, and timely?

As an example, the information flow model described above may be implemented to facilitate the inflow and outflow of information linking several functional areas of an organization, such as the design department, manufacturing department, marketing department, and customer relations department. The lack of communication among functional departments has been blamed for many of the organizational problems in industry. The use of a standard information flow model can help alleviate many of the communication problems. The information flow model can be expanded to take into account the uncertainties that may occur in the project environment.

2.10.1 Cost and Value of Information

Too much information is as bad as too little information. Too much information can impede the progress of a project. The marginal benefit of information decreases as its size increases. However, the marginal cost of obtaining additional information may increase as the size of the information increases. Figure 2–3 shows the potential behaviors of the value and cost curves with respect to the size of information.

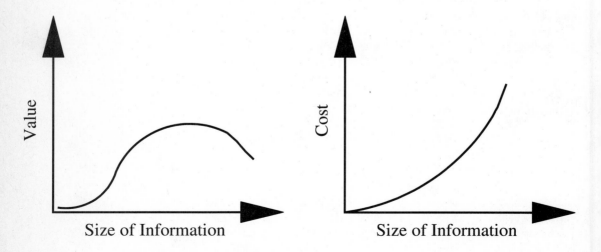

Figure 2-3. Value and Cost Curves for Project Information

The optimum size of information is determined by the point that represents the widest positive difference between the value of information and its cost. The costs associated with information can often be measured accurately. However, the benefits may be more difficult to document. The size of information may be measured in terms of a number of variables including number of pages of documentation, number of lines of code, and the size of the computer storage requirement. The amount of information presented for project management purposes should be condensed to cover only what is needed to make effective decisions. Information condensation may involve pruning the information that is already available or limiting what needs to be acquired.

The cost of information is composed of the cost of the resources required to generate the information. The required resources may include computer time, personnel hours, software, and so on. Unlike the value of information, which may be difficult to evaluate, the cost of information is somewhat more tractable. However, the development of accurate cost estimates prior to actual project execution is not trivial. The degree of accuracy required in estimating the cost of information depends on the intended uses of the estimates. Cost estimates may be used as general information for control purposes. Cost estimates may also be used as general guides for planning purposes or for developing standards. The bottom-up cost estimation approach is a popular method used in practice. This method proceeds by breaking the cost structure down into its component parts. The cost of each element is then established. The sum of these individual cost elements gives an estimate of the total cost of the information.

It is important to assess the value of project information relative to its cost before deciding to acquire it. Investments for information acquisition should be evaluated just like any other capital investment. The value of information is determined by how the information is used. In project management, information has two major uses. The first use relates to the need for information to run the daily operations of a project. Resource allocation, material procurement, replanning, rescheduling, hiring, and training are just a few of the daily functions for which information is needed. The second major use of information in project management relates to the need for information to make long-range project decisions. The value of information for such long-range decision making is even more difficult to estimate since the future cost of not having the information today is unknown.

The classical approach for determining the value of information is based on the value of perfect information. The expected value of perfect information is the maximum expected loss due to imperfect information. Using probability analysis or other appropriate quantitative methods, the project analyst can predict what a project outcome might be if certain information is available or not available. For example, if it is known for sure that it will rain on a certain day, a project manager might decide to alter the project schedule so that only nonweather sensitive tasks are planned on that particular day. The value of the perfect information about the weather would then be measured in terms of what loss could have been incurred if that information were not available. The loss may be in terms of lateness penalty, labor idle time, equipment damage, or ruined work.

An experienced project manager can accurately estimate the expected losses; and hence, the value of the perfect information about the weather. The cost of the same information may be estimated in terms of what it would cost to consult with a weather forecaster or the cost of buying a subscription to a special weather forecast channel on cable television. In Chapter 6, under statistical project control, we present a quantitative example of calculating the cost of perfect information.

2.11 THE TRIPLE C MODEL

The Triple C model presented by Badiru (1987) is an effective project planning tool. The model states that project management can be enhanced by implementing it within the integrated functions of

- Communication
- Cooperation
- Coordination

The model facilitates a systematic approach to project planning, organizing, scheduling, and control. The Triple C model is distinguished from the 3C approach commonly used in military operations. The military approach emphasizes personnel management in the hierarchy of command, control, and communication. This places communication as the last function. The Triple C, by contrast, suggests communication as the first and foremost function. The Triple C model can be implemented for project planning, scheduling, and control purposes. The model is shown graphically in Figure 2–4. It highlights what must be done and when. It can also help to identify the resources (personnel, equipment, facilities, etc.) required for each effort. If points out important questions such as

- Does each project participant know what the objective is?
- Does each participant know his or her role in achieving the objective?
- What obstacles may prevent a participant from playing his or her role effectively?

2.11.1 Communication

Communication makes working together possible. The communication function of project management involves making all those concerned become aware of project requirements and progress. Those who will be affected by the project directly or indirectly, as direct participants or as beneficiaries, should be informed as appropriate regarding the following:

- Scope of the project
- Personnel contribution required
- Expected cost and merits of the project
- Project organization and implementation plan
- Potential adverse effects if the project should fail
- Alternatives, if any, for achieving the project goal
- Potential direct and indirect benefits of the project

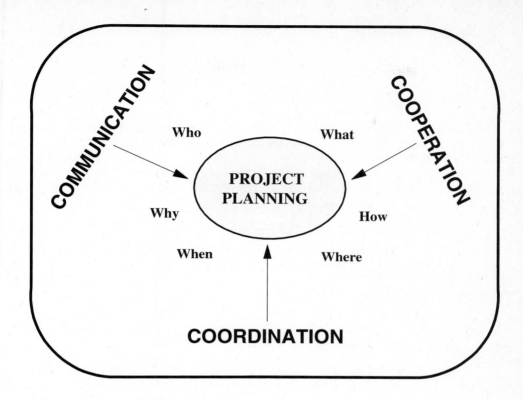

Figure 2-4.   Triple C Model for Project Planning

The communication channel must be kept open throughout the project life cycle. In addition to internal communication, appropriate external sources should also be consulted. The project manager must

- Exude commitment to the project
- Utilize the communication responsibility matrix
- Facilitate multichannel communication interfaces
- Identify internal and external communication needs
- Resolve organizational and communication hierarchies
- Encourage both formal and informal communication links

When clear communication is maintained between management and employees and among peers, many project problems can be averted. Project communication may be carried out in one or more of the following formats:

- One-to-many
- One-to-one
- Many-to-one

- Written and formal
- Written and informal
- Oral and formal
- Oral and informal
- Nonverbal gesture

Good communication is effected when what is implied is perceived as intended. Effective communications are vital to the success of any project. Despite the awareness that proper communications form the blueprint for project success, many organizations still fail in their communication functions. The study of communication is complex. Factors that influence the effectiveness of communication within a project organization structure include the following:

1. Personal perception. Each person perceives events on the basis of personal psychological, social, cultural, and experiential background. As a result, no two people can interpret a given event the same way. The nature of events is not always the critical aspect of a problem situation. Rather, the problem is often the different perceptions of the different people involved.

2. Psychological profile. The psychological makeup of each person determines personal reactions to events or words. Thus, individual needs and level of thinking will dictate how a message is interpreted.

3. Social environment. Communication problems sometimes arise because people have been conditioned by their prevailing social environment to interpret certain things in unique ways. Vocabulary, idioms, organizational status, social stereotypes, and economic situation are among the social factors that can thwart effective communication.

4. Cultural background. Cultural differences are among the most pervasive barriers to project communications, especially in today's multinational organizations. Language and cultural idiosyncrasies often determine how communication is approached and interpreted.

5. Semantic and syntactic factors. Semantic and syntactic barriers to communications usually occur in written documents. Semantic factors are those that relate to the intrinsic knowledge of the subject of the communication. Syntactic factors are those that relate to the form in which the communication is presented. The problems created by these factors become acute in situations where response, feedback, or reaction to the communication cannot be observed.

6. Organizational structure. Frequently, the organization structure in which a project is conducted has a direct influence on the flow of information and, consequently, on the effectiveness of communication. Organization hierarchy may determine how different personnel levels perceive a given communication.

7. Communication media. The method of transmitting a message may also affect the value ascribed to the message and, consequently, how it is interpreted or used. The common barriers to project communications are

- Inattentiveness
- Lack of organization
- Outstanding grudges
- Preconceived notions
- Ambiguous presentation
- Emotions and sentiments
- Lack of communication feedback.
- Sloppy and unprofessional presentation
- Lack of confidence in the communicator
- Lack of confidence by the communicator
- Low credibility of communicator
- Unnecessary technical jargon
- Too many people involved
- Untimely communication
- Arrogance or imposition
- Lack of focus

Some suggestions on improving the effectiveness of communication are presented next. The recommendations may be implemented as appropriate for any of the forms of communication listed earlier. The recommendations are for both the communicator and the audience.

1. Never assume that the integrity of the information sent will be preserved as the information passes through several communication channels. Information is generally filtered, condensed, or expanded by the receivers before relaying it to the next destination. When preparing a communication that needs to pass through several organization structures, one safeguard is to compose the original information in a concise form to minimize the need for recomposition.

2. Give the audience a central role in the discussion. A leading role can help make a person feel a part of the project effort and responsible for the project's success. He or she can then have a more constructive view of project communication.

3. Do homework and think through the intended accomplishment of the communication. This helps eliminate trivial and inconsequential communication efforts.

4. Carefully plan the organization of the ideas embodied in the communication. Use indexing or points of reference whenever possible. Grouping ideas into related chunks of information can be particularly effective. Present the short messages first. Short messages help create focus, maintain interest, and prepare the mind for the longer messages to follow.

5. Highlight why the communication is of interest and how it is intended to be used. Full attention should be given to the content of the message with regard to the prevailing project situation.

6. Elicit the support of those around you by integrating their ideas into the communication. The more people feel they have contributed to the issue, the more expeditious they are in soliciting the cooperation of others. The effect of the multiplicative rule can quickly garner support for the communication purpose.

7. Be responsive to the feelings of others. It takes two to communicate. Anticipate and appreciate the reactions of members of the audience. Recognize their operational circumstances and present your message in a form they can relate to.

8. Accept constructive criticism. Nobody is infallible. Use criticism as a springboard to higher communication performance.

9. Exhibit interest in the issue in order to arouse the interest of your audience. Avoid delivering your messages as a matter of a routine organizational requirement.

10. Obtain and furnish feedback promptly. Clarify vague points with examples.

11. Communicate at the appropriate time, at the right place, to the right people.

12. Reinforce words with positive action. Never promise what cannot be delivered. Value your credibility.

13. Maintain eye contact in oral communication and read the facial expressions of your audience to obtain real-time feedback.

14. Concentrate on listening as much as speaking. Evaluate both the implicit and explicit meanings of statements.

15. Document communication transactions for future references.

16. Avoid asking questions that can be answered yes or no. Use relevant questions to focus the attention of the audience. Use questions that make people reflect upon their words, such as, "How do you think this will work?" compared to "Do you think this will work?"

17. Avoid patronizing the audience. Respect their judgment and knowledge.

18. Speak and write in a controlled tempo. Avoid emotionally charged voice inflections.

19. Create an atmosphere for formal and informal exchanges of ideas.

20. Summarize the objectives of the communication and how they will be achieved.

Figure 2–5 shows an example of a design of a communication responsibility matrix. A communication responsibility matrix shows the linking of sources of communication and targets of communication. Cells within the matrix indicate the subject of the desired communication. There should be at least one filled cell in each row and each column of the matrix. This assures that each individual of a department has at least one communication source or target associated with him or her. With a communication responsibility matrix, a clear understanding of what needs to be communicated to whom can be developed.

Communication in a project environment can take any of several forms. The specific needs of a project may dictate the most appropriate mode. Three popular computer communication modes are discussed next in the context of communicating data and information for project management.

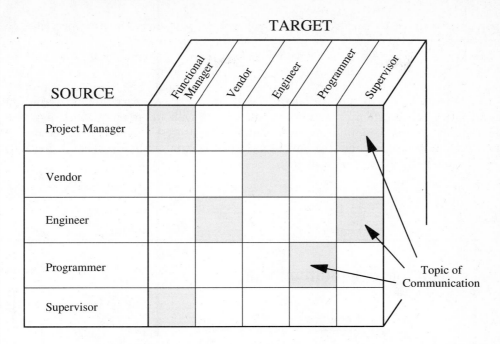

Figure 2-5. Triple C Communication Matrix

Simplex communication.   This is a unidirectional communication arrangement in which one project entity initiates communication to another entity or individual within the project environment. The entity addressed in the communication does not have a mechanism or capability for responding to the communication. An extreme example of this is a one-way, top-down communication from top management to the project personnel. In this case, the personnel have no communication access or input to top management. A budget-related example is a case where top management allocates budget to a project without requesting and reviewing the actual needs of the project. Simplex communication is common in authoritarian organizations.

Half-duplex communication.   This is a bidirectional communication arrangement whereby one project entity can communicate with another entity and receive a response within a certain time lag. Both entities can communicate with each other but not at the same time. An example of half-duplex communication is a project organization that permits communication with top management without a direct meeting. Each communicator must wait for a response from the target of the communication. Request and allocation without a budget meeting is another example of half-duplex data communication in project management.

Full-duplex communication.   This involves a communication arrangement that permits a dialogue between the communicating entities. Both individuals or entities can communicate with each other at the same time or face to face. As long as there is no clash of words, this appears to be the most receptive communication mode. It allows participative project planning in which each project personnel has an opportunity to contribute to the planning process.

Figure 2–6 presents a graphical representation of the communication modes discussed above. Each member of a project team needs to recognize the nature of the prevailing communication mode in the project. Management must evaluate the prevailing communication structure and attempt to modify it if necessary to enhance project functions. An evaluation of who is to communicate with whom about what may help improve the project data/information communication process. A communication matrix may include notations about the desired modes of communication between individuals and groups in the project environment.

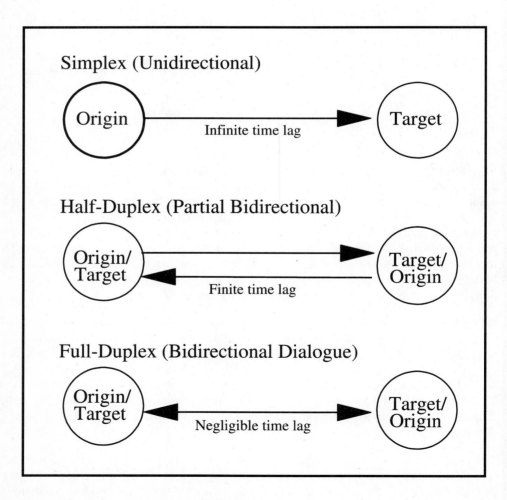

Figure 2-6.   Project Communication Modes

2.11.2 Cooperation

The cooperation of the project personnel must be explicitly elicited. Merely voicing a consent for a project is not enough assurance of full cooperation. The participants and beneficiaries of the project must be convinced of the merits of the project. Some of the factors that influence cooperation in a project environment include personnel requirements, resource requirements, budget limitations, past experiences, conflicting priorities, and lack of uniform organizational support. A structured approach to seeking cooperation should clarify the following:

- Cooperative efforts required
- Precedents for future projects
- Implication of lack of cooperation
- Criticality of cooperation to project success
- Organizational impact of cooperation
- Time frame involved in the project
- Rewards of good cooperation

Cooperation is a basic virtue of human interaction. More projects fail due to a lack of cooperation and commitment than any other project factors. To secure and retain the cooperation of project participants, you must elicit a positive first reaction to the project. The most positive aspects of a project should be the first items of project communication. For project management, there are different types of cooperation that should be understood.

Functional cooperation.   This is cooperation induced by the nature of the functional relationship between two groups. The two groups may be required to perform related functions that can only be accomplished through mutual cooperation.

Social cooperation.   This is the type of cooperation effected by the social relationship between two groups. The prevailing social relationship motivates cooperation that may be useful in getting project work done.

Legal cooperation.   Legal cooperation is the type of cooperation that is imposed through some authoritative requirement. In this case, the participants may have no choice other than to cooperate.

Administrative cooperation.   This is cooperation brought on by administrative requirements that make it imperative that two groups work together on a common goal.

Associative cooperation.   This type of cooperation may also be referred to as collegiality. The level of cooperation is determined by the association that exists between two groups.

Proximity cooperation.   Cooperation due to the fact that two groups are geographically close is referred to as proximity cooperation. Being close makes it imperative that the two groups work together.

Dependency cooperation. This is cooperation caused by the fact that one group depends on another group for some important aspect. Such dependency is usually of a mutual two-way nature. One group depends on the other for one thing while the latter group depends on the former for some other thing.

Imposed cooperation. In this type of cooperation, external agents must be employed to induced cooperation between two groups. This is applicable for cases where the two groups have no natural reason to cooperate. This is where the approaches presented earlier for seeking cooperation can become very useful.

Lateral cooperation. Lateral cooperation involves cooperation with peers and immediate associates. Lateral cooperation is often easy to achieve because existing lateral relationships create a conducive environment for project cooperation.

Vertical cooperation. Vertical or hierarchical cooperation refers to cooperation that is implied by the hierarchical structure of the project. For example, subordinates are expected to cooperate with their vertical superiors.

Whichever type of cooperation is available in a project environment, the cooperative forces should be channeled toward achieving project goals. Documentation of the prevailing level of cooperation is useful for winning further support for a project. Clarification of project priorities will facilitate personnel cooperation. Relative priorities of multiple projects should be specified so that a project that is of high priority to one segment of an organization is also of high priority to all groups within the organization. Some guidelines for securing cooperation for most projects are

- Establish achievable goals for the project.
- Clearly outline the individual commitments required.
- Integrate project priorities with existing priorities.
- Eliminate the fear of job loss due to industrialization.
- Anticipate and eliminate potential sources of conflict.
- Use an open-door policy to address project grievances.
- Remove skepticism by documenting the merits of the project.

Commitment. Cooperation must be supported with commitment. To cooperate is to support the ideas of a project. To commit is to willingly and actively participate in project efforts again and again through the thick and thin of the project. Provision of resources is one way that management can express commitment to a project.

Triple C + Commitment = Project Success

2.11.3 Coordination

After the communication and cooperation functions have successfully been initiated, the efforts of the project personnel must be coordinated. Coordination facilitates harmonious organization of project efforts. The construction of a responsibility chart can be very helpful at this stage. A responsibility chart is a matrix consisting of columns of individual or functional departments and rows of required actions. Cells within the matrix are filled with relationship codes that indicate who is responsible for what. Table 2–1 illustrates an example of a responsibility matrix for the planning for a seminar program. The matrix helps avoid neglecting crucial communication requirements and obligations. It can help resolve questions such as

- Who is to do what?
- How long will it take?
- Who is to inform whom of what?
- Whose approval is needed for what?
- Who is responsible for which results?
- What personnel interfaces are required?
- What support is needed from whom and when?

Table 2-1   Example of Project Responsibility Matrix

	Person responsible							Status of task					
TASKS	PULAT	FOOTE	BADIRU	DIRECTOR	LISA	JEAN	JANE	JAN 31	FEB 15	FEB 28	MAR 08	MAR 15	MAR 21
1. Brainstorming meeting	R	R	R	R				D					
2. Identify speakers				R					O				
3. Select seminar location	I	R	R						O	O			
4. Select banquet location	R	R								D			
5. Prepare publicity materials			C	R	I	I	R						
6. Draft brochures			C	R			R						
7. Develop schedule				R									
8. Arrange for visual aids				R									
9. Coordinate activities				R									
10. Periodic review of tasks	R	R	R	S	I								
11. Monitor progress of program	C	R	R										
12. Review program progress	R												
13. Closing arrangements	R												
14. Post-program review and evaluation	R	R	R	R									

Responsibility Code:
R = Responsible
I = Inform
S = Support
C = Consult

Task Code:
D = Done
O = On Track
B = Delayed

2.11.4 Resolving Project Conflicts with Triple C

When implemented as an integrated process, the Triple C model can help avoid conflicts in a project. When conflicts do develop, it can help in resolving the conflicts. Several sources of conflicts can exist in large projects. Some of these are discussed next.

Schedule conflict. Conflicts can develop because of improper timing or sequencing of project tasks. This is particularly common in large multiple projects. Procrastination can lead to having too much to do at once, thereby creating a clash of project functions and discord among project team members. Inaccurate estimates of time requirements may lead to infeasible activity schedules. Project coordination can help avoid schedule conflicts.

Cost conflict. Project cost may not be generally acceptable to the clients of a project. This will lead to project conflict. Even if the initial cost of the project is acceptable, a lack of cost control during project implementation can lead to conflicts. Poor budget allocation approaches and the lack of a financial feasibility study will cause cost conflicts later on in a project. Communication and coordination can help prevent most of the adverse effects of cost conflicts.

Performance conflict. If clear performance requirements are not established, performance conflicts will develop. Lack of clearly defined performance standards can lead each person to evaluate his or her own performance based on personal value judgments. In order to uniformly evaluate quality of work and monitor project progress, performance standards should be established by using the Triple C approach.

Management conflict. There must be a two-way alliance between management and the project team. The views of management should be understood by the team. The views of the team should be appreciated by management. If this does not happen, management conflicts will develop. A lack of a two-way interaction can lead to strikes and industrial actions which can be detrimental to project objectives. The Triple C approach can help create a conducive dialogue environment between management and the project team.

Technical conflict. If the technical basis of a project is not sound, technical conflicts will develop. New industrial projects are particularly prone to technical conflicts because of their significant dependence on technology. Lack of a comprehensive technical feasibility study will lead to technical conflicts. Performance requirements and systems specifications can be integrated through the Triple C approach to avoid technical conflicts.

Priority conflict. Priority conflicts can develop if project objectives are not defined properly and applied uniformly across a project. Lack of a direct project definition can lead each project member to define his or her own goals which may be in conflict with the intended goal of a project. Lack of consistency of the project mission is another potential source of priority conflicts. Overassignment of responsibilities with no guidelines for relative significance levels can also lead to priority conflicts. Communication can help defuse priority conflicts.

Resource conflict. Resource allocation problems are a major source of conflict in project management. Competition for resources, including personnel, tools, hardware, software, and so on, can lead to disruptive clashes among project members. The Triple C approach can help secure resource cooperation.

Power conflict. Project politics lead to a power play which can adversely affect the progress of a project. Project authority and project power should be clearly delineated. Project authority is the control that a person has by virtue of his or her functional post. Project power relates to the clout and influence which a person can exercise due to connections within the administrative structure. People with popular personalities can often wield a lot of project power in spite of low or nonexistent project authority. The Triple C model can facilitate a positive marriage of project authority and power to the benefit of project goals. This will help define clear leadership for a project.

Personality conflict. Personality conflict is a common problem in projects involving a large group of people. The larger a project, the larger the size of the management team needed to keep things running. Unfortunately, the larger management team creates an opportunity for personality conflicts. Communication and cooperation can help defuse personality conflicts.

In summary, conflict resolution through Triple C can be achieved by observing the following guidelines:

1. Confront the conflict and identify the underlying causes.
2. Be cooperative and receptive to negotiation as a mechanism for resolving conflicts.
3. Distinguish between proactive, inactive, and reactive behaviors in a conflict situation.
4. Use communication to defuse internal strife and competition.
5. Recognize that short-term compromise can lead to long-term gains.
6. Use coordination to work toward a unified goal.
7. Use communication and cooperation to turn a competitor into a collaborator.

2.11.5 The Abilene Paradox

A classic example of conflict in project planning is illustrated by the *Abilene Paradox* (Harvey 1974). The text of the paradox, as presented by Harvey, is summarized below.[1]

It was a July afternoon in Coleman, a tiny Texas town. It was a hot afternoon. The wind was blowing fine-grained West Texas topsoil through the house. Despite the harsh weather, the afternoon was still tolerable and potentially enjoyable. There was a fan blowing on the back porch; there was cold lemonade; and finally, there was entertainment: dominoes. Perfect for the conditions. The game required little more physical exertion than an occasional mumbled comment, "Shuffle 'em," and an unhurried movement of the arm to

[1] *Reprinted, by permission of publisher, from ORGANIZATIONAL DYNAMICS, Summer/1974 ©1974. American Management Association, New York. All rights reserved.*

place the spots in the appropriate position on the table. All in all, it had the makings of an agreeable Sunday afternoon in Coleman until Jerry's father-in-law suddenly said, "Let's get in the car and go to Abilene and have dinner at the cafeteria."

Jerry thought, "What, go to Abilene? Fifty-three miles? In this dust storm and heat? And in an un-airconditioned 1958 Buick?" But Jerry's wife chimed in with, "Sounds like a great idea. I'd like to go. How about you, Jerry?" Since Jerry's own preferences were obviously out of step with the rest, he replied, "Sounds good to me," and added, "I just hope your mother wants to go."

"Of course I want to go," said Jerry's mother-in-law. "I haven't been to Abilene in a long time." So into the car and off to Abilene they went. Jerry's predictions were fulfilled. The heat was brutal. The group was coated with a fine layer of dust that was cemented with perspiration by the time they arrived. The food at the cafeteria provided first-rate testimonial material for antacid commercials.

Some four hours and 106 miles later, they returned to Coleman, hot and exhausted. They sat in front of the fan for a long time in silence. Then, both to be sociable and to break the silence, Jerry said, "It was a great trip, wasn't it?" No one spoke. Finally, his father-in-law said, with some irritation, "Well, to tell the truth, I really didn't enjoy it much and would rather have stayed here. I just went along because the three of you were so enthusiastic about going. I wouldn't have gone if you all hadn't pressured me into it."

Jerry couldn't believe what he just heard. "What do you mean 'you all'?" he said. "Don't put me in the 'you all' group. I was delighted to be doing what we were doing. I didn't want to go. I only went to satisfy the rest of you. You're the culprits." Jerry's wife looked shocked. "Don't call me a culprit. You and Daddy and Mama were the ones who wanted to go. I just went along to be sociable and to keep you happy. I would have had to be crazy to want to go out in heat like that."

Her father entered the conversation abruptly. "Hell!" he said. He proceeded to expand on what was already absolutely clear. "Listen, I never wanted to go to Abilene. I just thought you might be bored. You visit so seldom I wanted to be sure you enjoyed it. I would have preferred to play another game of dominoes and eat the leftovers in the icebox."

After the outburst of recrimination, they all sat back in silence. There they were, four reasonably sensible people who, of their own volition, had just taken a 106-mile trip across a godforsaken desert in a furnace-like temperature through a cloud-like dust storm to eat unpalatable food at a hole-in-the-wall cafeteria in Abilene, when none of them had really wanted to go. In fact, to be more accurate, they'd done just the opposite of what they wanted to do. The whole situation simply didn't make sense. It was a paradox of agreement.

This example illustrates a problem that can be found in many organizations or project environments. Organizations often take actions that totally contradict their stated goals and objectives. They do the opposite of what they really want to do. For most organizations, the adverse effects of such diversion, measured in terms of human distress and economic loss, can be immense. A family group that experiences the Abilene paradox would soon get over the distress, but for an organization engaged in a competitive market, the distress may last a very long time. Six specific symptoms of the paradox are identified by Harvey (1974).

1. Organization members agree privately, as individuals, as to the nature of the situation or problem facing the organization.

2. Organization members agree privately, as individuals, as to the steps that would be required to cope with the situation or solve the problem they face.

3. Organization members fail to accurately communicate their desires and/or beliefs to one another. In fact, they do just the opposite and, thereby, lead one another into misinterpreting the intentions of others. They misperceive the collective reality. Members often communicate inaccurate data (e.g., "Yes, I agree"; "I see no problem with that"; "I support it") to other members of the organization. No one wants to be the lone dissenting voice in the group.

4. With such invalid and inaccurate information, organization members make collective decisions that lead them to take actions contrary to what they want to do and, thereby, arrive at results that are counterproductive to the organization's intent and purposes. For example, the Abilene group went to Abilene when it preferred to do something else.

5. As a result of taking actions that are counterproductive, organization members experience frustration, anger, irritation, and dissatisfaction with their organization. They form subgroups with supposedly trusted individuals and blame other subgroups for the organization's problems.

6. The cycle of the Abilene paradox repeats itself with increasing intensity if the organization members do not learn to manage their agreement.

Readers interested in further exposition of the Abilene paradox and similar examples should refer to Harvey (1974). The authors have witnessed many project situations where, in private conversations, individuals express their discontent about a project and yet fail to repeat their statements in a group setting. Consequently, other members are never aware of the dissenting opinions. In large organizations, the Triple C model can help in managing communication, cooperation, and coordination functions to avoid the Abilene paradox. The lessons to be learned from proper approaches to project planning can help avoid unwilling trips to Abilene.

12 EXERCISES

2.1. Develop a definition for the Triple C model of project management.

2.2. List and discuss five major functions in project planning.

2.3. Describe the role of the project manager in project planning.

2.4. Develop a project planning model for the construction of a new manufacturing facility. The facility will produce products to be sold in the international market.

2.5. List some of the factors that can impede the flow of information for project planning purposes. How can these factors be controlled?

2.6. Discuss the impact of hygiene factors and motivators in project planning.

2.7. Discuss how the concepts of MBO and MBE can be integrated for project planning purposes.

2.8. Discuss some of the criteria that can be used to assess the technical feasibility of a project.

2.9. Prepare the technical section of a project proposal dealing with environmental impact assessment for a new industry. Select a specific industry type for your response.

2.10. Prepare the management section of a project proposal dealing with environmental impact assessment for a new industry. Use the same industry type selected in Exercise 2–9.

3

Project Organization

"We trained hard, but it seemed that every time we were beginning to form into teams we would be reorganized. I was to learn later in life that we tend to meet any new situation by reorganizing; and what a wonderful method it can be for creating the illusion of progress while producing confusion, inefficiency and demoralization."—Petronius Arbiter, 210 B.C.

This chapter presents alternate organizational structures for projects. Structures presented include a matrix structure, functional structure, and product structure. The complexity of organizing multinational projects is addressed. A checklist of the factors to consider when organizing international projects is presented. This chapter also introduces new forms of organization structures for project management. It is recommended that a project be implemented in a hierarchical fashion such that each stage of the project builds on the one before it. The preceding chapter addressed project planning. This chapter addresses project organization. The transition of the project effort, so far, is represented as shown below. We will continue to build on this transition incrementally throughout this book.

SET GOALS → PLAN → ORGANIZE

3.1 ORGANIZATIONAL BREAKDOWN STRUCTURE

There are three major categories of breakdown structures. The *project breakdown structure (PBS)* is a project-specific structure that refers to the breakdown of the tasks to be performed in a specific project. This is often referred to as the work breakdown structure (WBS). The *organizational structure (OS)* refers to the organization of the environment within which a project is to be carried out. The *organizational breakdown structure (OBS)* refers to the identification and organization of the resources required to carry out the activities associated with a project. OBS is a model that provides a way of organizing resources into groups for better management within an organization, which may be a company, a project, or a division of a large enterprise. OBS can be used to keep track of resource allocation and specific work assignments. There is a strong interdependency between OBS and WBS. A good breakdown of work helps in estimating resource requirements more accurately. Good team organization is essential for the success of OBS. For example, a mixed team comprised of individuals with different technical backgrounds is required for technology-based projects. OBS team requirements are

- Background of the team leader: experience, education, technical knowledge.
- Scope of work to be accomplished: hours, skills, locations.
- Estimate of the number and type of personnel required.
- Equipment and work space requirements.
- Reporting procedures.

The focus of this chapter is alternate organizational structures. Further details on resource allocation, organization, and management are presented in Chapter 5.

.2 SELECTING AN ORGANIZATION STRUCTURE

After project planning, the next step is the selection of the organizational structure for the project. This involves the selection of an organizational structure that shows the management line and responsibilities of the project personnel. Any of several approaches may be utilized. Before selecting an organizational structure, the project team should assess the nature of the job to be performed and its requirements. The structure may be defined in terms of functional specializations, departmental proximity, standard management boundaries, operational relationships, or product requirements.

Large and complex projects should be based on well-designed structures that permit effective information and decision processes. The primary function of an organizational design is to facilitate effective information flow. Traditional organization models consist of the decision-making, bureaucracy, social, and systems structures. The decision-making structure handles the policies and general directions of the overall organization. The bureaucracy is concerned with the administrative processes. Some of the administrative functions may not be directly relevant to the main goal of the organization, but they are, nonetheless, deemed necessary. Bureaucratic processes are potential sources of delays for public projects because of government involvement. The social structure facilitates amiable interactions among the personnel. Such interpersonal relationships are essential for the group effort needed to achieve company objectives. The systems structure can better be described as the link among the various synergistic segments of the organization.

Many organizations still use the traditional or classical organization structure. A traditional organization is often utilized in service-oriented companies. The structure is sometimes referred to as the pure functional structure because groups with similar functional responsibilities are clustered at the same level of the structure. Figure 3–1

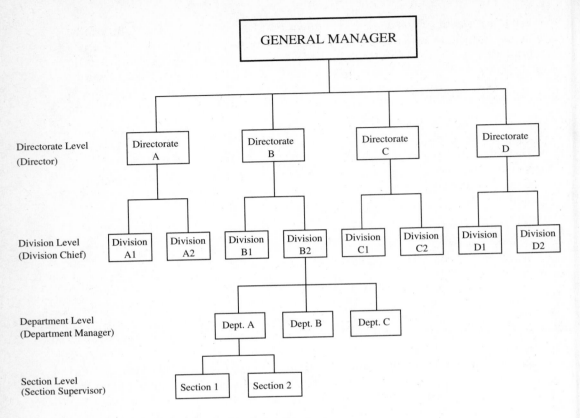

Figure 3-1. Traditional Organization Structure

illustrates the traditional organization structure. The positive characteristics of the structure are

- Availability of broad personnel base
- Identifiable technical line of control
- Grouping of specialists to share technical knowledge
- Collective line of responsibility
- Possible assignment of personnel to several different projects
- Clear hierarchy for supervision
- Continuity and consistency of functional disciplines
- Possibility for departmental policies, procedures, and missions

However, the traditional structure does have some negative characteristics.

- No one individual is directly responsible for the total project.
- Project-oriented planning may be impeded.
- There may not be clear line of reporting up from the lower levels.

- Coordination is complex.
- A higher level of cooperation is required between adjacent levels.
- The strongest functional group may wrongfully claim project authority.

3 FORMAL AND INFORMAL STRUCTURES

The formal organization structure represents the officially sanctioned structure of a functional area. The informal organization, on the other hand, develops when people organize themselves in an unofficial way to accomplish an objective that is in line with the overall project goals. The informal organization is often very subtle in that not everyone in the organization is aware of its existence. Both formal and informal organizations are practiced in every project environment. Even organizations with strict hierarchical structures, such as the military, still have some elements of informal organization.

3.3.1 Span of Control

The functional organization calls attention to the form and span of management that are suitable for company goals. The span of management (also known as the span of control) can be wide or narrow. In a narrow span, the functional relationships are streamlined with fewer subordinate units reporting to a single manager. Wide and narrow spans of control are illustrated in Figure 3–2. The wide span of management permits several subordinate units to report to the same boss. The span of control required for a project is influenced by a combination of the following:

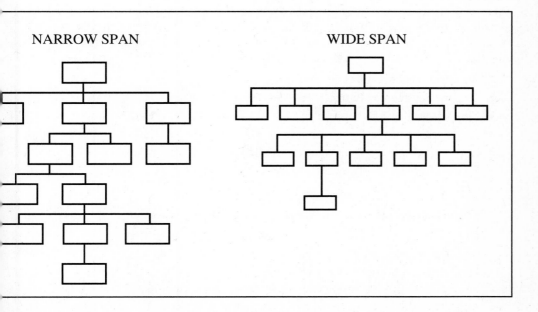

Figure 3-2.   Wide and Narrow Spans of Control

- Level of planning required
- Level of communication desired
- Effectiveness of delegating authority
- Dynamism and nature of the subordinate's job
- Competence of the subordinate in performing his or her job

Given a conducive project environment, the wide span of management can be very effective. From a motivational point of view, workers tend to have a better identification with upper management since there are fewer hierarchical steps to go through to reach the top. More professional growth is possible because workers assume more responsibilities. In addition, the wide span of management is more economical because of the absence of extra layers of supervision. However, the narrow span of management does have its own appeal in situations where there are several mutually exclusive skill levels in the organization.

3.4 FUNCTIONAL ORGANIZATION

The most common type of formal organization is known as the functional organization, whereby people are organized into groups dedicated to particular functions. Depending on the size and the type of auxiliary activities involved in a project, several minor, but supporting, functional units can be developed for a project.

Projects that are organized along functional lines are normally resident in a specific department or area of specialization. The project home office or headquarters is located in the specific functional department. For example, projects that involve manufacturing operations may be under the control of the vice-president of manufacturing while a project involving new technology may be assigned to the vice-president for advanced systems. Figure 3–3 shows examples of projects that are functionally organized. The advantages of a functional organization structure are

- Improved accountability
- Discernible line of control
- Flexibility in manpower utilization
- Enhanced comradeship of technical staff
- Improved productivity of specially skilled personnel
- Potential for staff advancement along functional path
- Use of home office can serve as a refuge for project problems

The disadvantages of a functional organization structure are

- Divided attention between project goals and regular functions
- Conflict between project objectives and regular functions
- Poor coordination of similar project responsibilities
- Unreceptive attitude by the surrogate department
- Multiple layers of management
- Lack of concentrated effort

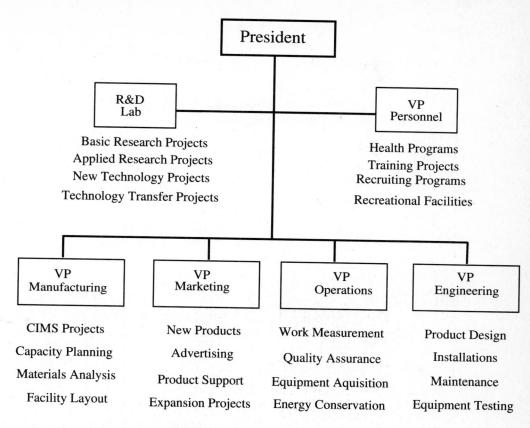

Figure 3-3.   Functional Organization Structure

It is difficult to separate the project environment from the traditional functional environment. There must be an integration. The project management approach affects the functional management approach and vice versa. The questions in Table 3–1 illustrate a comparison of the views of the functional and project environments. Since tasks are the basic components of a project and tasks are the major focus of functional endeavors, they form the basis for the integration of the project and functional environments.

Table 3-1   Functional versus Project Views

Project concerns	Functional concerns
What is the project to be done?	How will the task be done?
When will the project be done?	Where will the task be done?
Why will the project be done?	Who will do the task?
What resources are available?	How do functional inputs affect the project?
What is the project status?	How does the project affect the organization?

3.5 PRODUCT ORGANIZATION

Another approach to organizing a project is to use the end product or goal of the project as the determining factor for personnel structure. This is often referred to as the pure project organization or, simply, project organization. The project is set up as a unique entity within the parent organization. It has its own dedicated technical staff and administration. It is linked to the rest of the system through progress reports, organizational policies, procedures, and funding. The interface between product-organized projects and other elements of the organization may be strict or liberal depending on the organization. An example of a pure project organization is shown in Figure 3–4. The project staff is assembled by assigning personnel from different functional areas.

The product organization is common in large project-oriented organizations or organizations that have multiple product lines. Unlike the functional structure, the product organization decentralizes functions. It creates a unit consisting of specialized skills around a given project or product. Sometimes referred to as team, task force, or product group, the product organization is common in public, research, and manufacturing organizations where specially organized and designated groups are assigned specific functions. A major advantage of the product organization is that it gives the project members a feeling of dedication to and identification with a particular goal.

A possible shortcoming of the product organization is the requirement that the product group be sufficiently funded to be able to stand alone without sharing resources or personnel with other functional groups or programs. The product group may be viewed as an ad hoc unit that is formed for the purpose of a specific goal. The

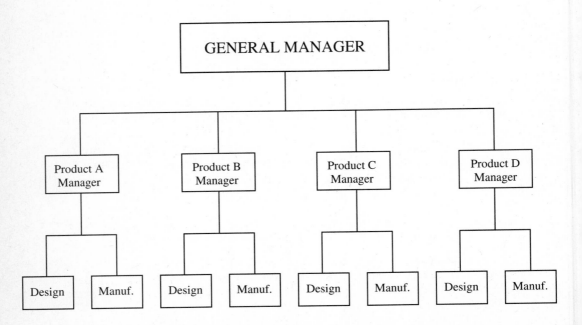

Figure 3-4. Product Organization Structure

personnel involved in the project are dedicated to the particular mission at hand. At the conclusion of the mission, they may be reassigned to other projects. The product organization can facilitate the most diverse and flexible grouping of project participants and permits highly dedicated attention to the project at hand. The advantages of the product organization structure are

- Simplicity of structure
- Unity of project purpose
- Localization of project failures
- Condensed and focused communication lines
- Full authority in the project manager
- Quicker decisions due to centralized authority
- Skill development due to project specialization
- Improved motivation, commitment, and concentration
- Flexibility in determining time, cost, performance trade-offs
- Accountability of project team to one boss (project manager)
- Individual acquisition and maintenance of expertise on a given project

The disadvantages are

- Narrow view of project personnel (as opposed to global organization view)
- Mutually exclusive allocation of resources (one man to one project)
- Duplication of efforts on different but similar projects
- Monopoly of organizational resources
- Concern about life after the project
- Reduced skill diversification

.6 MATRIX ORGANIZATION STRUCTURE

The matrix organization is a popular choice of management professionals. A matrix organization exists where there are multiple managerial accountability and responsibility for a job function. It attempts to combine the advantages of the traditional structure and the product organization structure. In pure product organization, technology utilization and resource sharing are limited because there is no single group responsible for overall project planning. In the traditional organization structure, time and schedule efficiency are sacrificed. Matrix organization can be defined as follows:

> *Matrix organization is a structure of management that facilitates maximum resource utilization and increased performance within time, cost, and performance constraints.*

There are usually two chains of command: horizontal and vertical. The horizontal line deals with the functional line of responsibility while the vertical line deals with the project line of responsibility. The project manager has total responsibility and accountability for project success. The functional managers have the responsibility to achieve and maintain high technical performance of the project. An example of a project organized under the matrix model is given in Figure 3–5.

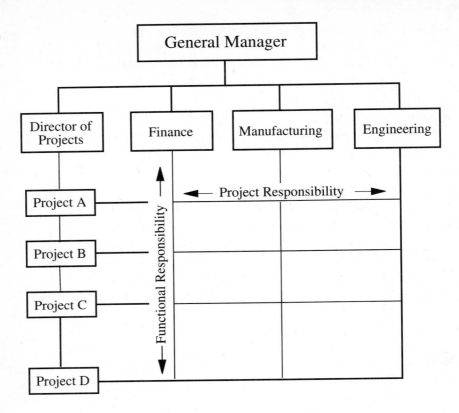

Figure 3-5.   Matrix Organization Structure

The project that is organized under a matrix structure may relate to specific problems, marketing issues, product quality improvement, and so on. The project line in the matrix is usually of a temporary nature while the functional line is more permanent. The matrix organization is dynamic. Its actual structure is determined by the prevailing activities of the project.

The matrix organization has the following advantages:

- Good team interaction
- Consolidation of objectives
- Multilateral flow of information
- Lateral mobility for job advancement
- Opportunity to work on a variety of projects
- Efficient sharing and utilization of resources
- Reduced project cost due to sharing of personnel
- Continuity of functions after project completion
- Stimulating interactions with other functional teams

- Cooperation of functional lines to support project efforts
- Home office for personnel after project completion
- Equal availability of company knowledge base to all projects

The disadvantages of matrix organization are

- Slow matrix response time for fast-paced projects
- Independent operation of each project organization
- High overhead cost due to additional lines of command
- Potential conflict of project priorities
- Problems of multiple bosses
- Complexity of the structure

Despite its disadvantages, the matrix organization is widely used in practice. Its numerous advantages seem to outweigh the disadvantages. In addition, the problems of the matrix organization structure can be overcome with a good project planning, which can set the tone for a smooth organization structure. Stuckenbruck (1979) presents more details on the matrix organization structure.

Matrix organization is a collaborative effort between product and functional organization structures. It permits both vertical and horizontal flows of information. The matrix model is sometimes called a multiple-boss organization. It is a model that is becoming increasingly popular as the need for information sharing increases. For example, large technical projects require the integration of specialties from different functional areas. Under matrix organization, projects are permitted to share critical resources as well as management expertise. Several project situations are suitable for implementing a matrix organization.

1. When the primary outputs of an organization are numerous, complex, and resource-critical.
2. When a complicated design calls for innovation and widespread expertise.
3. When expensive, sophisticated, and scarce technologies are needed in designing, building, and testing products.
4. When emergency response and flexibility are required for a project.

Traditionally, industrial projects are conducted in serial functional implementations such as R&D, engineering, manufacturing, and marketing. At each stage, unique specifications and work patterns may be used without consulting the preceding and succeeding phases. The consequence is that the end product may not possess the original intended characteristics. For example, the first project in the series might involve the production of one component while the subsequent projects might involve the production of other components. The composite product may not achieve the desired performance because the components were not designed and produced from a unified point of view. In today's interdependent market-oriented projects, such lack of a unified design will lead to overall project failure.

The major appeal of matrix organization is that it attempts to provide synergy within groups in an organization. This synergy can be realized if certain ground rules are observed when implementing a matrix organization. Some of those rules are

1. There must be an individual who is devoted full time to the project.
2. There must be both horizontal and vertical channels for communication.
3. There must be a quick access and conflict resolution strategy between managers.
4. All managers must have an input into the planning process.
5. Functional and project managers must be willing to negotiate and commit resources.

3.7 MIXED ORGANIZATION STRUCTURE

Another approach for organizing a project is to adopt a combined implementation of the functional, product, and matrix structures. This permits the different structures to coexist simultaneously in the same project. In an industrial project, for example, the project of designing a new product may be organized using a matrix structure while the subproject of designing the production line may be organized along functional lines. The mixed model facilitates flexibility in meeting special problem situations. The structure can adapt to the prevailing needs of the project or the needs of the overall organization. However, a disadvantage is the difficulty in identifying the lines of responsibility within a given project. There is a wide array of mixed organizations based on the matrix and other structures. These range from a single product/project manager with strong dependence on the functional organization, to a large product/project organization with little dependence on the functional organization. The functional personnel may be located in the project management office or in a separate geographical location. They may be fully dedicated to a single project manager or they may serve many project managers. In the next section, we present new ideas for unique organization structures that cater to specific project situations.

3.8 NEW ORGANIZATION STRUCTURES

New organization structures are needed to address unique organizational or project needs. Some of the structures presented below will, no doubt, be of a temporary nature.

3.8.1 Bubble Organization

The bubble organization, which can also be called the *blob organization*, is a structure that allows functional teams to rally around a central project goal. This may be suitable for grass-roots movement among society groups canvassing for a national need. The bubble structure will, most often, be temporary in nature. It disorganizes as soon as its goal is accomplished or it is deemed no longer worthwhile. Figure 3–6 illustrates the bubble organization structure.

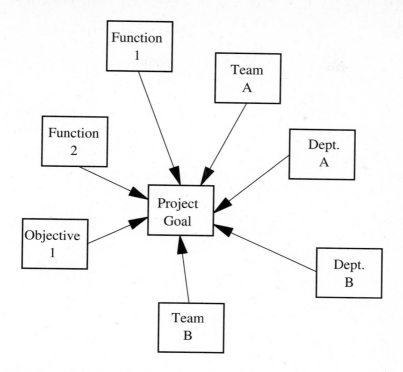

Figure 3-6.   Bubble Organization Structure

3.8.2 Market Organization

As world markets expand, it is important to be more responsive to the changing market forms. The market organization structure permits a project to adapt to market conditions. The market organization structure is illustrated in Figure 3–7.

3.8.3 Chronological Organization

The chronological organization is suitable for projects where time sequence is very essential in organizing tasks. This is for time-critical sets of tasks. A training program is suitable for the use of a chronological organization structure. Figure 3–8 presents an illustration of the chronological organization structure.

3.8.4 Sequential Organization

The sequential organization is similar to the chronological structure except that magnitude or quality of output rather than time is the basis for organizing the project. The quality of the output at each stage of the organization structure is needed to carry out the functions of the organization sequentially. A value-added production facility is an example of a system that is suitable for a sequential organization structure. Figure 3–9 presents the sequential organization structure.

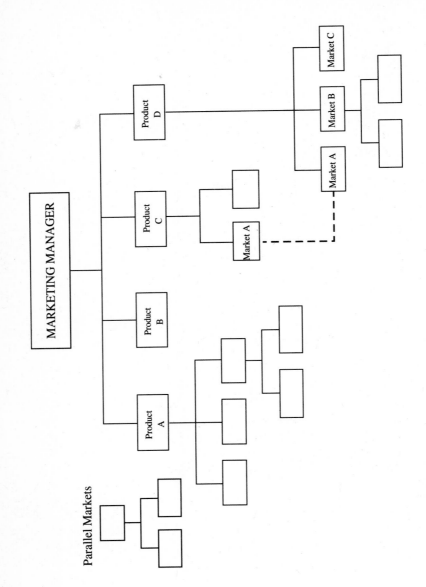

Figure 3-7. Market Organization Structure

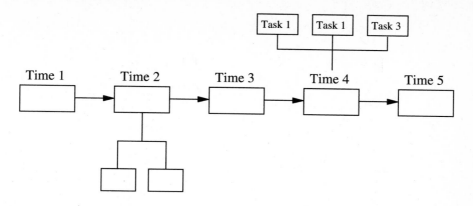

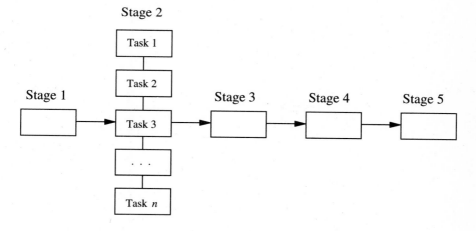

Figure 3-8. Chronological Organization Structure

Figure 3-9. Sequential Organization Structure

3.8.5 Military Organization

The military organization follows a strict hierarchical structure. It discourages informal lines of communication or responsibility. The name does not necessarily refer to the traditional military structure or configuration, but as in the military command structure, the block at the top of the military organization structure is notably more powerful than the lower blocks. A military organization structure is presented in Figure 3–10.

3.8.6 Political Organization

A political organization structure can be viewed as a rotary type of structure that is dynamic with respect to a time cycle. It has a very large base. The large base is collectively more powerful than the few blocks at the top. This may also be referred to as a democratic organization structure. The political organization structure is shown in Figure 3–11.

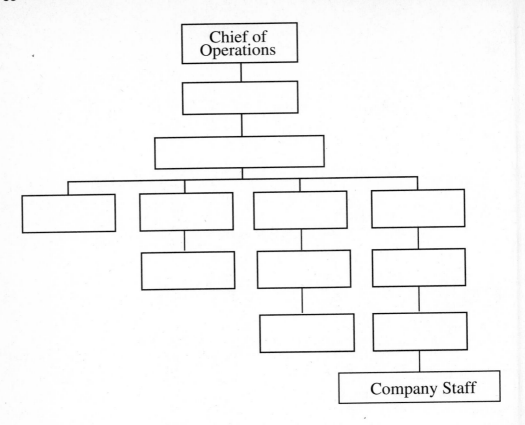

Figure 3-10.   Military Organization Structure

3.8.7 Autocratic Organization

An autocratic organization may be viewed as the reverse of the political organization structure. There is a single block at the top that is infinitely more powerful than the rest of the organization. Despite the large number of blocks at the lower levels of the structure, the top block remains almightily powerful. A major difference between the military and the autocratic organization structures is that there is a higher prospect that the block at the top of the military structure can be replaced. An autocratic organization structure is presented in Figure 3–12.

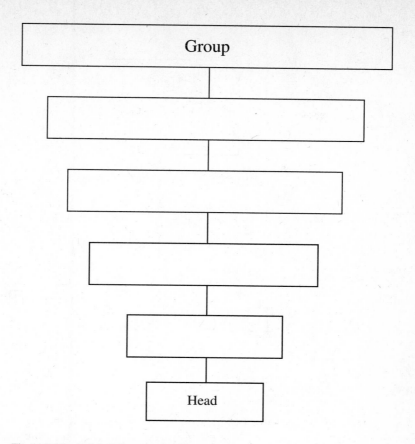

Figure 3-11. Political Organization Structure

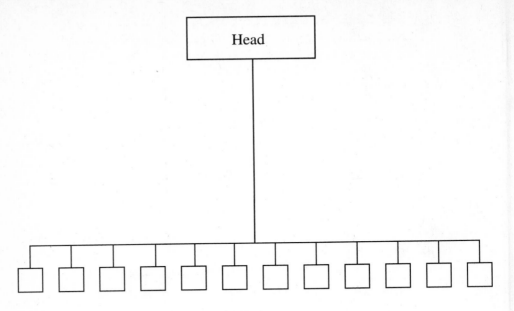

Figure 3-12.   Autocratic Organization Structure

3.9 PROJECT TRANSFER

Project and organizational transfers are important aspects of project management. The organizational structure that is in effect during a project can influence the transfer of the final product of a project. The organic (internal) organizational structure and the external organizational structure must be linked by a discernible transfer path as shown in Figure 3–13.

   Figures 3–14 and 3–15 show potential models for project transfer. Transfers may involve a complete project or components of a project. Figure 3–14 shows how products, ideas, concepts, and decisions move from one project environment to another. The receiving organization (referred to as the *transfer target*) uses the transferred elements to generate new products, ideas, concepts, and decisions, which follow a reverse transfer

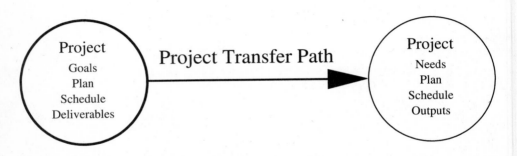

Figure 3-13.   Internal and External Organizational Linkage

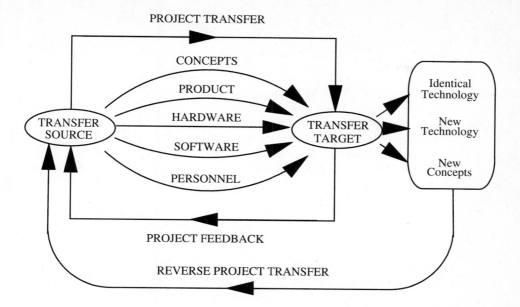

Figure 3-14.   Project Transfer Model

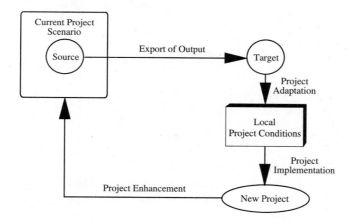

Figure 3-15.   Project Adaptation Loop

path to the *transfer source*. Thereby, both project environments operate on a symbiotic basis, with each contributing something to the other. Figure 3–15 shows a specific project adaptation process.

Project transfer can be achieved in various forms. Badiru (1991a) presents three transfer modes that represent basic strategies for getting one product from one point to another.

1. Transfer of complete project products. In this case, a fully developed product is transferred from one project to another project. Very little product development effort is carried out at the receiving point. However, information about the operations of the product is fed back to the source so that necessary product enhancements can be pursued. Hence, the recipient generates product information which serves as a resource for future work of the transfer source.

2. Transfer of procedures and guidelines. In this transfer mode, procedures and guidelines are transferred from one project to another. The project blueprints are implemented locally to generate the desired services and products. The use of local raw materials and personnel is encouraged for the local operations. Under this mode, the implementation of the transferred procedures can generate new operating procedures that can be fed back to enhance the original project. With this symbiotic arrangement, a loop system is created whereby both the transferring and the receiving organizations obtain useful benefits.

3. Transfer of project concepts, theories, and ideas. This strategy involves the transfer of the basic concepts, theories, and ideas associated with a project. The transferred elements can then be enhanced, modified, or customized within local constraints to generate new project outputs. The local modifications and enhancements have the potential to initiate an identical project, a new related project, or a new set of project concepts, theories, and ideas. These derived products may then be transferred back to the transfer source.

The important questions to ask about project transfer include the following:

- What exactly is being transferred?
- Who is receiving the transferred elements?
- What is the cost of the transfer?
- How is this project similar to previous projects?
- How is this project different from previous projects?
- Are the goals of the projects similar?
- What is expected from the transferred project?
- Is in-house skill adequate to use the transferred project?
- Is the prevailing management culture receptive to the new project?
- Is the current infrastructure capable of supporting the project?
- What modifications to the project will be necessary?

The selection of an appropriate transfer mode is particularly important for projects that cross national boundaries as discussed in the next section.

3.10 ORGANIZING MULTINATIONAL PROJECTS

Projects that cross national boundaries either in concept or in implementation have unique characteristics that create project management problems. In multinational projects, individual organizational policies are not enough to govern operations. Factors that normally influence these projects include

- Territorial laws and regulations
- Geographical segregation and restricted access
- Time differences
- Different scientific standards of measure
- Trade agreements
- Different government and political ideologies
- Different social, cultural, and labor practices
- Different stages of industrialization
- National security concerns
- Protection of proprietary technology information
- Strategic military implications
- Traditional national allies and adversaries
- Taxes, duties, and other import/export charges
- Foreign currency exchange rates
- National extradition/protection agreements
- Paperwork, permits, and restrictions
- Health, weather, and environmental considerations
- Poor, slow, or incompatible communication links
- Different native languages

International communication is perhaps one of the most difficult to deal with among all the factors listed above. The task of international transfer of technology and mutual project support takes on critical dimensions because of differences in the structures, objectives, and interests of the different countries involved. One common communication problem is that information destined for another country may have to pass through several levels of approval before reaching the point of use. The information is subject to all types of distortions and perils in its arduous journey. The integrity of the information may not be preserved as it is passed from one point to another. When implementing international projects, the following considerations should be reviewed:

1. Product
   - Type of product expected
   - Portability of the product
   - Product maintenance
   - Required training
   - Availability of spare parts
   - Feasibility for intended use
   - Local versus overseas productions
2. Technology
   - Local availability of required technology
   - Import/export restrictions on the technology
   - Implementation requirements
   - Supporting technologies

- Adaptation to local situations
- Lag between development and applications
- Operational approvals required

3. Political and social environment
   - Leadership and consistency of national policies
   - Political and social stability
   - Management views
   - Cultural adaptations
   - Bureaucracies
   - Structures of formal and informal organizations
   - Acceptance of foreigners and formation of relationships
   - Immigration laws
   - Ethnicity
   - General economic situation
   - Religious situations
   - Population pressures
   - Local development plans
   - Decision-making bureaucracies

4. Labor
   - Union regulations
   - Wage structures
   - Personnel dedication, loyalty, and motivation
   - Educational background and opportunities
   - Previous experience
   - Management relationships
   - Economic condition and level of contentment
   - Interests, attitudes, personalities, and leisure activities
   - Taxation policies
   - Logistics of employee relocation
   - Productivity consciousness
   - Demarcation of private and business activities
   - Local communication practices and facilities
   - Local customs
   - Language barriers

5. Market
   - Market needs
   - Inflation
   - Stability
   - Variety and availability of products
   - Cash, credit, and billing requirements
   - Exchange rates
   - National budget and gross national product
   - Competition and size of market
   - Transportation facilities

6. Plant and residential amenities
- Location
- Structural condition
- Accessibility
- Facilities available
- Proximity to business centers
- Topography
- Basic amenities (water, light, sewage, etc.)

7. Financial services
- Banking
- International money transfer
- Currency strength and stability
- Local sources of capital
- Interest rates
- Efficiency in conducting transactions
- Investment laws

Because of all these various factors, international project managers must undergo more extensive training than their conventional counterparts. A foreign manager in an international project must be open-minded, flexible, adaptive, and able to learn quickly. Since he or she will be working in an unfamiliar combination of social, cultural, political, and religious settings, he or she must have a keen sense of awareness and should be unassuming and responsive to local practices. An evaluation of the factors presented above in the context of the specific countries involved should help the international project manager to be better prepared for his or her expanded role. The complexity of an organizational structure for a multinational intercontinental project is illustrated by the example in Figure 3–16. The following case study illustrates the importance of a robust organizational structure for multinational projects.

3.10.1 Multinational Project Case Study *

This case study (Bahouth 1987) involves a large-scale international project which consists of the engineering, procurement, and construction of seven liquid gas tanks together with their ancillary system and control building in a Middle East country. The use and capacity of the tanks are as follows:

1. Three LNG tanks of 80,000 cubic meters each
2. Two LPG propane tanks of 50,000 cubic meters each
3. Two LPG butane tanks of 50,000 cubic meters each

The ancillary systems which are part of the project include

1. A propane and butane vapor recovery system
2. A low pressure flare system
3. An offsite control room

*The specifics of the project and the companies involved have been fictionalized.

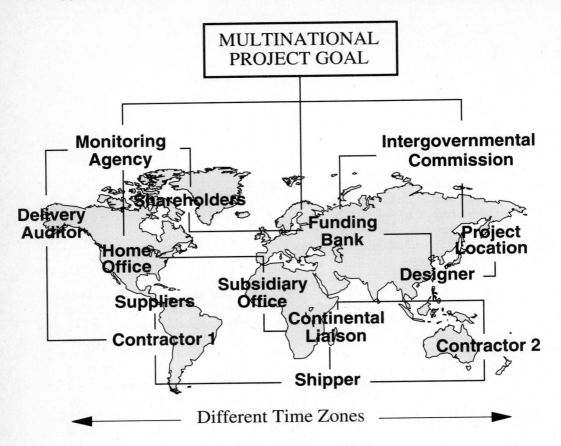

Figure 3-16. Multinational Intercontinental Organizational Structure

Project setting.   The construction is to be done on DAX Island in the Persian
Gulf (Middle East), approximately 300 miles from the city of Abu Dhabi (the capital
of the United Arab Emirates). DAX is a small island of two square miles and has a
population of 5,000, all of whom work in the only two plants on the island. Due to the
large project size, an additional 3,000 people will be working and living on the island
during the construction phase, which is a considerable increase in the population of the
island. The island has salty soil and cannot sustain any vegetation. Also, there isn't
any source for fresh water except for desalinated sea water. Desalination is a very slow
process and can cause considerable delays when it comes to pouring concrete.

Technical requirements.   Each of the seven tanks will consist of two separate
and structurally independent liquid containers: a primary inner metallic container and
a secondary outer concrete container. Each of the containers will be constructed of
material suitable for the low temperature liquid.

Each container should be capable of holding the required volume of the stored
liquid for an indefinite period without any deterioration of the container or its surround-
ings. The secondary concrete container should be capable of withstanding the effect

of fire exposure from the adjacent tanks and an external impact of 21 tons traveling at high speed without loss of structural integrity. Adequate insulation should be provided between the primary and secondary tanks to limit the heat in-leak. All the tanks should be provided with the necessary pumping and piping systems for the receipt, storage, and loading of the LNG and PLG from the nearby plant to the different tankers. A blast-proof offsite control room should be provided. All storage and loading operations should be controlled from this control room. The project total cost is estimated at US $600 million. The detailed project network consists of around 48,000 activities.

Organizational relationships.   GASC Company is the owner of the project. Their main offices are located in Abu Dhabi city. GASC is an operating company which has never managed any engineering or construction project. GASC turned to its mother company, NOCC, for help and signed an agreement with them for the management of this project.

NOCC is the national oil company of the Emirate of Abu Dhabi with main offices in Abu Dhabi city. It is completely owned by the government of Abu Dhabi and plays an extremely important role in the national economy. It is headed by a general manager responsible for the eight different directorates, one of which is the projects directorate. One of the divisions of the projects directorate is the gas projects division, headed by a manager. The gas projects division is selected to manage the project.

GASC is one of several operating companies whose majority of shares are owned by NOCC. NOCC's share in GASC is 60 percent and the remaining shares are owned by TOTAC of France, SHELC of Holland, PBC of Great Britain, and MITSC of Japan. While it is considered convenient to have the major owning company manage the project for GASC, such an arrangement has drawbacks, mainly the fact that the client company (who should have the final say in its own project) is a subsidiary of the hired management company.

The gas projects division is staffed with a skeleton staff. Although very capable, this staff can only oversee the management of the project and cannot perform all the actual engineering, procurement, and construction management. The Houston-based KELLC company is selected to provide the services under the direct supervision of the gas projects division. KELLC's scope of work includes the basic design of the tanks, the detailed design of all the piping and ancillary systems, the procurement of all free issue materials, and the management of construction. All the engineering and procurement activities are to be performed out of KELLC's regional office in London. KELLC, a well-known process engineering firm, has limited experience with concrete tanks. Their selection was conditional on their acceptance to hire the Belgian civil engineering firm TRACT as a consultant to help them in the critical civil engineering problems. NOCC will also hire the American-based consulting firm DMRC to do the soil investigation and testing work.

The construction work is packaged into 15 small contracts and 1 large (75 percent of the total construction work) contract. All the small contracts were awarded to local construction companies while the large main contract, which included the tanks, piping, and ancillary systems, was awarded to the Chicago-based CBAIC company. CBAIC will have to open three new offices. One office will be located in London, next to KELLC's regional office, during the engineering phase. Another office will be in Abu

Dhabi city for the construction management. A third office will be on DAX island for the construction operations. CBAIC is a reputable tanking contractor. However, its experience in concrete tanks is quite limited. They will have to hire the French civil engineering firm SBC as a subcontractor. SBC's experience in low temperature concrete is limited. They will have to hire the specialized Belgian firm CBC as a consultant.

Safety and environmental considerations.

The project's safety requirements are very high. The safety of those living on such a small island in case of an accident is of major concern to the owning company. To ensure that the required quality and safety standards are achieved, NOCC will hire the French third party inspection company BVC as a consultant.

Since the engineering office is in London, the French and the Belgian engineers are to commute to London as necessary to provide their inputs to the project. This will continue for the whole engineering phase which is expected to last two years.

Procurement activities will be handled out of KELLC's London office. Materials and equipment are to be delivered to the construction site. Supplies will be purchased in the open market at the most competitive prices. Steel will be purchased from Japan and Belgium; pipe from Germany, France and Japan; valves from Sweden and France; pumps from the United States; compressors from Switzerland and Japan; and vessels from Italy. A total of around 600 purchase orders will be issued.

Analysis of project scenario.

The organizational setup has to be quite flexible and change in accordance with the project requirements. Since it is expected to be a fast-track project, the most active period will be the second half of the engineering phase, which corresponds to the first half of the construction phase. The organizational setup for the project stretches across national boundaries, practices, and regulations. It is, thus, very important that efforts of the project staff be closely coordinated.

The existing scenario suggests that all the planning, organizing, and staffing functions are fixed. However, the existing organizational setup and overall project plan pose some possible problems. The possible problems associated with the existing setup are discussed along with suggestions on how they could be alleviated by applying sound management principles to the planning, organizing, and staffing functions. Management should especially consider and forecast possible future problems and make provisions for handling the problems.

Understanding the scope and limitations of the project is vital due to the fact that it is a multinational high-tech project and of such large size. Special management practices suitable for high-tech and multinational projects should be applied.

Sources of possible problems and solutions.

The core of possible problems associated with this project lies in the complex organizational structure and the fact that it is a large multinational high-tech project. Below are listed the problem sources, associated possible problems that could arise, and recommended solutions or alternatives to alleviate the problems.

1. GASC turned to its mother company NOCC for help and subcontracted with NOCC for management of the project. There could be several drawbacks associated with this situation. NOCC may take the attitude of overcontrolling since GASC is its subsidiary and begin to generate and implement decisions without contacting GASC for approval. For example, NOCC may overlook the performance standards specified in the blueprints and set new performance standards that are lower than those specified by GASC. If NOCC feels that their way of managing jobs is best and since GASC has to come to them for help, NOCC may feel that making their own changes can only benefit the project. However, this will only create a poor final product.

GASC knows the expectations of the project and has planned for that in the blueprints and specifications. However, GASC is playing no part in the internal decision making as the project evolves. This is going to produce conflicts which will slow down the project and create many more scheduling problems, financial problems, and other problems that were not planned for initially. If GASC carefully covers the planning function, they will foresee the potential problems and realize that their participation in decisions is of utmost importance for the project to be streamlined and for the final system to perform its intended function efficiently. Thus, GASC could either hire a more qualified internal staff with more expertise in management or they could take a team of their top managers to coordinate with NOCC on the project to ensure things will go as planned.

2. Too many companies are involved in the project. This results in several levels of responsibility. The situation is compounded by the fact that the companies are so globally widespread.

Too many levels of responsibility create unnecessary red tape and can cause much delay in the project's phases along with employee frustration and disinterest. First, NOCC is not equipped to handle such a large-scale project and, thus, should be eliminated from the organizational setup. This decision should have been made in the planning stages as mentioned previously. However, in the organizing stage, NOCC should also be eliminated because of its poor managing abilities which are clearly shown in the poorly planned organizational setup.

The organizational structure does not give a logical design of the interfaces needed among personnel in order to assure a dedicated pursuit of the common goals in association with the project nor has it been approached carefully so as to facilitate coordination. Although this large project requires much expertise in various fields, a better approach would be to simplify the levels of responsibilities by contracting with fewer companies that are more specialized. The benefits of this approach outweigh the costs. If the foundation of responsibility level and organizational hierarchy is not set up adequately from the start, then the directing and controlling functions will be limited and project success may not be possible.

3. Several communication problems could arise due to the fact that the project is high tech and multinational. The difference in cultural backgrounds and the complex organizational structure of this project will influence how communication is approached and how it will be perceived by different levels of personnel. The management should form an environment where a formal and informal exchange of ideas can take place. Due to the numerous levels of responsibility and cultural and language differences among countries, any form of communication that must pass through these several levels should be concisely formatted. This will ensure that the original information is clearly communicated and not be altered as it passes through the various levels. Clear communication is very important in preventing project problems.

The difference in times zones presents several communication obstacles to overcome. Thus, to help resolve these problems, a global time frame should be determined during which all the project offices in the following cities or countries can communicate on any given day (24 hours) by teleconference: London, Rome, Chicago, Abu Dhabi, Tokyo, Geneva, Bonn, Paris, Sweden, and Belgium. In approaching the problem of determining the time window, the two locations with the greatest time difference (Japan and Chicago) and the major project offices (London, Abu Dhabi, and DAX) should be jointly considered. The optimal time window should be one hour each day in which Chicago can communicate from 6:00 A.M. to 7:00 A.M. and Japan from 9:00 P.M. to 10:00 P.M. Thus, this time frame will give the major project offices a practical communication time during their daily business hours. This is important since most project interactions will originate from these offices.

Another problem in communication is the language barriers among the various countries. To ensure that information is streamlined and correctly perceived, interpreters who are fluent in these various foreign languages should be employed at the various project offices. Effective communication will ensure that coordination will be increased within all the units, thus enhancing cooperation and creating satisfaction among the members of the project.

4. A considerable increase in island population during the construction phase will have to be planned for. An additional 3,000 people will increase DAX's population by 60 percent. It is known that the island is currently starved of natural resources. Thus, high priority should be given to conserve these resources. Upper management should ensure that sufficient facilities be provided for the employees without polluting or destroying the natural environment.

5. Ordering supplies from so many different countries creates problems in logistics. Due to geographical diversity of the incoming supplies, delays and backlogs are likely to occur. A sound inventory system should be implemented to minimize any logistic problems or delays so that the project deadlines are accomplished.

Authors' digest. In dealing with multinational high-tech projects, the management culture must be versatile and congenial. Several factors must be considered including how to control the uncertainty in high-tech operations and how to streamline information among the various personnel. Multinational projects require management practices designed specifically to meet the unique challenges. Several factors to consider and specific suggestions for this project are presented next.

Strategic Planning. Since a frequent problem with technology is the extension of useful life well beyond the period of obsolescence, it is vital that NOCC define the long-range purpose and useful life of the project so that replacement of the system can be planned for financially.

Cost/Benefit Analysis. The bottom line in most operations is combined profit/ benefit and performance. NOCC should analyze the cost of the high technology required versus the benefits to see if this project could be profitable.

Technology Assessment. The technology itself should be assessed in detail. GASC should determine whether there are alternative technologies which can perform the required objective of the project.

Acquisition. In the planning stage, NOCC should evaluate the policy of procuring and implementing the high-tech system. NOCC should evaluate the hardware and software components of the project, communications site selection and installation, and development and implementation along with training and post-implementation evaluation of the project.

Trade agreements. Several contracts on procurement of supplies will need to be closely regulated to ensure all governmental policies of the various countries are strictly followed.

Different Labor Practices. Labor practices vary from country to country in terms of the number of hours of work per day, the number of days to work per week, unions, relationships between personnel, payment practices, and so on. A compromise should be developed among the various workers and contracting companies to address these various factors.

Different Governmental and Political Ideologies. Differences among countries on politics could create several conflicts especially if top management allows political ambitions to influence their actions toward work on the project. Thus, these differences in governments must be discussed and accounted for in the initial planning stage of the project.

Information Management. Information is a critical resource in managing high-tech operations. The organizational efficiency of the project will be affected by the quality of information generated and how the information is used. A multinational project manager may be bogged down with the problem of receiving too much information. Sophisticated data processing and information handling mechanisms will be needed to govern the mass of information that may be associated with the project.

In coordinating the activities of each phase of the project, a central division should be set up to monitor all the activities that go on and keep all the members of the project informed of the current situation. The organizational structure should also be streamlined to make use of the systems approach in which project teams interact both horizontally and vertically.

Undertaking such a vast multinational high-tech project requires a management with special expertise and capabilities. The project manager should have broad authority over all elements of the project and engage in all necessary managerial and technical actions required to complete the project successfully. Also, the project manager should have appropriate authority in making technical and design decisions and should be able to control funds, schedule, and quality of product.

Due to the large size of this international project, a team of managers is recommended in order to successfully control and direct the project. Cooperation, communication, and coordination are vital to the interactions among the managers. The management team should be experienced and well educated in management principles as applied to high-tech and multinational projects in order for harmony to exist throughout the decision-making processes. The case study presented here gives insight to a real-world situation in which documented management principles can be applied for optimal project outcome.

A conducive organization structure must be formulated. Since there are interdependencies and subsidiary relationships among many of the companies involved in the project, the conventional organization structures may not be adequate. A sophisticated form of the mixed organization structure will, thus, be required. In summary, the potential problems that may adversely affect the project organization are

- Shortage of food supply
- Security concern (high tension zone)
- Potential cost overruns
- Potential for catastrophic accidents
- Foreign rules and regulations
- Trade restrictions
- Communication delays
- Language barriers
- Lack of fresh water

Recommendations.   The following recommendations are offered for the case study project:

- Simplify the levels of responsibilities by contracting with fewer companies.
- Minimize the interdependencies of companies by using nonsubsidiary companies.
- Predetermine communication time zones that are compatible with the timing of project decisions.
- Reevaluate the level of technology required for the project.
- Arrange for special favorable trade exceptions to facilitate movement of personnel and materials.

3.11 EXERCISES

3.1. Suggest a suitable organization structure for an industrial development project. Justify your recommendation with a discussion.

3.2. Discuss how a poor organization structure can lead to project failure.

3.3. Which organization structure was used on the last project that you worked on?

3.4. Discuss how the matrix structure can be implemented in an organization that traditionally adheres to strict hierarchical structure.

3.5. Develop a model for a mixed organization structure using matrix, product, and functional structures.

3.6. Draw a market organization structure for a new product to be marketed worldwide. Select a specific product type to answer this exercise.

4

Project Scheduling

Project scheduling is the time-phased sequencing of activities subject to precedence relationships, time constraints, and resource limitations to accomplish specific objectives. This chapter presents the techniques of network analysis for project scheduling. The computational approaches to project network analysis using PERT (program evaluation and review technique), CPM (critical path method), and PDM (precedence diagramming method) are presented. Several graphical variations of Gantt charts are presented. This stage of the project management process can be represented as

$$\text{SET GOALS} \rightarrow \text{PLAN} \rightarrow \text{ORGANIZE} \rightarrow \text{SCHEDULE}$$

1 FUNDAMENTALS OF NETWORK ANALYSIS

Project scheduling is distinguished from job shop, flow shop, and other production sequencing problems because of the unique nature of many of the activities that make up a project. In production scheduling, the scheduling problem follows a standard procedure that determines the characteristics of production operations. A scheduling technique that works for one production run may be expected to work equally effectively for succeeding and identical production runs. By contrast, projects usually involve one-time endeavors that may not be duplicated in identical circumstances. In some cases, it may be possible to duplicate the concepts of the whole project or a portion of it.

Several techniques have been developed for the purpose of planning, scheduling, and controlling projects. The available scheduling models and solution approaches can be categorized as follows:

Project scheduling
    Resources unconstrained
        Critical path analysis

          Time-cost trade-off problem
      Resources constrained
         Heuristic techniques
         Mathematical programming techniques

Project schedules may be complex, unpredictable, and dynamic. Complexity may be due to interdependencies of activities, multiple resource requirements, multiple concurrent events, conflicting objectives, technical constraints, and schedule conflicts. Unpredictability may be due to equipment breakdowns, raw material inconsistency (delivery and quality), operator performance, labor absenteeism, and unexpected events. Dynamism may be due to resource variability, changes in work orders, and resource substitutions. We define *predictive scheduling* as a scheduling approach that attempts to anticipate the potential causes of schedule problems. These problems are corrected by contingency plans. We define *reactive scheduling* as a scheduling approach that reacts to problems that develop in the scheduling environment.

The most widely used scheduling aids involve network techniques, two of which are the critical path method (CPM) and the program evaluation and review technique (PERT). Network analysis procedures originated from the traditional Gantt chart, or bar chart, developed by Henry L. Gantt during World War I. There have been several mathematical techniques for scheduling activities especially where resource constraints are a major factor. Unfortunately, the mathematical formulations are not generally practical due to the complexity involved in implementing them realistically for large projects. Even computer implementations of the mathematical techniques sometimes become too cumbersome for real-time managerial applications. It should be recalled that the people responsible for project schedules are the managers who, justifiably, prefer simple and quick decision aids. To a project scheduler, a complex mathematical procedure constitutes an impediment rather than an aid in the scheduling routine. Nonetheless, the premise of the mathematical formulations rests on their applicability to small projects consisting of very few activities. Many of the techniques have been evaluated, applied, and reported in the literature.

A more practical approach to scheduling is the use of heuristics. If the circumstances of a problem satisfy the underlying assumptions, a good heuristic will yield schedules that are feasible enough to work with. A major factor in heuristic scheduling is to select a heuristic whose assumptions are widely applicable. A wide variety of scheduling heuristics exists for a wide variety of special cases. Whitehouse and Brown (1979), Bedworth and Bailey (1982), Thesen (1976), Elsayed (1982), Wiest (1977), Kurtulus and Davis (1982), Cooper (1976), Holloway et al. (1979), and Badiru (1988a) present practical scheduling heuristics. The procedure for using heuristics to schedule projects involves prioritizing activities in the assignment of resources and time slots in the project schedule. Many of the available priority rules consider activity durations and resource requirements in the scheduling process. Some of the common scheduling heuristics are presented in the next chapter.

If all activities are assigned priorities at the beginning and then scheduled, the scheduling heuristic is referred to as a *serial method*. If priorities are assigned to the set of activities eligible for scheduling at a given instant and the schedule is developed concurrently, then the scheduling heuristic is referred to as a *parallel method*. In the serial method, the relative priorities of activities remain fixed. In the parallel method, the priorities change with the current composition of activities.

The network of activities contained in a project provides the basis for scheduling the project. The critical path method (CPM) and the program evaluation and review technique (PERT) are the two most popular techniques for project network analysis. The precedence diagramming method (PDM) has gained in popularity in recent years because of the move toward concurrent engineering in manufacturing operations. A project network is the graphical representation of the contents and objectives of the project. The basic project network analysis is typically implemented in three phases: network planning phase, network scheduling phase, and network control phase.

Network planning is sometimes referred to as activity planning. This involves the identification of the relevant activities for the project. The required activities and their precedence relationships are determined. Precedence requirements may be determined on the basis of technological, procedural, or imposed constraints. The activities are then represented in the form of a network diagram. The two popular models for network drawing are the *activity-on-arrow (AOA)* and the *activity-on-node (AON)* conventions. In the AOA approach, arrows are used to represent activities, while nodes represent starting and ending points of activities. In the AON approach, nodes represent activities, while arrows represent precedence relationships. Section 4.7 explains AOA and AON graphs in detail. Time, cost, and resource requirement estimates are developed for each activity during the network planning phase. The estimates may be based on historical records, time standards, forecasting, regression functions, or other quantitative models.

Network scheduling is performed by using forward pass and backward pass computational procedures. These computations give the earliest and latest starting and finishing times for each activity. The amount of slack or float associated with each activity is determined. The activity path with the minimum slack in the network is used to determine the critical activities. This path also determines the duration of the project. Resource allocation, and time-cost trade-offs are other functions performed during network scheduling.

Network control involves tracking the progress of a project on the basis of the network schedule and taking corrective actions when needed. An evaluation of actual performance versus expected performance determines deficiencies in the project progress. The advantages of project network analysis are as follows:

- Advantages for communication
  Clarifies project objectives.
  Establishes the specifications for project performance.
  Provides a starting point for more detailed task analysis.
  Presents a documentation of the project plan.
  Serves as a visual communication tool.
- Advantages for control
  Presents a measure for evaluating project performance.
  Helps determine what corrective actions are needed.
  Gives a clear message of what is expected.
  Encourages team interactions.

- Advantages for team interaction
  Offers a mechanism for a quick introduction to the project.
  Specifies functional interfaces on the project.
  Facilitates ease of application.

Figure 4–1 shows the graphical representation for an AON network. The components of the network are explained next.

1. Node: A node is a circular representation of an activity.
2. Arrow: An arrow is a line connecting two nodes and having an arrowhead at one end. The arrow implies that the activity at the tail of the arrow precedes the one at the head of the arrow.
3. Activity: An activity is a time-consuming effort required to perform a part of the overall project. An activity is represented by a node in the AON system or by an arrow in the AOA system. The job the activity represents may be indicated by a short phrase or symbol inside the node or along the arrow.
4. Restriction: A restriction is a precedence relationship which establishes the sequence of activities. When one activity must be completed before another activity can begin, the first is said to be a predecessor of the second.
5. Dummy: A dummy is used to indicate one event of a significant nature (e.g., milestone). It is denoted by a dashed circle and treated as an activity with zero time duration. A dummy is not required in the AON method. However, it may be included for convenience, network clarification, or to represent a milestone in the progress of the project.
6. Predecessor activity: A predecessor activity is one which immediately precedes the one being considered. In Figure 4–1a, A is a predecessor of B and C.
7. Successor activity: A successor activity is one which immediately follows the one being considered. In Figure 4–1a, activities B and C are successors to A.
8. Descendent activity: A descendent activity is any activity restricted by the one under consideration. In Figure 4–1a, activities B, C, and D are all descendents of activity A.
9. Antecedent activity: An antecedent activity is any activity which must precede the one being considered. Activities A and B are antecedents of D. Activity A is antecedent of B, and A has no antecedent.
10. Merge point: A merge point (see Figure 4–1b) exists when two or more activities are predecessors to a single activity. All activities preceding the merge point must be completed before the merge activity can commence.
11. Burst point: A burst point (see Figure 4–1c) exists when two or more activities have a common predecessor. None of the activities emanating from the same predecessor activity can be started until the burst point activity is completed.
12. Precedence diagram: A precedence diagram (see Figure 4–1d) is a graphical representation of the activities making up a project and the precedence requirements needed to complete the project. Time is conventionally shown to be from left to right, but no attempt is made to make the size of the nodes or arrows proportional to time.

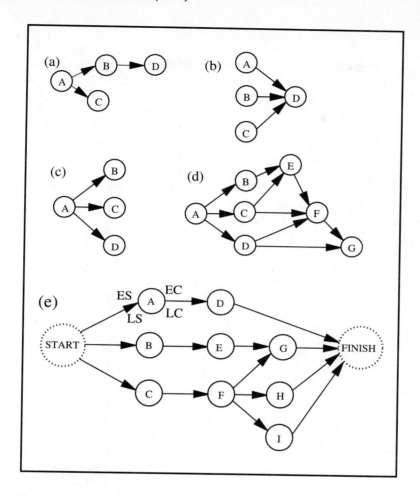

Figure 4-1.   Graphical Representation of an AON Network

2 CRITICAL PATH METHOD (CPM)

Precedence relationships in a critical path method (CPM) network fall into three major categories:

1. Technical precedence
2. Procedural precedence
3. Imposed precedence

Technical precedence requirements are caused by the technical relationships among activities in a project. For example, in conventional construction, walls must be erected before the roof can be installed. Procedural precedence requirements are determined by policies and procedures. Such policies and procedures are often subjective, with no concrete justification. Imposed precedence requirements can be classified as resource-imposed, state-imposed, or environment-imposed. For example, resource shortages may require that one task be before another. The current status of a project (e.g., percent completion) may determine that one activity be performed before another. The environment of a project, for example, weather changes or the effects of concurrent projects, may determine the precedence relationships of the activities in a project.

The primary goal of a CPM analysis of a project is the determination of the *critical path*. The critical path determines the minimum completion time for a project. The computational analysis involves *forward pass* and *backward pass* procedures. The forward pass determines the earliest start time and the earliest completion time for each activity in the network. The backward pass determines the latest start time and the latest completion time for each activity. Conventional network logic is always drawn from left to right. If this convention is followed, there is no need to use arrows to indicate the directional flow in the activity network. The notations used for activity A in the network are explained as follows:

A :	Activity identification
ES :	Earliest starting time
EC :	Earliest completion time
LS :	Latest starting time
LC :	Latest completion time
t :	Activity duration

During the forward pass analysis of the network, it is assumed that each activity will begin at its earliest starting time. An activity can begin as soon as the last of its predecessors is finished. The completion of the forward pass determines the earliest completion time of the project. The backward pass analysis is the reverse of the forward pass analysis. The project begins at its latest completion time and ends at the latest starting time of the first activity in the project network. The rules for implementing the forward pass and backward pass analyses in CPM are presented below. These rules are implemented iteratively until the ES, EC, LS, and LC have been calculated for all nodes in the activity network.

Rule 1
Unless otherwise stated, the starting time of a project is set equal to time 0. That is, the first node, *node 1*, in the network diagram has an earliest start time of 0. Thus,

$$ES(1) = 0$$

If a desired starting time, t_0, is specified, then

$$ES(1) = t_0$$

Rule 2
The earliest start time (ES) for any node (activity j) is equal to the maximum of the earliest completion times (EC) of the immediate predecessors of the node. That is,

$$ES(i) = \underset{j \in P(i)}{\text{Max}}\{EC(j)\}$$

where $P(i) = \{$set of immediate predecessors of activity $i\}$.

Rule 3
The earliest completion time (EC) of activity i is the activity's earliest start time plus its estimated time, t_i. That is,

$$EC(i) = ES(i) + t_i$$

Rule 4
The earliest completion time of a project is equal to the earliest completion time of the very last node, *node n*, in the project network. That is,

$$EC(Project) = EC(n)$$

Rule 5
Unless the latest completion time (LC) of a project is explicitly specified, it is set equal to the earliest completion time of the project. This is called the *zero project slack convention*. That is,

$$LC(Project) = EC(Project)$$

Rule 6
If a desired deadline, T_p, is specified for the project, then

$$LC(Project) = T_p$$

It should be noted that a latest completion time or deadline may sometimes be specified for a project on the basis of contractual agreements.

Rule 7
The latest completion time (LC) for activity j is the smallest of the latest start times of the activity's immediate successors. That is,

$$LC(j) = \underset{i \in S(j)}{\text{Min}}$$

where $S(j) = \{$immediate successors of activity $j\}$

Rule 8
The latest start time for activity j is the latest completion time minus the activity time. That is,

$$LS(j) = LC(j) - t_i$$

4.2.1 CPM Example

Table 4–1 presents the data for a simple project network. This network and extensions of it will be used for computational examples in this chapter and subsequent chapters. The AON network for the example is given in Figure 4–2. Dummy activities are included in the network to designate single starting and ending points for the network.

Table 4–1   Data for Sample Project for CPM Analysis

Activity	Predecessor	Duration (days)
A	—	2
B	—	6
C	—	4
D	A	3
E	C	5
F	A	4
G	B, D, E	2

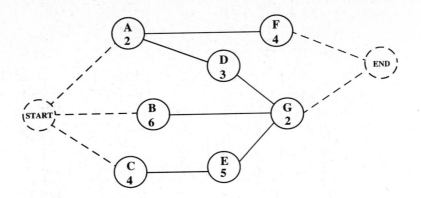

Figure 4-2. Example of Activity Network

Forward pass. The forward pass calculations are shown in Figure 4–3. Zero is entered as the ES for the initial node. Since the initial node for the example is a dummy node, its duration is 0. Thus, the EC for the starting node is equal to its ES. The ES values for the immediate successors of the starting node are set equal to the EC of the start node and the resulting EC values are computed. Each node is treated as the start node for its successor or successors. However, if an activity has more than one predecessor, the maximum of the ECS of the preceding activities is used as the activity's starting time. This happens in the case of activity G, whose ES is determined as Max{6, 5, 9} = 9. The earliest project completion time for the example is 11 days. Note that this is the maximum of the immediately preceding earliest completion time: Max{6, 11} = 11. Since the dummy ending node has no duration, its earliest completion time is set equal to its earliest start time of 11 days.

Backward pass. The backward pass computations establish the latest start time (LS) and latest completion time (LC) for each node in the network. The results

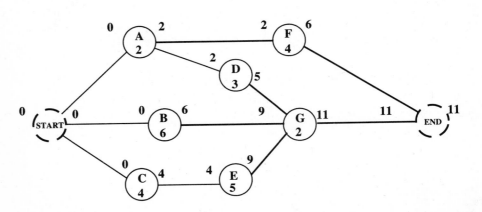

Figure 4-3. Forward Pass Analysis for CPM Example

of the backward pass computations are shown in Figure 4-4. Since no deadline is specified, the latest completion time of the project is set equal to the earliest completion time. By backtracking and using the network analysis rules presented earlier, the latest completion and latest start times are determined for each node. Note that in the case of activity A with two immediate successors, the latest completion time is determined as the minimum of the immediately succeeding latest start times. That is, Min{6, 7} = 6. A similar situation occurs for the dummy starting node. In that case, the latest completion time of the dummy start node is Min{0, 3, 4} = 0. Since this dummy node has no duration, the latest starting time of the project is set equal to the node's latest completion time. Thus, the project starts at time 0 and is expected to be completed by time 11.

Within a project network, there are usually several possible paths and a number of activities that must be performed sequentially and some activities that may be performed concurrently. If an activity has ES and EC times that are not equal, then the actual start and completion times of that activity may be flexible. The amount of flexibility an activity possesses is called a slack time. The slack time is used to determine the critical activities in the network as discussed next.

4.2.2 Determination of Critical Activities

The critical path is defined as the path with the least slack in the network diagram. All the activities on the critical path are said to be critical activities. These activities can create bottlenecks in the network if they are delayed. The critical path is also the longest path in the network diagram. In some networks, particularly large ones, it is possible to have multiple critical paths or a critical path subnetwork. If there is a large number of paths in the network, it may be very difficult to visually identify all the critical paths. The slack time of an activity is also referred to as its *float*. There are four basic types of activity slack:

- Total slack (TS). Total slack is defined as the amount of time an activity may be delayed from its earliest starting time without delaying the latest completion time

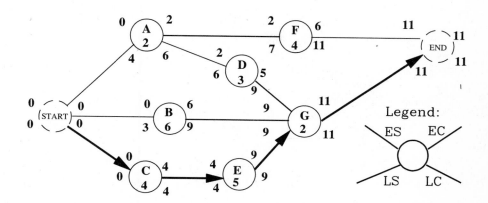

Figure 4-4. Backward Pass Analysis for CPM Example

of the project. The total slack of activity j is the difference between the latest completion time and the earliest completion time of the activity, or the difference between the latest starting time and the earliest starting time of the activity.

$$TS(j) = LC(j) - EC(j) \text{ or } TS(j) = LS(j) - ES(j)$$

Total slack is the measure that is used to determine the critical activities in a project network. The critical activities are identified as those having the minimum total slack in the network diagram. If there is only one critical path in the network, then all the critical activities will be on that one path.

- Free slack (FS). Free slack is the amount of time an activity may be delayed from its earliest starting time without delaying the starting time of any of its immediate successors. Activity free slack is calculated as the difference between the minimum earliest starting time of the activity's successors and the earliest completion time of the activity.

$$FS(j) = \underset{i \in S(j)}{\text{Min}} \{ES(i)\text{'s} - EC(j)\}$$

- Interfering slack (IS). Interfering slack or interfering float is the amount of time by which an activity interferes with (or obstructs) its successors when its total slack is fully used. This is rarely used in practice. The interfering float is computed as the difference between the total slack and the free slack.

$$IS(j) = TS(j) - FS(j)$$

- Independent float (IF). Independent float or independent slack is the amount of float that an activity will always have regardless of the completion times of its predecessors or the starting times of its successors. Independent float is computed as

$$IF = \underset{j \in S(k),\ i \in P(k)}{\text{Max}} \left\{ 0, \left(\underset{j \in S(k)}{\text{Min}} ES_j - \underset{i \in P(k)}{\text{Max}} LC_i - t_k \right) \right\}$$

where ES_j is the earliest starting time of the succeeding activity, LC_i is the latest completion time of the preceding activity, and t is the duration of the activity whose independent float is being calculated. Independent float takes a pessimistic view of the situation of an activity. It evaluates the situation whereby the activity is pressured from either side, that is, when its predecessors are delayed as late as possible while its successors are to be started as early as possible. Independent float is useful for conservative planning purposes, but it is not used much in practice. Despite its low level of use, independent float does have practical implications for better project management. Activities can be buffered with independent floats as a way to handle contingencies.

For Figure 4–4 the total slack and the free slack for activity A are calculated, respectively, as

$$TS = 6 - 2 = 4 \text{ days}$$
$$FS = \text{Min}\{2, 2\} - 2 = 2 - 2 = 0$$

Similarly, the total slack and the free slack for activity F are

$$TS = 11 - 6 = 5\,\text{days}$$
$$FS = \text{Min}\{11\} - 6 = 11 - 6 = 5\,\text{days}$$

Table 4–2 presents a tabulation of the results of the CPM example. The table contains the earliest and latest times for each activity as well as the total and free slacks. The results indicate that the minimum total slack in the network is 0. Thus, activities C, E, and G are identified as the critical activities. The critical path is highlighted in Figure 4–4 and consists of the following sequence of activities:

$$\text{START} \rightarrow \text{C} \rightarrow \text{E} \rightarrow \text{G} \rightarrow \text{END}$$

The total slack for the overall project itself is equal to the total slack observed on the critical path. The minimum slack in most networks will be zero since the ending LC is set equal to the ending EC. If a deadline is specified for a project, then we would set the project's latest completion time to the specified deadline. In that case, the minimum total slack in the network would be given by

$$TS_{\text{Min}} = (\text{project deadline}) - EC \text{ of the last node}$$

This minimum total slack will then appear as the total slack for each activity on the critical path. If a specified deadline is lower than the EC at the finish node, then the project will start out with a negative slack. That means that it will be behind schedule before it even starts. It may then become necessary to expedite some activities (i.e., crashing) in order to overcome the negative slack. Figure 4–5 shows an example with a specified project deadline. In this case, the deadline of 18 days comes after the earliest completion time of the last node in the network.

Using forward pass to determine the critical path.   The critical path in CPM analysis can be determined from the forward pass only. This can be helpful in cases where it is desired to quickly identify the critical activities without performing all the other calculations needed to obtain the latest starting times, the latest completion times, and total slacks. The steps for determining the critical path from the forward pass only are as follows:

1. Complete the forward pass in the usual manner.

Table 4–2   Result of CPM Analysis for Sample Project

Activity	Duration	ES	EC	LS	LC	TS	FS	Critical
A	2	0	2	4	6	4	0	—
B	6	0	6	3	9	3	3	—
C	4	0	4	0	4	0	0	Critical
D	3	2	5	6	9	4	4	—
E	5	4	9	4	9	0	0	Critical
F	4	2	6	7	11	5	5	—
G	2	9	11	9	11	0	0	Critical

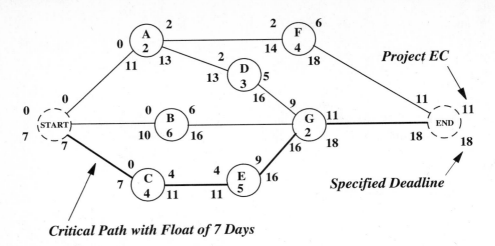

Critical Path with Float of 7 Days

Figure 4-5. CPM Network with Deadline

2. Identify the last node in the network as a critical activity.
3. If activity i is an immediate predecessor of activity j, which is determined as a critical activity, then check EC_i and ES_j. If $EC_i = ES_j$, then label activity i as a critical activity. When all immediate predecessors of activity j are considered, mark activity j.
4. Continue the backtracking from each unmarked critical activity until the project starting node is reached. Note that if there is a single starting node or a single ending node in the network, then that node will always be on the critical path.

4.2.3 Subcritical Paths

In a large project network, there may be paths that are near critical. Such paths require almost as much attention as the critical path since they have a high potential of becoming critical when changes occur in the network. Analysis of subcritical paths may help in the classification of tasks into ABC categories on the basis of Pareto analysis. Pareto analysis separates the "vital" few activities from the "trivial" many activities. This permits a more efficient allocation of resources. The principle of Pareto analysis originated from the work of Italian economist Vilfredo Pareto (1848–1923). In his studies, Pareto discovered that most of the wealth in his country was held by a few individuals.

For project control purposes, the Pareto principle states that 80% of the bottlenecks are caused by only 20% of the tasks. This principle is applicable to many management processes. For example, in cost analysis, one can infer that 80% of the total cost is associated with only 20% of the cost items. Similarly, 20% of an automobile's parts cause 80% of the maintenance problems. In personnel management, about 20% of employees account for about 80% of the absenteeism. For critical path analysis, 20% of the network activities will take up 80% of our control efforts. The ABC classification based on Pareto analysis divides items into three priority categories: A (most important),

B (moderately important), and C (least important). Appropriate percentages (e.g., 20%, 25%, 55%) are assigned to the categories.

With Pareto analysis, attention can be shifted from focusing only on the critical path to managing critical and near-critical tasks. The level of criticality of each path may be assessed by the following procedure:

Step 1: Sort activities in increasing order of total slack.

Step 2: Partition the sorted activities into groups based on the magnitudes of their total slacks.

Step 3: Sort the activities within each group in increasing order of their earliest starting times.

Step 4: Assign the highest level of criticality to the first group of activities (e.g., 100%). This first group represents the usual critical path.

Step 5: Calculate the relative criticality indices for the other groups in decreasing order of criticality.

Define the following variables:

α_1 = the minimum total slack in the network
α_2 = the maximum total slack in the network
β = total slack for the path whose criticality is to be calculated.

Compute the path's criticality as

$$\lambda = \frac{\alpha_2 - \beta}{\alpha_2 - \alpha_1}(100\%)$$

This procedure yields relative criticality levels between 0% and 100%. Table 4–3 presents a hypothetical example of path criticality indices. The criticality level may be converted to a scale between 1 (least critical) and 10 (most critical) by the expression

$$\lambda' = 1 + 0.09\lambda$$

Table 4-3 Analysis of Subcritical Paths

Path number	Activities on path	Total slack	λ	λ'
1	A, C, G, H	0	100%	10
2	B, D, E	1	97.56	9.78
3	F, I	5	87.81	8.90
4	J, K, L	9	78.05	8.03
5	O, P, Q, R	10	75.61	7.81
6	M, S, T	25	39.02	4.51
7	N, AA, BB, U	30	26.83	3.42
8	V, W, X	32	21.95	2.98
9	Y, CC, EE	35	17.14	2.54
10	DD, Z, FF	41	0	1.00

4.3 GANTT CHARTS

When the results of a CPM analysis are fitted to a calendar time, the project plan becomes a schedule. The Gantt chart is one of the most widely used tools for presenting a project schedule. A Gantt chart can show the planned and actual progress of activities. The time scale is indicated along the horizontal axis, while horizontal bars or lines representing activities are ordered along the vertical axis. As a project progresses, markers are made on the activity bars to indicate actual work accomplished. Gantt charts must be updated periodically to indicate project status. Figure 4–6 presents the Gantt chart for our illustrative example using the earliest starting (ES) times from Table 4–2. Figure 4–7 presents the Gantt chart for the example based on the latest starting (LS) times. Critical activities are indicated by the shaded bars.

Figure 4–6 shows that the starting time of activity F can be delayed from day 2 until day 7 (i.e., $TS = 5$) without delaying the overall project. Likewise, A, D, or both may be delayed by a combined total of four days ($TS = 4$) without delaying the overall project. If all the 4 days of slack are used up by A, then D cannot be delayed. If A is delayed by 1 day, then D can be delayed by up to 3 days without causing a delay of G, which determines project completion. The Gantt chart also indicates that activity B may be delayed by up to 3 days without affecting the project completion time.

In Figure 4–7, the activities are scheduled by their latest completion times. This represents the extreme case where activity slack times are fully used. No activity in this schedule can be delayed without delaying the project. In Figure 4–7, only one activity is scheduled over the first three days. This may be compared to the schedule in Figure 4–6, which has three starting activities. The schedule in Figure 4–7 may be useful if there is a situation that permits only a few activities to be scheduled in the

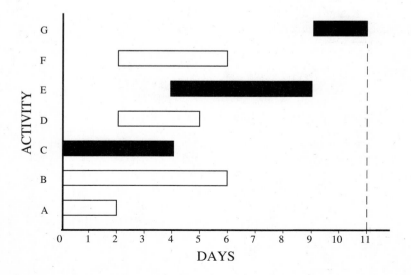

Figure 4-6.   Gantt Chart Based on Earliest Starting Times

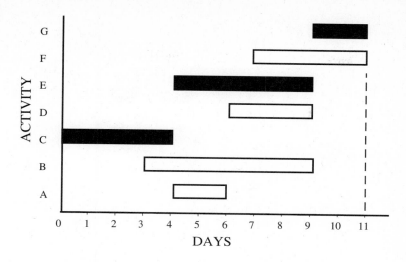

Figure 4-7.   Gantt Chart Based on Latest Starting Times

early stages of the project. Such situations may involve shortage of project personnel, lack of initial budget, time for project initiation, time for personnel training, allowance for learning period, or general resource constraints. Scheduling of activities based on ES times indicates an optimistic view. Scheduling on the basis of LS times represents a pessimistic approach.

4.3.1 Gantt chart Variations

The basic Gantt chart does not show the precedence relationships among activities. The chart can be modified to show these relationships by linking appropriate bars, as shown in Figure 4–8. However, the linked bars become cluttered and confusing for large networks. Figure 4–9 shows a Gantt chart which presents a comparison of planned and actual schedules. Note that two tasks are in progress at the current time indicated in the figure. One of the ongoing tasks is an unplanned task. Figure 4–10 shows a Gantt chart on which important milestones have been indicated. Figure 4–11 shows a Gantt chart in which bars represent a combination of related tasks. Tasks may be combined for scheduling purposes or for conveying functional relationships required on a project. Figure 4–12 presents a Gantt chart of project phases. Each phase is further divided into parts. Figure 4–13 shows a multiple-projects Gantt chart. Multiple-projects charts are useful for evaluating resource allocation strategies. Resource loading over multiple projects may be needed for capital budgeting and cash flow analysis decisions. Figure 4–14 shows a project slippage chart that is useful for project tracking and control. Other variations of the basic Gantt chart may be developed for specific needs.

4.3.2 Activity Crashing and Schedule Compression

Schedule compression refers to reducing the length of a project network. This is often accomplished by crashing activities. *Crashing*, sometimes referred to as expediting,

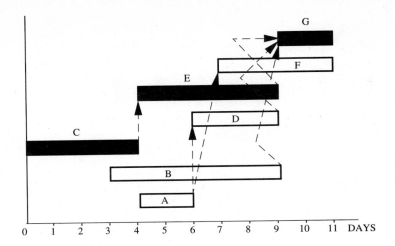

Figure 4-8. Linked Bars in Gantt Chart

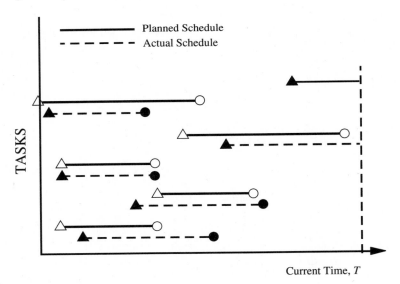

Figure 4-9. Progress-Monitoring Gantt Chart

reduces activity durations, thereby reducing project duration. Crashing is done as a trade-off between shorter task duration and higher task cost. It must be determined whether the total cost savings realized from reducing the project duration is enough to justify the higher costs associated with reducing individual task durations. If there is a delay penalty associated with a project, it may be possible to reduce the total project cost even though individual task costs are increased by crashing. If the cost savings on a delay penalty are higher than the incremental cost of reducing the project duration, then crashing is justified. Under conventional crashing, the further the duration of a project is

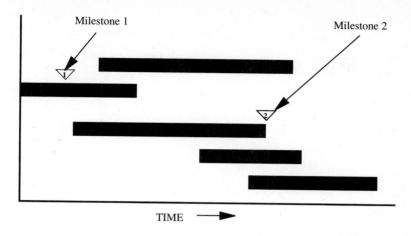

Figure 4-10.   Milestone Gantt Chart

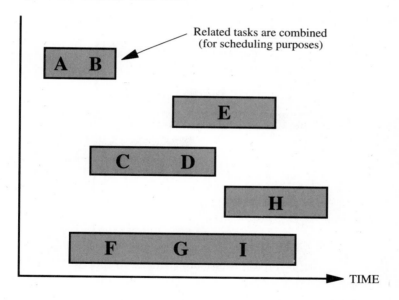

Figure 4-11.   Task-Combination Gantt Chart

compressed, the higher the total cost of the project. The objective is to determine at what point to terminate further crashing in a network. *Normal task duration* refers to the time required to perform a task under normal circumstances. *Crash task duration* refers to the reduced time required to perform a task when additional resources are allocated to it.

If each activity is assigned a range of time and cost estimates, then several combinations of time and cost values will be associated with the overall project. Iterative procedures are used to determine the best time or cost combination for a project. Time-cost trade-off analysis may be conducted, for example, to determine the marginal cost of reducing the duration of the project by one time unit. Table 4–4 presents an exten-

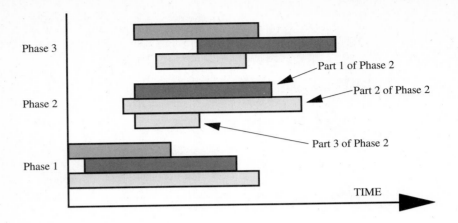

Figure 4-12.   Phase-Based Gantt Chart

Figure 4-13.   Multiple-Projects Gantt Chart

sion of the data for the example problem to include normal and crash times as well as normal and crash costs for each activity. The normal duration of the project is 11 days, as seen earlier, and the normal cost is $2,775.

If all the activities are reduced to their respective crash durations, the total crash cost of the project will be $3,545. In that case, the crash time is found by CPM analysis to be 7 days. The CPM network for the fully crashed project is shown in Figure 4–15. Note that activities C, E, and G remain critical. Sometimes, the crashing of activities may result in additional critical paths. The Gantt chart in Figure 4–16 shows a schedule of the crashed project using the ES times. In practice, one would not crash all activities in a network. Rather, some heuristic would be used to determine which activity should be crashed and by how much. One approach is to crash only the critical activities or those activities with the best ratios of incremental cost versus time reduction. The last column in Table 4–4 presents the respective ratios for the activities in our example. The crashing ratios are computed as

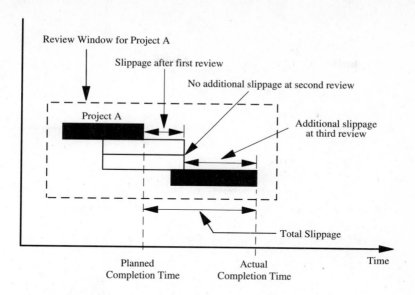

Figure 4-14. Project-Slippage-Tracking Gantt Chart

Table 4-4 Normal and Crash Time and Cost Data

Activity	Normal duration	Normal cost	Crash duration	Crash cost	Crashing ratio
A	2	$210	2	$210	0
B	6	400	4	600	100
C	4	500	3	750	250
D	3	540	2	600	60
E	5	750	3	950	100
F	4	275	3	310	35
G	2	100	1	125	25
		$2,775		$3,545	

$$r = \frac{\text{crash cost} - \text{normal cost}}{\text{normal duration} - \text{crash duration}}$$

This method of computing the crashing ratio gives crashing priority to the activity with the lowest cost slope. It is a commonly used approach to expediting in CPM networks.

Activity G offers the lowest cost per unit time reduction of $25. If our approach is to crash only one activity at a time, we may decide to crash activity G first and evaluate the increase in project cost versus the reduction in project duration. The process can then be repeated for the next best candidate for crashing, which in this case is activity F. The project completion time is not reduced any further since activity F is not a critical activity. After F has been crashed, activity D can then be crashed. This approach is

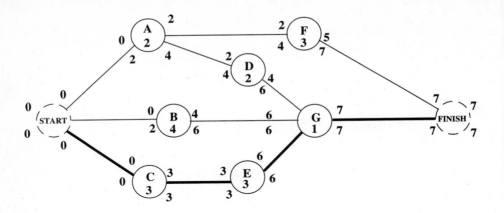

Figure 4-15. Example of Fully Crashed CPM Network

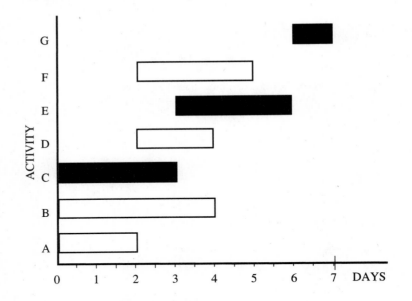

Figure 4-16. Gantt Chart of Fully Crashed CPM Network

repeated iteratively in order of activity preference until no further reduction in project duration can be achieved or until the total project cost exceeds a specified limit.

A more comprehensive analysis is to evaluate all possible combinations of the activities that can be crashed. However, such a complete enumeration would be prohibitive, since there would be a total of 2^c crashed networks to evaluate, where c is the number of activities that can be crashed out of the n activities in the network ($c \leq n$). For our example, only 6 out of the 7 activities in the network can be crashed. Thus, a complete enumeration will involve $2^6 = 64$ alternate networks. Table 4–5 shows 7 of the 64 crashing options. Activity G, which offers the best crashing ratio, reduces the project duration by only 1 day. Even though activities F, D, and B are crashed by a total of 4 days at an incremental cost of $295, they do not generate any reduction in project duration. Activity E is crashed by 2 days and it generates a reduction of 2 days in project duration. Activity C, which is crashed by 1 day, generates a further reduction of 1 day in the project duration. It should be noted that the activities which generate reductions in project duration are the ones that were identified earlier as the critical activities.

Figure 4–17 shows the crashed project duration versus the crashing options and a plot of the total project cost after crashing. As more activities are crashed, the project duration decreases while the total project cost increases. If full enumeration were performed, the plot would contain additional points between the minimum possible project duration of 7 days (fully crashed) and the normal project duration of 11 days (no crashing). Similarly, the plot for total project cost would contain additional points between the normal cost of $2,775 and the crash cost of $3,545.

In general, there may be more than one critical path, so one needs to check for the set of critical activities with the least total crashing ratio in order to minimize the total crashing cost. Also, one needs to update the critical paths every time a set of activities is crashed because new activities may become critical in the meantime. For the network given in Figure 4–15, the path C−E−G is the only critical path throughout $7 \leq T \leq 11$. Therefore, one need not consider crashing other jobs since the incurred cost will not affect the project completion time. There are 12 possible ways one can crash activities C, G, and E in order to reduce the project time.

Table 4–6 defines possible strategies and crashing costs for durations of $7 \leq T \leq 11$. Again, the strategies involve only critical arcs (activities), since crashing a noncritical arc is clearly fruitless. Figure 4–18 is a plot of the strategies with respect to cost and project duration values. The optimal strategy for each T value is the strategy with the minimum cost. Optimal strategies are connected in Figure 4–18. This piecewise linear and convex curve is referred to as the time-cost trade-off curve. Chapter 7 outlines an algorithm for a time-cost trade-off problem.

Several other approaches exist for determining which activities to crash in a project network. Two alternate approaches are presented below for computing the crashing ratio, r. The first one directly uses the criticality of an activity to determine its crashing ratio while the second one uses a computational expression as shown below:

Table 4–5 Selected Crashing Options for CPM Example

Option number	Activities crashed	Network duration	Time reduction	Incremental cost	Total cost
1.	None	11	—	—	$2,775
2.	G	10	1	$25	2,800
3.	G, F	10	0	35	2,835
4.	G, F, D	10	0	60	2,895
5.	G, F, D, B	10	0	200	3,095
6.	G, F, D, B, E	8	2	200	3,295
7.	G, F, D, B, E, C	7	1	250	3,545

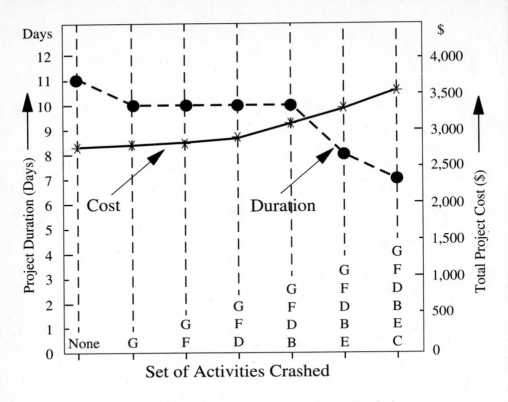

Figure 4-17. Plot of Duration and Cost as a Function of Crashing Options

Table 4-6 Project Compression Strategies

Project duration	Crashing strategy	Description of crashing	Total cost
$T = 11$	S_1	Activities at normal duration	$2,775
$T = 10$	S_2	Crash G by 1 unit	2,800
	S_3	Crash C by 1 unit	3,025
	S_4	Crash E by 1 unit	2,875
$T = 9$	S_5	Crash G and C by 1 unit	3,050
	S_6	Crash G and E by 1 unit	2,900
	S_7	Crash C and E by 1 unit	3,125
	S_8	Crash E by 2 units	2,975
$T = 8$	S_9	Crash G, C, and E by 1 unit	3,150
	S_{10}	Crash G by 1 unit, E by 2 units	3,000
	S_{11}	Crash C by 1 unit, E by 2 units	3,225
$T = 7$	S_{12}	Crash G and C by 1 unit, and E by 2 units	3,500

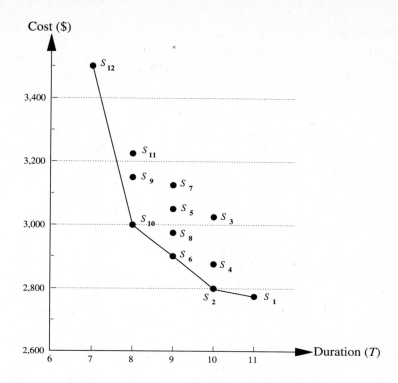

Figure 4-18.   Time-Cost Plots for the Strategies in Table 4–6

$$r = \text{criticality index}$$

$$r = \frac{\text{crash cost} - \text{normal cost}}{(\text{normal duration} - \text{crash duration})(\text{criticality index})}$$

The first approach gives crashing priority to the activity with the highest probability of being on the critical path. In deterministic networks, this refers to the critical activities. In stochastic networks, an activity is expected to fall on the critical path only a percentage of the time. The second approach is a combination of the approach used for the illustrative example and the criticality index approach. It reflects the process of selecting the least-cost expected value. The denominator of the expression represents the expected number of days by which the critical path can be shortened. For different project networks, different crashing approaches should be considered, and the one that best fits the nature of the network should be selected. Other approaches to project compression are presented by Elmaghraby and Pulat (1979), Mustafa and Murphree (1989), and Yau and Ritchie (1990).

4.4 PERT NETWORK ANALYSIS

The program evaluation and review technique (PERT) is an extension of CPM which incorporates variabilities in activity durations into project network analysis. PERT has been used extensively and successfully in practice.

In real life, activities are often prone to uncertainties which determine the actual durations of the activities. In CPM, activity durations are assumed to be deterministic. In PERT, the potential uncertainties in activity durations are accounted for by using three time estimates for each activity. The three time estimates represent the spread of the estimated activity duration. The greater the uncertainty of an activity, the wider the range of the estimates.

4.4.1 PERT Estimates and Formulas

PERT uses the three time estimates and simple equations to compute the expected duration and variance for each activity. The PERT formulas are based on a simplification of the expressions for the mean and variance of a beta distribution. The approximation formula for the mean is a simple weighted average of the three time estimates, with the end points assumed to be equally likely and the mode four times as likely. The approximation formula for PERT is based on the recognition that most of the observations from a distribution will lie within plus or minus three standard deviations, or a spread of six standard deviations. This leads to the simple method of setting the PERT formula for standard deviation equal to one-sixth of the estimated duration range. While there is no theoretical validation for these approximation approaches, the PERT formulas do facilitate ease of use. The formulas are presented below:

$$t_e = \frac{a + 4m + b}{6}$$

$$s^2 = \frac{(b - a)^2}{36}$$

where

a = optimistic time estimate
m = most likely time estimate
b = pessimistic time estimate ($a < m < b$)
t_e = expected time for the activity
s^2 = variance of the duration of the activity

After obtaining the estimate of the duration for each activity, the network analysis is carried out in the same manner previously illustrated for the CPM approach. The major steps in PERT analysis are as follows:

1. Obtain three time estimates a, m, and b for each activity.
2. Compute the expected duration for each activity by using the formula for t_e.

3. Compute the variance of the duration of each activity from the formula for s^2. It should be noted that CPM analysis cannot calculate variance of activity duration, since it uses a single time estimate for each activity.

4. Compute the expected project duration, T_e. As in the case of CPM, the duration of a project in PERT analysis is the sum of the durations of the activities on the critical path.

5. Compute the variance of the project duration as the sum of the variances of the activities on the critical path. The variance of the project duration is denoted by S^2. It should be recalled that CPM cannot compute the variance of the project duration, since variances of activity durations are not computed.

6. If there are two or more critical paths in the network, choose the one with the largest variance to determine the project duration and the variance of the project duration. Thus, PERT is pessimistic with respect to the variance of project duration when there are multiple critical paths in the project network. For some networks, it may be necessary to perform a mean-variance analysis to determine the relative importance of the multiple paths by plotting the expected project duration versus the path duration variance.

7. If desired, compute the probability of completing the project within a specified time period. This is not possible under CPM.

In practice, a question often arises as to how to obtain good estimates of a, m, and b. Several approaches can be used in obtaining the required time estimates for PERT. Some of the approaches are

- Estimates furnished by an experienced person
- Estimates extracted from standard time data
- Estimates obtained from historical data
- Estimates obtained from simple regression and/or forecasting
- Estimates generated by simulation
- Estimates derived from heuristic assumptions
- Estimates dictated by customer requirements

Several researchers including Golenko-Ginzburg (1988, 1989a, 1989b), Donaldson (1965), Johnson and Schou (1990), Littlefield and Randolph (1987), Sasieni (1986), Troutt (1989), Farnum and Stanton (1987), Gallagher (1987), Grubbs (1962), MacCrimmon and Ryavec (1964), McBride and McClelland (1967), and Welsh (1965) have addressed the deficiencies in the PERT estimation procedure.

The pitfall of using estimates furnished by an individual is that they may be inconsistent, since they are limited by the experience and personal bias of the person providing them. Individuals responsible for furnishing time estimates are usually not experts in estimation, and they generally have difficulty in providing accurate PERT time estimates. There is often a tendency to select values of a, m, and b that are optimistically skewed. This is because a conservatively large value is typically assigned to b by inexperienced individuals.

The use of time standards, on the other hand, may not reflect the changes occurring in the current operating environment due to new technology, work simplification, new personnel, and so on. The use of historical data and forecasting is very popular because estimates can be verified and validated by actual records. In the case of regression and forecasting, there is the danger of extrapolation beyond the data range used for fitting the regression and forecasting models. If the sample size in a historical data set is sufficient and the data can be assumed to reasonably represent prevailing operating conditions, the three PERT estimates can be computed as follows:

$$\hat{a} = \bar{t} - kR$$
$$\hat{m} = \bar{t}$$
$$\hat{b} = \bar{t} + kR$$

where

R = range of the sample data
$\bar{t}$ = arithmetic average of the sample data
$k = 3/d_2$
d_2 = an adjustment factor for estimating the standard deviation of a population

If $kR > \bar{t}$, then set $a = 0$ and $b = 2\bar{t}$. The factor d_2 is widely tabulated in the quality control literature as a function of the number of sample points, n. Selected values of d_2 are presented below.

n	5	10	15	20	25	30	40	50	75	100
d_2	2.326	3.078	3.472	3.735	3.931	4.086	4.322	4.498	4.806	5.015

In practice, probability distributions of activity times can be determined from historical data. The procedure involves three steps:

1. Appropriate organization of the historical data into histograms.
2. Determination of a distribution that reasonably fits the shape of the histogram.
3. Testing the goodness-of-fit of the hypothesized distribution by using an appropriate statistical model. The chi-square test and the Kolmogrov-Smirnov (K-S) test are two popular methods for testing goodness-of-fit. Most statistical texts present the details of how to carry out goodness-of-fit tests.

4.4.2 The Beta Distribution

PERT analysis assumes that the probabilistic properties of activity duration can be modeled by the beta probability density function. The beta distribution is defined by two end points and two shape parameters. The beta distribution was chosen by the original developers of PERT as a reasonable distribution to model activity times because it has finite end points and can assume a variety of shapes based on different shape parameters. While the true distribution of activity time will rarely ever be known, the beta distribution serves as an acceptable model. Figure 4–19 shows examples of alternate

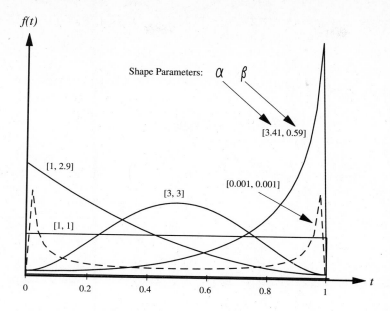

Figure 4-19. Alternate Shapes of the Beta Distribution

shapes of the standard beta distribution between 0 and 1. The uniform distribution between 0 and 1 is a special case of the beta distribution with both shape parameters equal to one.

Many analytical studies of the beta distribution and the PERT approach have been presented in the literature. Interested readers should consult MacCrimmon and Ryavec (1964), Grubbs (1962), McBride and McClelland (1967), Van Slyke (1963), Sasieni (1986), Welsh (1965), Farnum and Stanton (1987), Troutt (1989), Gallagher (1987), and Golenko-Ginzburg (1988, 1989a). The standard beta distribution is defined over the interval 0 to 1, while the general beta distribution is defined over any interval a to b. The general beta probability density function is given by

$$f(t) = \frac{\Gamma(\alpha + \beta)}{\Gamma(\alpha)\Gamma(\beta)} \cdot \frac{1}{(b-a)^{\alpha+\beta-1}} \cdot (t-a)^{\alpha-1}(b-t)^{\beta-1}$$

$$\text{for } a \leq t \leq b \quad \text{and} \quad \alpha > 0, \ \beta > 0$$

where

a = lower end point of the distribution
b = upper end point of the distribution
α, β are the shape parameters for the distribution.

The mean, variance, and mode of the general beta distribution are defined as

$$\mu = a + (b - a)\frac{\alpha}{\alpha + \beta}$$

$$\sigma^2 = (b - a)^2 \frac{\alpha\beta}{(\alpha + \beta + 1)(\alpha + \beta)^2}$$

$$m = \frac{\alpha(\beta - 1) + b(\alpha - 1)}{\alpha + \beta - 2}$$

The general beta distribution can be transformed into a standardized distribution by changing its domain from $[a, b]$ to the unit interval, $[0,1]$. This is accomplished by using the relationship $t_g = a + (b - a)t_s$, where t_s is the standard beta random variable between 0 and 1. This yields the standardized beta distribution, given by

$$f(t) = \frac{\Gamma(\alpha + \beta)}{\Gamma(\alpha)\Gamma(\beta)} t^{\alpha-1}(1 - t)^{\beta-1}; \qquad 0 < t < 1; \ \alpha, \beta > 0$$

$$= 0; \qquad\qquad\qquad\qquad\qquad \text{elsewhere}$$

with mean, variance, and mode defined as

$$\mu = \frac{\alpha}{\alpha + \beta}$$

$$\sigma^2 = \frac{\alpha\beta}{(\alpha + \beta + 1)(\alpha + \beta)^2}$$

$$m = \frac{a(\beta - 1) + b(\alpha - 1)}{\alpha + \beta - 2}$$

4.4.3 Triangular Distribution

The triangular probability density function has been used as an alternative to the beta distribution for modeling activity times. The triangular density has three essential parameters: minimum value (a), mode (m), and maximum (b). The triangular density function is defined mathematically as

$$f(t) = \frac{2(t - a)}{(m - a)(b - a)}; \qquad a \leq t \leq m$$

$$= \frac{2(b - t)}{(b - m)(b - a)}; \qquad m \leq t \leq b$$

with mean and variance defined, respectively, as

$$\mu = \frac{a + m + b}{3}$$

$$\sigma^2 = \frac{a(a - m) + b(b - a) + m(m - b)}{18}$$

Figure 4–20 presents a graphical representation of the triangular density function. The three time estimates of **PERT** can be inserted into the expression for the mean

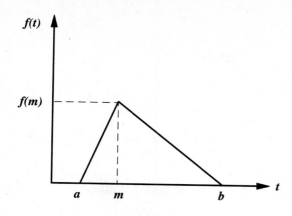

Figure 4-20. Triangular Probability
Density Function

of the triangular distribution to obtain an estimate of the expected activity duration. Note that in the conventional PERT formula, the mode (m) is assumed to carry four times as much weight as either a or b when calculating the expected activity duration. By contrast, under the triangular distribution, the three time estimates are assumed to carry equal weights.

4.4.4 Uniform Distribution

For cases where only two time estimates instead of three are to be used for network analysis, the uniform density function may be assumed for the activity times. This is acceptable for situations where the extreme limits of an activity duration can be estimated and it can be assumed that the intermediate values are equally likely to occur. The uniform distribution is defined mathematically as

$$f(t) = \frac{1}{b-a}; \qquad a \leq t \leq b$$
$$= 0; \qquad\qquad \text{otherwise}$$

with mean and variance defined, respectively, as

$$\mu = \frac{a+b}{2}$$
$$\sigma^2 = \frac{(b-a)^2}{12}$$

Figure 4–21 presents a graphical representation of the uniform distribution. In the case of the uniform distribution, the expected activity duration is computed as the average of the upper and lower limits of the distribution. The appeal of using only two time estimates a and b is that the estimation error due to subjectivity can be reduced and the estimation task simplified. Even when a uniform distribution is not assumed, other statistical distributions can be modeled over the range of a to b.

Other distributions that have been explored for activity time modeling include the normal distribution, lognormal distribution, truncated exponential distribution, and Weibull distribution. Once the expected activity durations have been computed, the

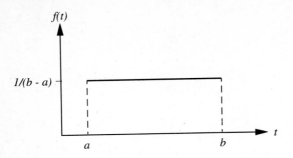

Figure 4-21. Uniform Probability Density Function

analysis of the activity network is carried out just as in the case of single-estimate CPM network analysis.

4.4.5 Distribution of Project Duration

Regardless of the distribution assumed for activity durations, the *central limit theorem* suggests that the distribution of the project duration will be approximately normally distributed. The theorem says that the distribution of averages obtained from any probability density function will be approximately normally distributed if the sample size is large and the averages are independent. In mathematical terms, the theorem is stated as follows:

Central limit theorem.   Let $X_1, X_2, \ldots, X_N$ be independent and identically distributed random variables. Then the sum of the random variables tends to be normally distributed for large values of N. The sum is defined as

$$T = X_1 + X_2 + \ldots + X_N$$

In activity network analysis, T represents the total project length as determined by the sum of the durations of the activities on the critical path. The mean and variance of T are expressed as

$$\mu = \sum_{i=1}^{N} E[X_i]$$

$$\sigma^2 = \sum_{i=1}^{N} V[X_i]$$

where

$E[X_i]$ = expected value of random variable X_i
$V[X_i]$ = variance of random variable X_i

When applying the central limit theorem to activity networks, it should be noted that the assumption of independent activity times may not always be satisfied. Because of precedence relationships and other interdependencies of activities, some activity durations may not be independent.

4.4.6 Probability Calculation

If the project duration T_e can be assumed to be approximately normally distributed based on the central limit theorem, then the probability of meeting a specified deadline T_d can be computed by finding the area under the standard normal curve to the left of T_d. Figure 4–22 shows an example of a normal distribution describing the project duration.

Using the familiar transformation formula below, a relationship between the standard normal random variable z and the project duration variable can be obtained:

$$z = \frac{T_d - T_e}{S}$$

where

T_d = specified deadline
T_e = expected project duration based on network analysis
S = standard deviation of the project duration

The probability of completing a project by the deadline T_d is then computed as

$$P(T \leq T_d) = P\left(z \leq \frac{T_d - T_e}{S}\right)$$

The probability is obtained from the standard normal table presented in the appendix A. The following example illustrates the procedure for probability calculations in PERT.

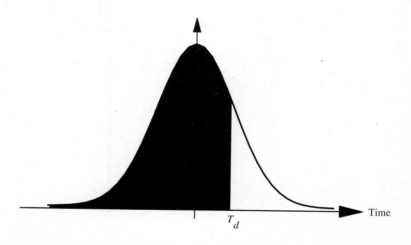

Figure 4-22.   Area under the Normal Curve

4.4.7 PERT Network Example

Suppose we have the project data presented in Table 4–7. The expected activity durations and variances as calculated by the PERT formulas are shown in the two right-hand columns of the table. Figure 4–23 shows the PERT network. Activities C, E, and G are shown to be critical, and the project completion time is 11 time units.

Table 4-7   PERT Project Data

Activity	Predecessors	a	m	b	t_e	s^2
A	—	1	2	4	2.17	0.2500
B	—	5	6	7	6.00	0.1111
C	—	2	4	5	3.83	0.2500
D	A	1	3	4	2.83	0.2500
E	C	4	5	7	5.17	0.2500
F	A	3	4	5	4.00	0.1111
G	B, D, E	1	2	3	2.00	0.1111

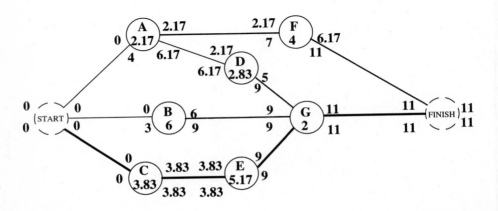

Figure 4-23.   PERT Network Example

The probability of completing the project on or before a deadline of 10 time units (i.e., $T_d = 10$) is calculated as

$$T_e = 11$$
$$S^2 = V[C] + V[E] + V[G]$$
$$= 0.25 + 0.25 + 0.1111$$
$$= 0.6111$$
$$S = \sqrt{0.6111}$$
$$= .7817$$
$$P(T \leq T_d) = P(T \leq 10)$$
$$= P\left(z \leq \frac{10 - T_e}{S}\right)$$
$$= P\left(z \leq \frac{10 - 11}{0.7817}\right)$$
$$= P(z \leq -1.2793)$$
$$= 1 - P(z \leq 1.2793)$$
$$= 1 - 0.8997$$
$$= 0.1003$$

Thus, there is just over 10% probability of finishing the project within 10 days. By contrast, the probability of finishing the project in 13 days is calculated as

$$P(T \leq 13) = P\left(z \leq \frac{13 - 11}{0.7817}\right)$$
$$= P(z \leq 2.5585)$$
$$= 0.9948$$

This implies that there is over 99% probability of finishing the project within 13 days. Note that the probability of finishing the project in exactly 13 days will be 0. An exercise at the end of this chapter requires the reader to show that $P(T = T_d) = 0$. If we desire the probability that the project can be completed within a certain lower limit (T_L) and a certain upper limit (T_U), the computation will proceed as follows: Let $T_L = 9$ and $T_U = 11.5$. Then,

$$P(T_L \leq T \leq T_U) = P(9 \leq T \leq 11.5)$$
$$= P(T \leq 11.5) - P(T \leq 9)$$
$$= P\left(z \leq \frac{11.5 - 11}{0.7817}\right) - P\left(z \leq \frac{9 - 11}{0.7817}\right)$$
$$= P(z \leq 0.6396) - P(z \leq -2.5585)$$
$$= P(z \leq 0.6396) - [1 - P(z \leq -2.5585)]$$
$$= 0.7389 - [1 - 0.9948]$$
$$= 0.7389 - 0.0052$$
$$= 0.7337$$

4.5 PRECEDENCE DIAGRAMMING METHOD

The precedence diagramming method (PDM) was developed in the early 1960s as an extension of the basic PERT/CPM network analysis (Crandall 1973; Wiest 1981; Moder et al. 1983; Harhalakis 1990). PDM permits mutually dependent activities to be performed partially in parallel instead of serially. The usual finish-to-start dependencies between activities are relaxed to allow activities to be overlapped. This facilitates schedule compression. An example is the requirement that concrete should be allowed to dry for a number of days before drilling holes for handrails. That is, drilling cannot start until so many days after the completion of concrete work. This is a finish-to-start constraint. The time between the finishing time of the first activity and the starting time of the second activity is called the *lead-lag* requirement between the two activities. Figure 4–24 shows the graphical representation of the basic lead-lag relationships between activity A and activity B. The terminology presented in Figure 4–24 is explained as follows.

SS_{AB} (start-to-start) lead: This specifies that activity B cannot start until activity A has been in progress for at least SS time units.

FF_{AB} (finish-to-finish) lead: This specifies that activity B cannot finish until at least FF time units after the completion of activity A.

FS_{AB} (finish-to-start) lead: This specifies that activity B cannot start until at least FS time units after the completion of activity A. Note that PERT/CPM approaches use $FS_{AB} = 0$ for network analysis.

SF_{AB} (start-to-finish) lead: This specifies that there must be at least SF time units between the start of activity A and the completion of activity B.

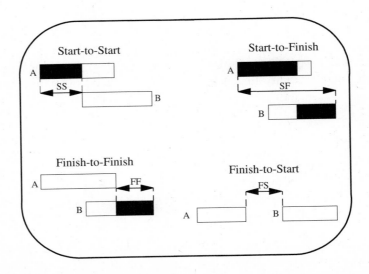

Figure 4-24. Lead-Lag Relationships in PDM

The leads or lags may, alternately, be expressed in percentages rather than time units. For example, we may specify that 25 percent of the work content of activity A must be completed before activity B can start. If percentage of work completed is used for determining lead-lag constraints, then a reliable procedure must be used for estimating the percent completion. If the project work is broken up properly using work breakdown structure (WBS), it will be much easier to estimate percent completion by evaluating the work completed at the elementary task levels. The lead-lag relationships may also be specified in terms of *at most* relationships instead of *at least* relationships. For example, we may have at most a FF lag requirement between the finishing time of one activity and the finishing time of another activity. Splitting activities often simplifies the implementation of PDM, as will be shown later with some examples. Some of the factors that will determine whether or not an activity can be split are technical limitations affecting splitting of a task, morale of the person working on the split task, setup times required to restart split tasks, difficulty involved in managing resources for split tasks, loss of consistency of work, and management policy about splitting jobs.

Figure 4–25 presents a simple CPM network consisting of three activities. The activities are to be performed serially and each has an expected duration of 10 days. The conventional CPM network analysis indicates that the duration of the network is 30 days. The earliest times and the latest times are as shown in the figure.

The Gantt chart for the example is shown in Figure 4–26. For comparison, Figure 4–27 shows the same network but with some lead-lag constraints. For example, there is an SS constraint of 2 days and an FF constraint of 2 days between activities A and B. Thus, activity B can start as early as 2 days after activity A starts, but it cannot finish until 2 days after the completion of A. In other words, *at least* 2 days must be between the starting times of A and B. Likewise, *at least* 2 days must separate the finishing time of A and the finishing time of B. A similar precedence relationship exists between activity B and activity C. The earliest and latest times obtained by considering the lag constraints are indicated in Figure 4–27.

The calculations show that if B is started just 2 days after A is started, it can be completed as soon as 12 days as opposed to the 20 days obtained in the case of conventional CPM. Similarly, activity C is completed at time 14, which is considerably less than the 30 days calculated by conventional CPM. The lead-lag constraints allow us to compress or overlap activities. Depending on the nature of the tasks involved, an activity does not have to wait until its predecessor finishes before it can start. Figure 4–28 shows the Gantt chart for the example incorporating the lead-lag constraints. It should be noted that a portion of a succeeding activity can be performed simultaneously with a portion of the preceding activity.

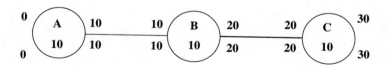

Figure 4-25. Serial Activities in CPM Network

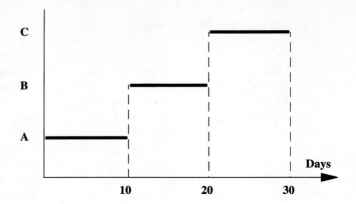

Figure 4-26. Gantt Chart of Serial Activities in CPM Example

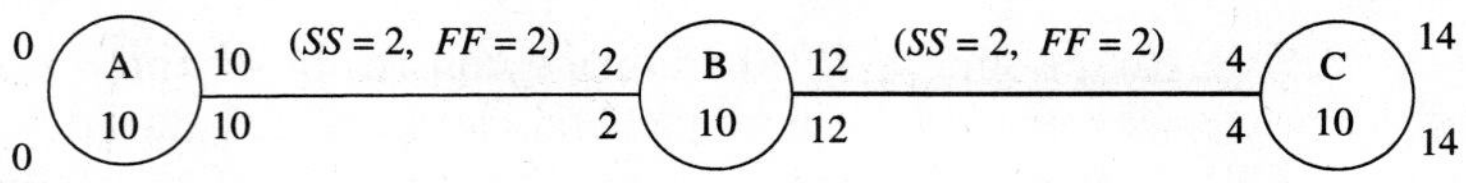

Figure 4-27. PDM Network Example

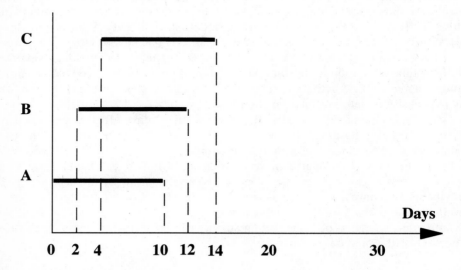

Figure 4-28. Gantt Chart for PDM Example

A portion of an activity that overlaps with a portion of another activity may be viewed as a distinct portion of the required work. Thus, partial completion of an activity may be evaluated. Figure 4–29 shows how each of the three activities is partitioned into contiguous parts. Even though there is no physical break or termination of work in any activity, the distinct parts (beginning and ending) can still be identified. This means that there is no physical splitting of the work content of any activity. The distinct parts are determined on the basis of the amount of work that must be completed before or after another activity, as dictated by the lead-lag relationships. In Figure 4–29, activity A is partitioned into the parts A_1 and A_2. The duration of A_1 is 2 days because there is an $SS = 2$ relationship between activity A and activity B. Since the original duration of A is 10 days, the duration of A_2 is then calculated to be $10 - 2 = 8$ days.

Likewise, activity B is partitioned into the parts B_1, B_2, and B_3. The duration of B_1 is 2 days because there is an $SS = 2$ relationship between activity B and activity C. The duration of B_3 is also 2 days because there is an $FF = 2$ relationship between activity A and activity B. Since the original duration of B is 10 days, the duration of B_2 is calculated to be $10-(2+2) = 6$ days. In a similar fashion, activity C is partitioned into C_1 and C_2. The duration of C_2 is 2 days because there is an $FF = 2$ relationship between activity B and activity C. Since the original duration of C is 10 days, the duration of C_1 is then calculated to be $10 - 2 = 8$ days. Figure 4–30 shows a conventional CPM network drawn for the three activities after they are partitioned into distinct parts. The conventional forward and backward passes reveal that all the activity parts are on the critical path. This makes sense, since the original three activities are performed serially and no physical splitting of activities has been performed. Note that there are three critical paths in Figure 4–30, each with a length of 14 days. It should also be noted that the distinct parts of each activity are performed contiguously.

Figure 4–31 shows an alternate example of three serial activities. The conventional CPM analysis shows that the duration of the network is 30 days. When lead-lag constraints are introduced into the network as shown in Figure 4–32, the network duration is compressed to 18 days.

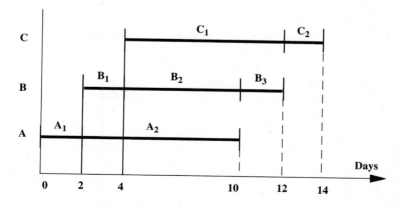

Figure 4-29. Partitioning of Activities in PDM Example

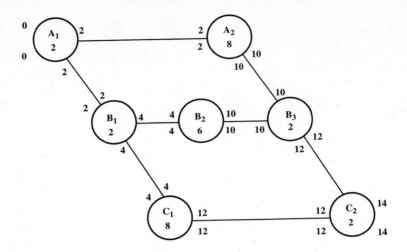

Figure 4-30.   CPM Network of Partitioned Activities

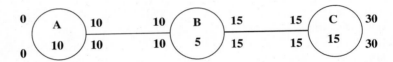

Figure 4-31.   Another CPM Example of Serial Activities

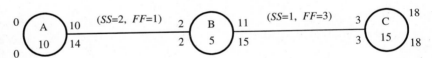

Figure 4-32.   Compressed PDM Network

In the forward pass computations in Figure 4–32, note that the earliest completion time of B is time 11, because there is an $FF = 1$ restriction between activity A and activity B. Since A finishes at time 10, B cannot finish until at least time 11. Even though the earliest starting time of B is time 2 and its duration is 5 days, its earliest completion time cannot be earlier than time 11. Also note that C can start as early as time 3 because there an $SS = 1$ relationship between B and C. Thus, given a duration of 15 days for C, the earliest completion time of the network is $3 + 15 = 18$ days. The difference between the earliest completion time of C and the earliest completion time of B is $18 - 11 = 7$ days, which satisfies the $FF = 3$ relationship between B and C.

In the backward pass, the latest completion time of B is 15 (i.e., $18 - 3 = 15$), since there is an $FF = 3$ relationship between activity B and activity C. The latest start time for B is time 2 (i.e., $3 - 1 = 2$), since there is an $SS = 1$ relationship between activity B and activity C. If we are not careful, we may erroneously set the latest start time of B to 10 (i.e., $15 - 5 = 10$). But that would violate the $SS = 1$ restriction between B and C. The latest completion time of A is found to be 14 (i.e., $15 - 1 = 14$), since there is an *FF* = 1 relationship between A and B. All the earliest times and latest times at each node must be evaluated to ensure that they conform to all the lead-lag constraints. When computing earliest start or earliest completion times, the smallest possible value that satisfies the lead-lag constraints should be used. By the same reasoning, when computing the latest start or latest completion times, the largest possible value that satisfies the lead-lag constraints should be used.

Manual evaluations of the lead-lag precedence network analysis can become very tedious for large networks. A computer program may be used to simplify the implementation of PDM. If manual analysis must be done for PDM computations, it is suggested that the network be partitioned into more manageable segments. The segments may then be linked after the computations are completed. The expanded CPM network in Figure 4–33 was developed on the basis of the precedence network in Figure 4–32. It is seen that activity A is partitioned into two parts, activity B is partitioned into three parts, and activity C is partitioned into two parts. The forward and backward passes show that only the first parts of activities A and B are on the critical path. Both parts of activity C are on the critical path.

Figure 4–34 shows the corresponding earliest-start Gantt chart for the expanded network. Looking at the earliest start times, one can see that activity B is physically split at the boundary of B_2 and B_3 in such a way that B_3 is separated from B_2 by 4 days. This implies that work on activity B is temporarily stopped at time 6 after B_2 is finished and is not started again until time 10. Note that despite the 4-day delay in starting B_3, the entire project is not delayed. This is because B_3, the last part of activity B, is not on the critical path. In fact, B_3 has a total slack of 4 days. In a situation like this, the duration of activity B can actually be increased from 5 days to 9 days without any adverse effect on the project duration. It should be recognized, however, that increasing the duration of an activity may have negative implications for project cost and personnel productivity.

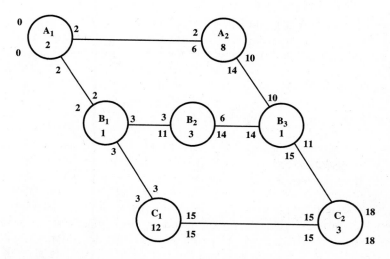

Figure 4-33. CPM Expansion of Second PDM Example

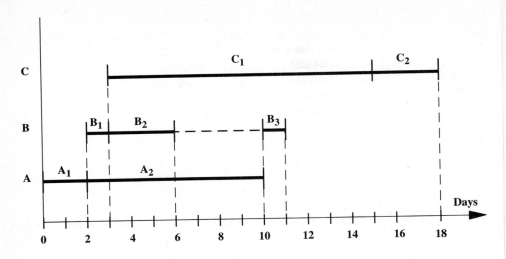

Figure 4-34. Compressed PDM Schedule Based on ES Times

If the physical splitting of activities is not permitted, then the best option available in Figure 4–34 is to stretch the duration of B_2 so as to fill up the gap from time 6 to time 10. An alternative is to delay the starting time of B_1 until time 4 so as to use up the 4-day slack right at the beginning of activity B. Unfortunately, delaying the starting time of B_1 by 4 days will delay the overall project by 4 days, since B_1 is on the critical path as shown in Figure 4–33. The project analyst will need to evaluate the appropriate trade-offs among splitting activities, delaying activities, increasing activity durations, and incurring higher project costs. The prevailing project scenario should be considered when making such trade-off decisions. Figure 4–35 shows the Gantt chart for the compressed PDM schedule based on latest start times. In this case, it will be necessary to split both activities A and B even though the total project duration remains the same at 18 days. If activity splitting is to be avoided, then we can increase the duration of activity A from 10 to 14 days and the duration of B from 5 to 13 days without adversely affecting the entire project duration. The important benefit of precedence diagramming is that the ability to overlap activities facilitates some flexibilities in manipulating individual activity times and compressing the project duration.

4.5.1 Anomalies in PDM Networks

Care must be exercised when working with PDM networks because of the potential for misuse or misinterpretation. Because of the lead and lag requirements, activities that do not have any slacks may appear to have generous slacks. Also, *reverse critical activities* may occur in PDM. Reverse critical activities are activities that can cause a decrease in project duration when their durations are increased. This may happen when the critical path enters the completion of an activity through a finish lead-lag constraint. Also, if a finish-to-finish dependency and a start-to-start dependency are connected to a reverse critical task, a reduction in the duration of the task may actually lead to an increase in the project duration. Figure 4–36 illustrates this anomalous situation. The

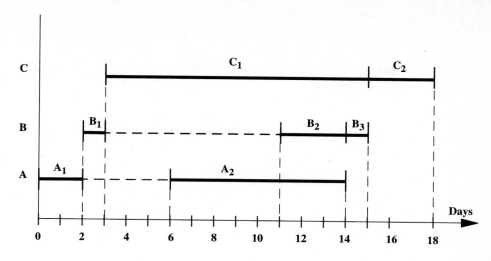

Figure 4-35. Compressed PDM Schedule Based on LS Times

finish-to-finish constraint between A and B requires that B should finish no earlier than 20 days. If the duration of task B is reduced from 10 days to 5 days, the start-to-start constraint between B and C forces the starting time of C to be shifted forward by 5 days, thereby resulting in a 5-day increase in the project duration.

The preceding anomalies can occur without being noticed in large PDM networks. One safeguard against their adverse effects is to make only one activity change at a time and document the resulting effect on the network structure and length. Wiest (1981) suggests using the following categorizations for the unusual characteristics of activities in PDM networks:

Normal critical (NC): This refers to an activity for which the project duration shifts in the same direction as the shift in the duration of the activity.

Reverse critical (RC): This refers to an activity for which the project duration shifts in the reverse direction to the shift in the duration of the activity.

Bicritical (BC): This refers to an activity for which the project duration increases as a result of any shift in the duration of the activity.

Start critical (SC): This refers to an activity for which the project duration shifts in the direction of the shift in the start time of the activity but is unaffected (neutral) by a shift in the overall duration of the activity.

Finish critical (FC): This refers to an activity for which the project duration shifts in the direction of the shift in the finish time of the activity but is unaffected (neutral) by a shift in the overall duration of the activity.

Mid-normal Critical (MNC): This refers to an activity whose mid portion is normal critical.

Mid-reverse Critical (MRC): This refers to an activity whose mid portion is reverse critical.

Mid-BiCritical (MBC): This refers to an activity whose mid portion is bicritical.

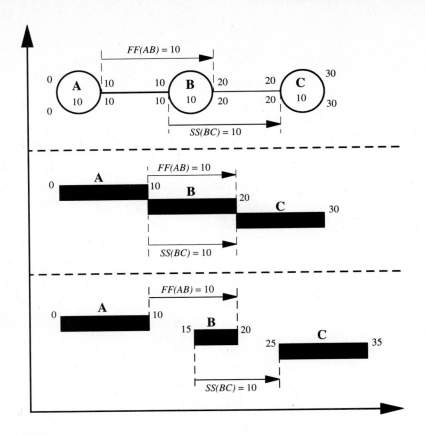

Figure 4-36. Reverse Critical Activity in PDM Network

4.6 COMPLEXITY OF PROJECT NETWORKS

The performance of a scheduling heuristic will be greatly influenced by the complexity of the project network. The more activities there are in the network and the more resource types are involved, the more complex the scheduling effort. Numerous analytical experiments (Badiru 1988c) have revealed the lack of consistency in heuristic performances. Some heuristics perform well for both small and large projects. Some perform well only for small projects. Still, some heuristics that perform well for certain types of small projects may not perform well for other projects of comparable size. The implicit network structure based on precedence relationships and path interconnections influences network complexity and, hence, the performance of scheduling heuristics. The complexity of a project network may indicate the degree of effort that has been devoted to planning the project. The better the planning for a project, the lower the complexity of the project network can be expected to be. This is because many of the redundant interrelationships among activities can be identified and eliminated through better planning.

There have been some attempts to quantify the complexity of project networks. Since the structures of projects vary from very simple to very complex, it is desirable to have a measure of how difficult it will be to schedule a project. Kaimann (1974) presented the following coefficients of network complexity (C):

For PERT networks,
$$C = (\text{number of activities})^2/(\text{number of events})$$
where an event is defined as an end point (or node) of an activity.

For precedence networks,
$$C = (\text{preceding work items})^2/(\text{work items})$$

The above expressions represent simple measures of the degree of interrelationship of the project network. Davies (1974) presented the following measure for network complexity:

$$C = 2(A - N + 1)/(N - 1)(N - 2)$$

where A is the number of activities and N is the number of nodes in the project network. T. J. R. Johnson (1967) presented a measure defined as the total activity density, D, to convey the complexity of a project network. The network density is defined as

$$D = \sum_{i=1}^{N} \text{Max}\{0, (p_i - s_i)\}$$

where N is the number of activities, p_i is the number of predecessor activities for activity i, and s_i is the number of successor activities for activity i. Davis (1975) presented an alternate measure of network complexity (C); a measure of network density (D); a measure of total work content for resource type j (w_j); an obstruction factor (O), which is a measure of the ratio of excess resource requirements to total work content; adjusted obstruction per period based on earliest start time schedule (O_{est}); adjusted obstruction per period based on latest start time schedule (O_{1st}); and a resource utilization factor (U). These measures are

$$C = \frac{\text{number of activities}}{\text{number of nodes}}$$

$$D = \frac{\text{sum of job durations}}{\text{sum of job durations} + \text{total free slack}}$$

$$w_j = \sum_{i=1}^{N} d_i r_{ij}$$
$$= \sum_{t=1}^{CP} r_{jt}$$

where

d_i = duration of job i
r_{ij} = per-period requirement of resource type j by job i
t = time period
N = number of jobs
CP = original critical path duration
r_{jt} = total resource requirements of resource type j in time period t

$$O = \sum_{j=1}^{M} O_j$$

$$= \sum_{j=1}^{M} \left(\frac{\sum_{t=1}^{CP} \text{Max}\{0, r_{jt} - A_j\}}{w_j} \right)$$

where:

O_j = the obstruction factor for resource type j
CP = original critical path duration
A_j = units of resource type j available per period
M = number of different resource types
w_j = total work content for resource type j
r_{jt} = total resource requirements of resource type j in time period t

$$O_{\text{est}} = \sum_{j=1}^{m} \left[\frac{\sum_{t=1}^{CP} \text{Max}\{0, r_{jt(\text{est})} - A_j\}}{(M)(CP)} \right]$$

where $r_{jt(\text{est})}$ is the total resource requirements of resource type j in time period t based on earliest start times.

$$O_{\text{est}} = \sum_{j=1}^{m} \left[\frac{\sum_{t=1}^{CP} \text{Max}\{0, r_{jt(\text{1st})} - A_j\}}{(M)(CP)} \right]$$

where $r_{jt(\text{1st})}$ is the total resource requirements of resource type j in time period t based on latest start times. The measures O_{est} and O_{1st} incorporate the calculation of excess resource requirements adjusted by the number of periods and the number of different resource types.

$$U = \underset{j}{\text{Max}}\{f_j\}$$

$$= \underset{j}{\text{Max}}\left\{\frac{w_j}{(CP)(A_j)}\right\}$$

where f_j is the resource utilization factor for resource type j. This measures the ratio of the total work content to the total work initially available. Badiru (1988a) defined another measure of the complexity of a project network as

$$\lambda = \frac{p}{d}\left[\left(1-\frac{1}{L}\right)\sum_{i=1}^{L}t_i + \sum_{j=1}^{R}\left(\frac{\sum_{i=1}^{L}t_i x_{ij}}{Z_j}\right)\right]$$

where:

λ = project network complexity
L = number of activities in the network
t_i = expected duration for activity i
R = number of resource types
x_{ij} = units of resource type j required by activity i
Z_j = maximum units of resource type j available
p = maximum number of immediate predecessors in the network
d = PERT duration of the project with no resource constraint

The terms in the expression for the complexity are explained as follows: The maximum number of immediate predecessors, p, is a multiplicative factor that increases the complexity and potential for bottlenecks in a project network. The $(1-1/L)$ term is a fractional measure (between 0.0 and 1.0) that indicates the time intensity or work content of the project. As L increases, the quantity $(1-1/L)$ increases, and a larger fraction of the total time requirement (sum of t_i) is charged to the network complexity. Conversely, as L decreases, the network complexity decreases proportionately with the total time requirement. The sum of $(t_i x_{ij})$ indicates the time-based consumption of a given resource type j relative to the maximum availability. The term is summed over all the different resource types. Having PERT duration in the denominator helps to express the complexity as a dimensionless quantity by canceling out the time units in the numerator. In addition, it gives the network complexity per unit of total project duration.

In addition to the approaches presented above, other project network complexity measures have been proposed in the literature (Patterson 1976; Elmaghraby and Herroelen 1980). There is always a debate as to whether or not the complexity of a project can be accurately quantified. There are several quantitative and qualitative factors with unknown interactions that are present in any project network. As a result, any measure of project complexity should be used as a relative measure of comparison rather than as an absolute indication of the difficulty involved in scheduling a given project. Since the performance of a scheduling approach can deteriorate sometimes with the increase in project size, a further comparison of the rules may be done on the basis of a collection of large projects. A major deficiency in the existing measures of project network complexity is that there is a lack of well-designed experiments to compare and verify the effectiveness of the measures. Also, there is usually no guideline as to whether a complexity measure should be used as an ordinal or a cardinal measure, as is illustrated in the following example.

4.6.1 Example of Complexity Computation

Table 4–8 presents a sample project for illustrating the network complexity measure presented by Badiru (1988a).

Using the formulation for network complexity presented by Badiru (1988a), we obtain

$$p = 1 \qquad L = 6 \qquad d = 6.33$$

$$\sum_{i=1}^{6} t_i = 13.5, \qquad \sum_{i=1}^{6} t_i x_{i1} = 22.5, \qquad \sum_{i=1}^{6} t_i x_{i2} = 6.3$$

$$\lambda = \frac{1}{6.33}\left[\left(\frac{6-1}{6}\right)(13.5) + \left(\frac{22.58}{5} + \frac{6.25}{2}\right)\right]$$
$$= 2.99$$

If the above complexity measure is to be used as an ordinal measure, then it must be used to compare and rank alternate project networks. For example, when planning a project, one may use the complexity measure to indicate the degree of simplification that has been achieved in each iteration of the project plan. Similarly, when evaluating project options, one may use the ordinal complexity measure to determine which network option will be easiest to manage. If the complexity measure is to be used as a cardinal measure, then a benchmark value must be developed. In other words, control limits will be needed to indicate when a project network can be classified as simple, medium, or complex. For Badiru's complexity measure, the following classification ranges are suggested:

Table 4-8 Data for Project Complexity Example

Activity number	PERT estimates $(a,\ m,\ b)$	Preceding activities	Required resources (x_{i1}, x_{i2})
1	1, 3, 5	—	1, 0
2	0.5, 1, 3	—	1, 1
3	1, 1, 2	—	1, 1
4	2, 3, 6	Activity 1	2, 0
5	1, 3, 4	Activity 2	1, 0
6	1.5, 2, 2	Activity 3	4, 2
			$Z_1 = 5,\ Z_2 = 2$

> Simple network: $0 \le \lambda \le 5.0$
>
> Medium network: $5.0 < \lambda \le 12.0$
>
> Complex network: $\lambda > 12.0$

The above ranges are based on the result of an experimental investigation involving 30 alternate projects of various degrees of network complexity. Of course, the ranges cannot be said to be universally applicable, because consistency of measurement cannot be assured from one project network to another. Users can always determine what ranges will best suit their needs. The complexity measure can then be used accordingly. Perhaps the greatest utility of a complexity measure is obtained when evaluating computer implementation of network analysis. This is addressed further by the discussion in the next section.

4.6.2 Evaluation of Solution Time

By using solution time as a performance measure, another comparison of the scheduling heuristics may be conducted. Computer processing time should be recorded for each scheduling rule under each test problem. The following procedure may then be used to perform the solution time analysis. We let τ_{mn} denote computer processing time for scheduling heuristic m for test problem n. The sum of the processing times over the set of test problems is expressed as

$$\Psi_n = \sum_m \tau_{mn}, \qquad n = 1, 2, \ldots, N; m = 1, 2, \ldots, M$$

where m, n, M, and N are as previously defined. Then,

$$\mu_n = \frac{\Psi_n}{M}$$

denotes the average time for scheduling project n, where M is the number of rules considered. The normalized solution time for heuristic m under test problem n can then be denoted as

$$\Omega_{mn} = \frac{\tau_{mn}}{\mu_n}$$

Each heuristic m is ranked on the basis of the sum of normalized solution times over all test problems. That is,

$$\theta_m = \sum_n \Omega_{mn}, m = 1, 2, \ldots, M; \; n = 1, 2, \ldots, N$$

It is obvious that the solution time of each scheduling heuristic depends on its computational complexity. The computations for some heuristics (e.g., highest number of immediate successors) do not require a prior analysis of the PERT network. For heuristics that consider several factors in the activity sequencing process, their computations will be more complex than other scheduling heuristics. The solution time analysis results should thus be coupled with schedule effectiveness in order to judge the overall acceptability of any given heuristic. Prior analysis and selection of the most effective scheduling heuristic for a given project can help minimize schedule changes and delays often encountered in impromptu scheduling practices.

4.6.3 Performance of Scheduling Heuristics

In addition to comparing scheduling heuristics on the basis of project durations, the following aggregate measures may also be used. The first one is an evaluation of the ratio of the minimum project duration observed to the project duration obtained under each heuristic. For each heuristic m, the ratio under each test problem n is computed as

$$\rho_{mn} = \frac{q_n}{PL_{mn}}, \qquad m = 1, 2, \ldots, M; n = 1, 2, \ldots, N$$

where

ρ_{mn} = efficiency ratio for heuristic m under test problem n
PL_{mn} = project duration for test problem n under heuristic m
q_n = $\text{Min}_m\{PL_{mn}\}$; minimum duration observed for test problem n
M = the number of scheduling heuristics considered
N = the number of test problems.

From the above definitions, the maximum value for the ratio is 1.0. Thus, it is alternately referred to as the rule efficiency ratio. The value q_n is, of course, not necessarily the global minimum project duration for test problem n. Rather, it represents the local minimum based on the particular scheduling heuristics considered. If the global minimum duration for a project is known (probably from an optimization model), then it should be used in the expression for ρ_{mn}. Rules can be compared on the basis of the absolute values for ρ_{mn} or on the basis of the sums of ρ_{mn}. The sums of ρ_{mn} over the index n are defined as

$$\Phi_m = \sum_{n=1}^{N} \rho_{mn}, \qquad m = 1, 2, \ldots, M$$

The use of the sums of ρ_{mn} is a practical approach to comparing scheduling heuristics. It is possible to have a scheduling rule that will consistently yield near minimum project durations for all test problems. On the other hand, there may be another rule that performs very well for some test problems while performing poorly for other problems. A weighted sum helps to average out the overall performance over several test problems. The other comparison measure involves the calculation of the percentage deviations from the observed minimum project duration. The deviations are computed as

$$S_{mn} = \left(\frac{PL_{mn} - q_n}{q_n} \right)(100), \qquad m = 1, 2, \ldots, M; n = 1, 2, \ldots, N$$

which denotes the percentage deviation from the minimum project duration for rule m under test problem n. A project analyst will need to consider the several factors discussed above when selecting and implementing scheduling heuristics. The potential variability in the work rates of resources is another complicating factor in the project scheduling problem. A methodology for analyzing resource work rates is presented in Chapter 5.

4.7 FORMULATION OF PROJECT GRAPH

As discussed previously, a project network can be represented in two different modes: a precedence (activity-on-node) diagram and an arrow (activity-on-arc) diagram. While the first mode is frequently used in practical situations, the second mode is often used in optimization. In this section, we present the formulation of project graph using AOA representation. This formulation will be used later in Chapter 7 under project optimization. The concept of precedence is the key to the construction of project networks. If activity a precedes activity b, then it is indicated as $a \leftarrow b$. The relationship is transitive. That is, if $a \leftarrow b$ and $b \leftarrow c$, then $a \leftarrow c$. Regardless of the representation mode, a project network has the following structural properties:

1. It is acyclic. In other words, there does not exist a directed path $i-k_1-k_2-k_3\ldots$ k_p-i meaning an activity precedes itself!
2. Each node should have at least one arc directed into the node and one arc directed out of the node with the exception of the start and finish nodes. The start node does not have any arc into it and the finished node has no arc out of it.
3. All of the nodes and arcs of the network have to be visited (that is realized) in order to complete the project.

Since project networks are acyclic, the existence of a cycle in the network due to an error in the network construction or due to an error in entering the data into a computer will lead to cycling in the procedures. More specifically, critical path calculations will not terminate. Therefore, one may need to detect the cycles in the network. One way to accomplish that is to number the nodes such that for each $\text{arc}(i,j)i < j$. If such a numbering is possible, then the network is acyclic. However, this may not be an easy task when thousands of nodes are involved. A depth-first search (dfs) procedure can be used to detect cycles in a project network.

4.7.1 Depth-First Search (DFS) Method

The procedure starts with all nodes unvisited and unmarked. The source node is visited first and its dfs number is set equal to one since it is the first visited node in the depth search. In general, the procedure locates an $\text{arc}(i,j)$ emanating from a visited node i going to an unvisited node j. Existence of such an arc leads to node j being visited and initiates a new search at node j. The dfs number of node j will be one more than the dfs number of node i. If no arc connects node i to unvisited nodes, then the search from i is completed, and the node is marked. The procedure then backs up to a node with the dfs number one less than the dfs number of node i and initiates a new search process. The procedure terminates when the source node is marked. Depth-first traversal of the network **G** leads to four different arc definitions. A *tree arc* is an arc which connects a visited node to an unvisited node. A *forward arc* connects a low-numbered node to a high-numbered node. A *cross arc* connects a visited node to a visited and marked node. A *back arc* connects a visited node to a visited but unmarked node. Existence of back arcs indicates that cycles exist in the network. The procedure for detecting cycles in a network is summarized below.

Procedure to detect cycles in a network G = (N, A) using the dfs method

1. Initialize all nodes as unvisited and unmarked. $A_1 = \varnothing$. Set the dfs number of node 1, $dfs(1) = 1$. Node 1 is now visited. Set $i = 1$. Go to step 2.
2. If all the arcs emanating from node i are explored, then go to step 3. Otherwise, let (i,j) be an unexplored arc connecting node i to node j. If node j is unvisited, then set $dfs(j) = dfs(i)+1$. Node j is now visited. Update arc set as $A_1 = A_1 + (i,j)$. Set $i = j$. Return to step 2. If node j is visited and marked then $A_1 = A_1 + (i,j)$. Return to step 2. If node j is visited and unmarked then stop. The network contains cycles. Arc(i,j) is an arc of the cycle. If the set of other arcs leading to a cycle needs to be detected, then return to step 2 without adding (i,j) to A_1 instead of stopping.
3. Mark node i. Stop, if node i is node 1. The set of arcs in $A - A_1$ forms cycles in the network. Otherwise, choose node k with $dfs(k) = dfs(i)-1$. Set $i = k$. Return to step 2.

The dfs method applied to the network in Figure 4–37 produces arc (5,2) as the back arc indicating that it causes a cycle in the network.

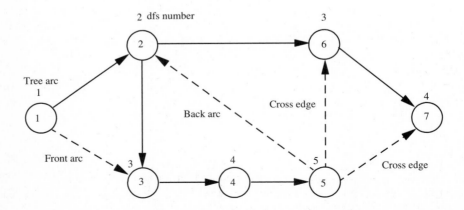

Figure 4-37. Illustration of Depth-First Search for Cycles in Network

4.7.2 Activity-On-Arc (AOA) Representation

In this mode, the nodes represent the realizations of some milestones (events) of the project and the arcs represent activities. An arc connecting nodes i and j represented by an arrow from node i to node j is referred to as arc(i,j) or activity (i,j). Node i, the immediate predecessor node of arc(i,j) is the start node for the activity and node j, the immediate successor node of arc(i,j), is the end node for the activity. This is shown in Figure 4–38.

Suppose we want to draw the activity network of the daily set of chores to be undertaken by a family of three. Table 4–9 outlines the tasks to be performed and the precedence relationship for the project. As shown in Figure 4–39, the project can be represented as a directed graph with 12 nodes and 15 arcs. Arcs (8,10) and (9,10) are dummy arcs which do not consume any resources, but they are needed to represent the precedence relationship.

Since activities i, j and k have the same immediate predecessor, activity h, and the same immediate successor, activity l, this necessitates three parallel arcs between nodes 7 and 10. An activity network allows only one arc between any two nodes and hence nodes 8 and 9 are drawn and connected to node 10 via dummy arcs to overcome the problem. In general, dummy arcs are needed when

1. A set of jobs has the same set of immediate predecessors and the same set of immediate successors;

Node   Node

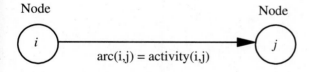

arc(i,j) = activity(i,j)

Figure 4-38.   A Generic AOA Element

Table 4-9   Sample Project for After-Work-Hours Chores

Activity	Activity description	Immediate predecessors
a	Dad, Mom, and son arrive home in the same car.	—
b	Dad and Mom change clothes.	a
c	Son watches TV.	a
d	Mom warms up the food.	b
e	Dad sets the table.	b
f	Son does homework.	c
g	Mom fixes salad.	d
h	The family eats dinner.	e, f, g
i	Dad loads the dishwasher.	h
j	Mom checks son's homework.	h
k	Son practices piano.	h
l	All go to son's basketball game.	i, j, k
m	All wash up and go to bed.	l

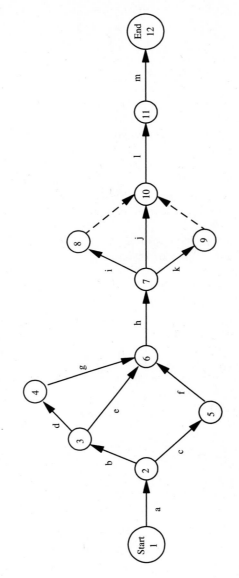

Figure 4-39. A Project Graph of the After-Work-Hours Chores

2. Two or more jobs have the same set of immediate predecessors, some of which are also immediate predecessors to other jobs;
3. Two or more jobs have the same set of immediate successors, some of which are immediate successors to other jobs.

As indicated earlier, AOA representation is assumed by most of the project optimization models. It is, therefore, desired that redundant dummy arcs be removed from the network in order to reduce the problem size. Redundant dummy arcs exist when they are the only immediate predecessors of a node, or the only immediate successors of a node. Also, if a dummy arc represents a precedence relationship which has already been represented by the project arc, then it is redundant. The inclusion of the redundant arcs in the network does not affect the optimization process but may affect the solution time and the memory requirements once the procedure is coded. One should be more careful not to omit any existing precedence relationship since failure to do so leads to incorrect decisions. Any cycles created signal an incorrect diagram and can be detected by using the depth-first search procedure explained earlier. Figure 4–40 is the activity-on-node network corresponding to the project given in Table 4–9.

In the AON mode, the nodes represent the activities and the arcs are drawn to indicate the precedence relationship. If node i is connected to node j through arc(i,j), then activity i is an immediate predecessor of activity j. The in degree (out degree) of a node is the number of arcs leading into (out of) the node. If the in degree of node j is 2 and the out degree is 3, then activity j has two immediate predecessors and three immediate successors. The dummy arcs are not needed in this mode. However, a dummy start node is needed when more than one activity does not have immediate predecessors. Similarly, a dummy end node is needed when two or more activities do not have immediate successors. The dummy start (end) node is connected to the set of initial (last) activities using dummy arcs. This representation mode is easier to construct and is similar to bar chart representations often used in industry.

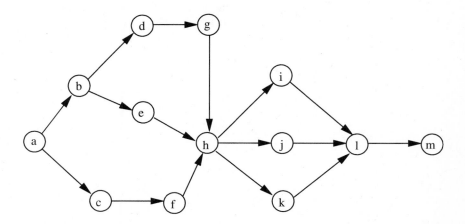

Figure 4-40.   Activity-on-Node Graph of the After-Work-Hours Chores

4.8 EXERCISES

4.1. Show that for any PERT project network, the probability of finishing the project in *exactly* x time units is zero.

4.2. Repeat the crashing exercise presented in Table 4–5 by crashing only the three critical activities, C, E, and G. What is the new project duration? What is the new total project cost?

4.3. The following historical data was compiled for the time (days) it takes to perform a task: 11, 18, 25, 30, 8, 42, 25, 30, 13, 21, 26, 35, 15, 23, 29, 36, 17, 25, 30, 21. Based on the given data, determine reasonable values for the PERT estimates a, m, and b for the task.

4.4. Answer the following questions using the tabulated PERT project data.

Activity	Predecessor	Duration (days) a, m, b
A	—	7, 9, 11
B	—	7, 8, 10
C	A	2, 4, 6
D	A	1, 3, 7
E	B	1, 3, 7
F	B	2, 4, 6

(a) Draw the AON CPM network for the project, perform the forward and backward computations, and determine the project duration.
(b) Draw the Gantt chart using the earliest start times.
(c) Draw the Gantt chart using the latest start times.
(d) Compare and discuss the relative differences, advantages, and disadvantages between the two Gantt charts in parts (b) and (c).

4.5. The CPM network below is subject to the following two simultaneous restrictions:
(a) The project cannot start until 5 days from time 0 (i.e., *ES* of the start node = 5).
(b) There is a deadline of 35 days (i.e., *LC* of the finish node = 35). Perform the forward and backward CPM computations and show ES, LS, EC, LC, TS, and FS for each activity. What is the TS for each activity on the critical path?

Activity	Predecessor	Duration (days)
A	—	2
B	A	4
C	B	6
D	A	3
E	B	2
F	E	1
G	C, D	2
H	G	3
I	H	4

4.6. Using the mathematical expressions for free slack (FS) and total slack (TS), show that FS cannot be larger than TS.

4.7. Show that even if $TS > 0$, FS may or may not be greater than 0.

4.8. Suppose five activities are on a CPM critical path and each activity has $TS = 2$. Determine the overall effect on the critical path length if two of the critical activities are delayed by 2 time units each. Discuss.

4.9. Give practical examples of SS, SF, FF, and FS precedence constraints.

4.10. Suppose two serial activities A and B are the only activities in a precedence diagramming network. Activity A precedes activity B. There is an SS restriction of 4 time units and an FF restriction of 3 time units between A and B.

(a) Determine the mathematical relationship between the duration of $A(t_A)$ and the duration of $B(t_B)$.

(b) Verify your response to part (a) by considering the following alternate cases:

Case 1:	$t_A = 0$
Case 2:	$0 < t_A <= 1$
Case 3:	$1 < t_A < 4$
Case 4:	$t_A = 4$
Case 5:	$t_A > 4$

4.11. Perform a PERT analysis of the following project and find the probability of finishing the project between 16 and 20 days inclusive.

Activity	Predecessors	Duration estimates (days) (a, m, b)
1	—	1, 4, 5
2	—	2, 3, 4
3	1	6, 10, 13
4	1	6, 6, 7
5	2	2, 2, 2
6	3	1, 2, 3
7	4, 5	5, 8, 9
8	2	12, 16, 19

4.12. Answer the following questions using the tabulated project data.

Project phase	Preceding phase	Normal time	Normal cost	Crash time	Crash cost	Variance of duration
A	—	40	$9,000	30	$12,000	10
B	A	53	15,000	50	15,300	9
C	A	60	7,500	30	6,000 *	600
D	A	35	20,000	30	22,000	25
E	C, D	28	12,000	20	15,000	40
F	B, E	30	6,000	27	7,000	30

* Crashing of this activity results in significant savings in direct labor cost.

(a) There is a penalty of $3,000 per day beyond the normal PERT duration. Perform a crashing analysis to compare total normal cost to total crash cost. First crash only the critical activities; then crash all activities.

(b) Find the probability that the total lateness penalty will be less than $3,000.

(c) Find the probability that the total lateness penalty will be $3,000 or less.

(d) Find the probability that the total lateness penalty will be greater than $3,000.

(e) Find the probability that the total lateness penalty will be less than $9,000.

4.13. Suppose it is known that the duration of a project has a triangular distribution with a lower limit of 24 months, an upper limit of 46 months, and a mode of 36 months. Find the probability that the project can be completed in 40 months or less.

4.14. Suppose it is known that the duration of a project is uniformly distributed between 24 months and 46 months. Find the probability that the project can be completed in 40 months or less.

4.15. You are given the following project data:

Activity	Predecessor	Estimated CPM duration, t (days)
A	—	6
B	—	5
C	—	2
D	A, B	4
E	B	5
F	C	6
G	B, F	1
H	D, E	7

If the PERT time estimates for each activity are defined as

$$a = t - 0.3t$$

$$m = t$$

$$b = t + \sqrt{t}$$

find a deadline, T_d, such that there is 0.85 probability of finishing the project on or before the deadline. *Note:* Carry all computations to four decimal places.

4.16. Suppose we are given the following data for a project:

Activity	Predecessor	PERT duration	Duration
A	—	4	0.5000
B	A	5	1.0000
C	A	7	0.2500
D	A	2	0.3000
E	C, D	8	0.3750
F	B, E	9	0.4375

There is no penalty if the project is completed on or before the normal PERT duration. However, a penalty of $3,000 is charged for each day that the project lasts beyond the normal PERT duration. Using standard probability approach, find the *expected dollar penalty* for this project. *Note:* Carry all intermediate computations to four decimal places.

4.17. Presented below is the data for a precedence diagram problem. The *SS*, *SF*, and *FF* relationships between pairs of activities are given.

Activity	Predecessor	Duration
A	—	5
B	A	10
C	A	4
D	B, C	15

SS(between A and B) = 1
FF(between A and B) = 8
SS(between A and C) = 2
SF(between A and C) = 8
SS(between B and D) = 5
SS(between C and D) = 7
FF(between C and D) = 10

(a) Perform the forward and backward computations *without* considering the precedence constraints. What is the project duration?

(b) Perform the forward and backward computations considering the precedence constraints. What is the project duration?

4.18. (a) What potential effect (s) does the crashing of a noncritical activity have on the critical path and the project duration?

(b) Use a hypothetical example of a CPM project network consisting of seven activities to support your answer above. First show the network computations without crashing. Identify the critical path. Then show the computations with the crashing of one noncritical activity.

4.19. Assume that you are the manager responsible for the project whose data is presented as follows:

Activity	Description	Duration (hours)	Preceding activities
A	Develop required material list	8	—
B	Procure pipe	200	A
C	Erect scaffold	12	—
D	Remove scaffold	4	I, M
E	Deactivate line	8	—
F	Prefabricate sections	40	B
G	Install new pipes	32	F, L
H	(Deleted from work plan)	—	—
I	Fit-up pipes and valves	8	G, K
J	Procure valves	225	A
K	Install valves	8	J, L
L	Remove old pipes and valves	35	C, E
M	Insulate	24	G, K
N	Pressure test	6	I
O	Cleanup and start-up	4	D, N

(a) Draw the CPM diagram for this project.

(b) Perform the forward and backward pass calculations on the network and indicate the project duration.

(c) List all the different paths in the network in *decreasing* order of criticality index in the tabulated form shown below. Define the criticality index as follows:

Let S_k = sum of the total slacks on path number k (i.e., sum of TS values for all activities on path k).

S_{max} = maximum value of S (i.e., Max$\{S_k\}$).

$C_k = [(S_{max} - S_k)/S_{max}] * 100\%$.

Define the most critical path as the one with $C_k = 100\%$.

Define the least critical path as the one with $C_k = 0\%$.

k	Activities in path	Sum of TS on path (S_k)	Path criticality index (C_k)
1.			100%
2.			
...	....	....	....
n			0%

4.20. Assume that we have the following precedence diagramming network data:

Activity	Predecessor	Duration
A	—	8
B	A	12
C	B	4
D	B	7
E	C, D	8
F	D	12
G	E, F	3

The following PDM restrictions are applicable to the network:

$SS_{AB} = 3$

$FF_{AB} = 12$

$FS_{BC} = 1$

$SS_{BD} = 6$

$SF_{DE} = 13$

$FS_{FG} = 3$

If the project start time is $t = 5$ and the project due date is $t = 52$, perform conventional CPM analysis of the network without considering the lead-lag PDM restrictions. Show both forward and backward passes. Identify the critical path. Tabulate the total slack (TS) and free slack (FS) for the activities in the network.

4.21. Repeat Exercise 4.20 and consider the lead-lag PDM restrictions. Draw the Gantt chart schedule that satisfies all the lead-lag constraints.

4.22. Draw an expanded CPM network for the Gantt chart in Exercise 4.21 considering the need to split activities.

4.23. Would you classify the CPM procedure as a management by objective (MBO) approach or a management by exception (MBE) approach? Discuss.

4.24. In activity scheduling, which of the following carries the higher priority?
   (i) Activity precedence constraint
   (ii) Resource constraint

4.25. For the project data in Exercise 4.20, assume that the given activity durations represent the most likely PERT estimates (m). Define the other PERT estimates a and b as follows:

$$a = m - 1$$
$$b = m + 1$$

Disregarding the PDM lead-lag constraints, find the probability of finishing the project within 51 time units.

4.26. The *median rule* of project control refers to the due date that has a 0.50 probability of being achieved. Suppose the duration, T, of a project follows a triangular distribution with end points a and b. If the mode of the distribution is closer to b than it is to a, find a general expression for the median (denoted by m_d) such that

$$P(T \leq m_d) = 0.50$$

4.27. Draw an AOA network for the project described in Exercise 4.19. Use as few dummy arcs as possible.

4.28. For the problem given in Exercise 4.12, assuming that the variance of duration is 0 for all the project phases, determine the time-cost trade-off curve by defining the strategies and calculating crashing cost for each strategy for $T_{\max} \leq T \leq T_{\max}$.

5

Resource Allocation

This chapter addresses resource allocation and management strategies. The differences between unconstrained PERT/CPM networks and resource-constrained networks are discussed. The resource loading and resource leveling strategies are presented. The resource idleness graph is introduced as a measure of the level of resource idleness in resource-constrained project schedules. Various resource allocation heuristics are presented. A procedure for calculating resource work rates to assess project productivity is presented. An example of a resource-constrained PDM network is presented. The chapter also illustrates the use of the new graphical tools, referred to as the critical resource diagram (CRD) and the resource schedule Gantt chart. The CRD is used to represent the interrelationships among resource units as they perform their respective tasks. It is also used to identify bottlenecked resources in a project network. The resource schedule Gantt chart indicates time intervals of allocation for specific resource types. The chapter concludes with examples of probabilistic evaluation of resource utilization levels. The stage of the project management process has now reached the following stage:

SET GOALS → PLAN → ORGANIZE → SCHEDULE →
ALLOCATE RESOURCES

5.1 RESOURCE ALLOCATION AND MANAGEMENT

Basic CPM and PERT approaches assume unlimited resource availability in project network analysis. In this chapter, both the time and resource requirements of activities are considered in developing network schedules. Projects are subject to three major constraints: time limitations, resource constraints, and performance requirements. Since these constraints are difficult to satisfy simultaneously, trade-offs must be made. The smaller the resource base, the longer the project schedule. The quality of work may also be adversely affected by poor resource allocation strategies.

Good planning, scheduling, and control strategies must be developed to determine what the next desired state of a project is, when the next state is expected to be reached, and how to move toward that next state. Resource availability as well as other internal and external factors will determine the nature of the progress of a project from one state to another. Network diagrams, Gantt charts, progress charts, and resource loading graphs are visual aids for resource allocation strategies. One of the first requirements for resource management is to determine what resources are required versus what resources are available. Table 5–1 shows a model of a resource availability data base. The data base is essential when planning resource loading strategies for resource-constrained projects.

.2 RESOURCE-CONSTRAINED SCHEDULING

A resource-constrained scheduling problem arises when the available resources are not enough to satisfy the requirements of activities that can be performed concurrently. To satisfy this constraint, *sequencing rules* (also called priority rules, activity urgency factors, scheduling rules, or scheduling heuristics) are used to determine which of the competing activities will have priority for resource allocation. Several optimum-yielding techniques are available for generating resource-constrained schedules. Unfortunately, the optimal techniques are not generally used in practice because of the complexity involved in implementing them for large projects.

Even using a computer to generate an optimal schedule is sometimes cumbersome because of the modeling requirements, the drudgery of lengthy data entry, and the combinatorial nature of interactions among activities. However, whenever possible, effort should be made in using these methods since they provide the best solution. Chapter 7 discusses optimization methods for resource allocation problems.

Most of the available mathematical techniques are based on integer programming that formulates the problem using 0 and 1 indicator variables. The variables indicate whether or not an activity is scheduled in specific time periods. Three of the common objectives in project network analysis are to minimize project duration, minimize total project cost, and maximize resource utilization. One or more of these objectives are attempted subject to one or more of the following constraints:

Table 5–1   Format for Resource Availability Data base

Resource ID	Brief description	Special skills	When available	Duration of availability	How many
Type 1	Manager	Planning	1/1/95	10 months	1
Type 2	Analyst	Scheduling	12/25/94	Indefinite	5
Type 3	Engineer	Design	Now	36 months	20
•	•	•	•	•	•
•	•	•	•	•	•
•	•	•	•	•	•
Type n − 1	Operator	Machining	Immediate	Indefinite	10
Type n	Programmer	Software tools	9/2/94	12 months	2

1. Limitation on resource availability
2. Precedence restrictions
3. Activity-splitting restrictions
4. Nonpreemption of activities
5. Project deadlines
6. Resource substitutions
7. Partial resource assignments
8. Mutually exclusive activities
9. Variable resource availability
10. Variable activity durations

Instead of using mathematical formulations, a scheduling heuristic uses logical rules to prioritize and assign resources to competing activities. Many scheduling rules have been developed in recent years. Some of the most frequently used ones are discussed in this chapter.

5.3 RESOURCE ALLOCATION EXAMPLE

Table 5–2 shows a project with resource constraints. There is one resource type (operators) in the project data and there are only 10 units of it available. The PERT estimates for the activity durations are expressed in terms of days. It is assumed that the resource units are reusable. Each resource unit is reallocated to a new activity at the completion of its previous assignment. Resource units can be idle if there are no eligible activities for scheduling or if enough units are not available to start a new activity. For simplification, it is assumed that the total units of resource required by an activity must be available before the activity can be scheduled. If partial resource allocation is allowed, then the work rate of the partial resources must be determined. A methodology for determining resource work rates is presented in a later section of this chapter.

The unconstrained PERT duration was found earlier to be 11 days. The resource limitations are considered when creating the Gantt chart for the resource-constrained schedule. For this example, we will use the *longest-duration-first* heuristic to prioritize

Table 5-2 PERT Project Data with Resource Requirements $Z_1 = 10$

Activity	Predecessor	PERT estimates a	m	b	Number of operators required
A	—	1	2	4	3
B	—	5	6	7	5
C	—	2	4	5	4
D	A	1	3	4	2
E	C	4	5	7	4
F	A	3	4	5	2
G	B, D, E	1	2	3	6

activities for resource allocation. Other possible heuristics are *shortest-duration-first*, *critical-activities-first*, *maximum-predecessors-first*, and so on. For very small project networks, many of the heuristics will yield identical schedules.

5.3.1 Longest-Duration-First

The initial step is to rank the activities in decreasing order of their PERT durations, t_e. This yields the following priority order:

$$B, E, F, C, D, A, G$$

At each scheduling instant, only the eligible activities are considered for resource allocation. Eligible activities are those whose preceding activities have been completed. Thus, even though activity B has the highest priority for resource allocation, it can compete for resources only if it has no pending predecessors. Referring to the PERT network shown in the preceding chapter, note that activities A, B, and C can start at time 0 since they all have no predecessors. These three activities require a total of 12 operators $(3 + 5 + 4)$ altogether, but we have only 10 operators available. So, a resource allocation decision must be made. We check our priority order and find that B and C have priority over A. So, we schedule B with 5 operators and C with 4 operators. The remaining 1 operator is not enough to meet the need of any of the remaining activities. The two scheduled activities are drawn on the Gantt chart as shown in Figure 5–1.

We have one operator idle from time 0 until time 3.83, when activity C finishes and releases 4 operators. At time 3.83, we have 5 operators available. Since activity E can start after activity C, it has to compete with activity A for resources. According to the established priority, E has priority over A so, 4 operators are assigned to E. The remaining 1 operator is not enough to perform activity A, so it has to wait and 1 operator remains idle. If E had required more operators than were available, activity A would have been able to get resources at time 3.83. No additional scheduling is done until time 6, when activity B finishes and releases 5 operators. So, we now have 6 operators available and there are no activities to compete with A for resources. Thus, activity A is finally scheduled at time 6 and we are left with 3 idle operators. Even though the 3 operators are enough to start either activity D or activity F, neither of these activities can start until activity A finishes, because of the precedence requirement.

When activity A finishes at time 8.17, F and D are scheduled in that order. Activity G is scheduled at time 11 and finishes at time 13 to complete the project. Figure 5–1 shows the complete project schedule. The numbers in parentheses in the figure are the resource requirements. Note that our assumption is that activity splitting and partial resource assignments are not allowed. An activity cannot start until all the units of resources required are available. In real project situations, this assumption may be relaxed so that partial resource assignments are permissible. If splitting and partial assignments are allowed, the scheduling process will still be the same except that more record keeping will be required to keep track of pending jobs.

Resource allocation may be affected by several factors including duration of availability, skill level required, cost, productivity level, and priority strategy. Ranking activities for resource allocation may be done under *parallel priority* or *serial priority*. In serial priority, the relative ranking of all activities is done at the beginning prior to

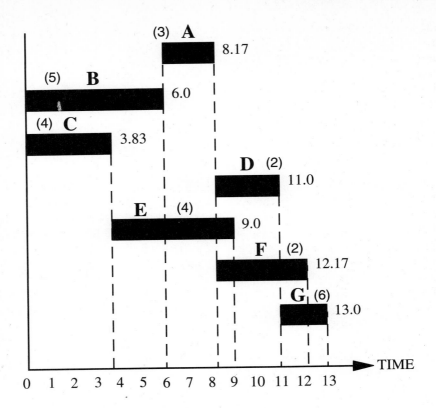

Figure 5-1. Resource-Constrained PERT Schedule

starting the scheduling process. The activities maintain their relative priority ranking throughout the scheduling process. In the parallel priority approach, relative ranking is done at each scheduling instant and it is done only for the activities that are eligible for scheduling at that instant. Thus, under the parallel priority approach, the relative ranking of an activity may change at any time, depending on which activities it is competing with for resources at that time. The illustrative example above uses the serial priority approach. If desired, any other resource allocation heuristic could be used for the scheduling example. Some of the available resource allocation rules are presented in the following section.

5.4 RESOURCE ALLOCATION HEURISTICS

Resource allocation heuristics facilitate scheduling large projects subject to resource limitations. Some heuristics are very simple and intuitive, while others require computer implementations. Several scheduling heuristics have been developed in recent years (Wiest 1967; Elsayed 1982; Slowinski 1980; Kurtulus and Davis 1982; Patterson 1984; Davis and Patterson 1975; Thesen 1976; Fisher 1980; Russell 1986; Cooper 1976; Holloway et al. 1979; Badiru 1988a). Many of these are widely applied to real projects.

Many project management software packages use proprietary resource allocation rules that are not transparent to the user. A good scheduling heuristic should be simple, unambiguous, easily understood, and easily executable by those who must use it. The heuristic must be flexible and capable of resolving schedule conflicts. When users trust and use a scheduling heuristic, then project scheduling becomes an effective communication tool in project management. Some of the most frequently used scheduling rules are presented next.

5.4.1 ACTIM (Activity-Time)

ACTIM (Whitehouse and Brown 1979) is one of the earlier activity sequencing rules. The rule was developed by George H. Brooks and used in his algorithm, Brooks's algorithm (Brooks and White 1965; Bedworth 1973). The original algorithm considered only the single-project, single-resource case, but it lends itself to extensions for the multiresource cases. The ACTIM scheduling heuristic represents the maximum time that an activity controls in the project network on any one path. This is represented as the length of time between when the activity finishes and when the project finishes. It is computed for each project activity by subtracting the activity's latest start time from the critical path time as shown below:

$$ACTIM = (\text{critical path time}) - (\text{activity latest start time})$$

5.4.2 ACTRES (Activity-Resource)

ACTRES is a scheduling heuristic proposed by Bedworth (1973). This is a combination of the activity time and resource requirements. It is computed as

$$ACTRES = (\text{activity time}) * (\text{resource requirement})$$

For multiple resources, the computation of ACTRES can be modified to account for various resource types. For this purpose, the resource requirement can be replaced by a scaled sum of resource requirements over different resource types.

5.4.3 TIMRES (Time-Resources)

TIMRES is another priority rule proposed by Bedworth (1973). It is composed of equally weighted portions of ACTIM and ACTRES. It is expressed as

$$TIMRES = 0.5(ACTIM) + 0.5(ACTRES)$$

5.4.4 GENRES

GENRES is a search technique proposed by Whitehouse and Brown (1979) as an extension of Brooks's algorithm (Brooks and White 1965). It is a modification of TIMRES with various weighted combinations of ACTIM and ACTRES. GENRES is implemented as a computer search technique whereby iterative weights (w) between 0 and 1 are used in the expression

$$GENRES = (w)(ACTIM) + (1 - w)(ACTRES)$$

5.4.5 ROT (Resource over Time)

ROT is a scheduling criterion proposed by Elsayed (1982). It is calculated as the resource requirement divided by the activity time as

$$ROT = \frac{\text{resource requirement}}{\text{activity time}}$$

The resource requirement can be replaced by the scaled sum of resource requirements in the case of multiple resource types with different units.

5.4.6 CAF (Composite Allocation Factor)

CAF is a comprehensive rule developed by Badiru (1988c). For each activity i, CAF is computed as a weighted and scaled sum of two components RAF (resource allocation factor) and SAF (stochastic activity duration factor) as

$$CAF_i = (w)RAF_i + (1 - w)SAF_i$$

where w is a weight between 0 and 1. RAF is defined for each activity i as

$$RAF_i = \frac{1}{t_i} \sum_{j=1}^{R} \frac{x_{ij}}{y_j}$$

where

         x_{ij} = number of resource type j units required by activity i
         $y_j = \text{Max}_j\{x_{ij}\}$, maximum units of resource type j required
         t_i = the expected duration of activity i
         R = the number of resource types

RAF is a measure of the expected resource consumption per unit time. In the case of multiple resource types, the different resource units are scaled by the y_j component in the formula for RAF. This yields dimensionless quantities that can be summed in the formula for RAF. The RAF formula yields real numbers that are expressed per unit time. To eliminate the time-based unit, the following scaling method is used:

$$\text{scaled } RAF_i = \frac{RAF_i}{\text{Max}\{RAF_i\}}(100)$$

The above scaling approach yields unitless values of RAF between 0 and 100 for the activities in the project. Resource-intensive activities have larger magnitudes of RAF and, therefore, require a higher priority in the scheduling process. To incorporate the stochastic nature of activity times in a project schedule, SAF is defined for each activity i as

$$SAF_i = t_i + \frac{s_i}{t_i}$$

where

t_i = expected duration for activity i

s_i = standard deviation of duration for activity i

s_i/t_i = coefficient of variation of the duration of activity i

It should be noted that the formula for SAF has one component (t_i) with units of time and one component s_i/t_i with no units. To facilitate the required arithmetic operation, each component is scaled as shown below:

$$\text{scaled } t_i = \frac{t_i}{\text{Max}\{t_i\}}(50)$$

$$\text{scaled } (s_i/t_i) = \frac{(s_i/t_i)}{\text{Max}\{s_i/t_i\}}(50)$$

When these scaled components are plugged into the formula for SAF, we automatically obtain unitless scaled SAF values that are on a scale of 0 to 100. However, the 100 weight will be observed only if the same activity has the highest scaled t_i value and the highest scaled s_i/t_i value at the same time. Similarly, the 0 weight will be observed only if the same activity has the lowest scaled t_i value and the lowest scaled s_i/t_i value at the same time. The scaled values of SAF and RAF are now inserted in the formula for CAF as

$$CAF_i = (w)\{\text{scaled } RAF_i\} + (1 - w)\{\text{scaled} SAF_i\}.$$

To ensure that the resulting CAF values range from 0 to 100, the following final scaling approach is applied:

$$\text{scaled } CAF_i = \frac{CAF_i}{\text{Max}\{CAF_i\}}(100)$$

It is on the basis of the magnitudes of CAF that an activity is assigned a priority for resource allocation in the project schedule. Activities with larger values of CAF have higher priorities for resource allocation. An activity that lasts longer, consumes more resources, and varies more in duration will have a larger magnitude of CAF.

5.4.7 RSM (Resource Scheduling Method)

RSM was developed by Brand, Meyer, and Shaffer (1964). The rule gives priority to the activity with the minimum value of d_{ij} calculated as

$$d_{ij} = \text{increase in project duration when activity } j \text{ follows activity } i$$
$$= \text{Max}\{0, (EC_i - LS_j)\}$$

where EC_i is the earliest completion time of activity i and LS_j is the latest start time of activity j. Competing activities are compared two at a time in the resource allocation process.

5.4.8 GRD (Greatest Resource Demand)

This rule gives priority to the activity with the largest total resource-unit requirements. The GRD measure is calculated as

$$g_j = d_j \sum_{i=1}^{n} r_{ij}$$

where

g_j = priority measure for activity j

d_j = duration of activity j

r_{ij} = units of resource type i required by activity j per period

n = number of resource types (Resource units are expressed in common units.)

5.4.9 GRU (Greatest Resource Utilization)

The GRU rule assigns priority to activities that, if scheduled, will result in maximum utilization of resources or minimum idle time. For large problems, computer procedures are often required to evaluate the various possible combinations of activities and the associated utilization levels.

5.4.10 Most Possible Jobs

This approach assigns priority in such a way that the greatest number of activities are scheduled in any period.

5.4.11 Other Scheduling Rules

- Most total successors
- Most critical activity
- Most immediate successors
- Any activity that will finish first
- Minimum activity latest start (Min LS)
- Maximum activity latest start (Max LS)
- Minimum activity earliest start (Min ES)
- Maximum activity latest completion (Max LC)
- Minimum activity earliest completion (Min EC)
- Maximum activity earliest completion (Max EC)
- Minimum activity latest completion (Min LC)
- Maximum activity earliest start (Max ES)
- Minimum activity total slack (Min TS)
- Maximum activity total slack (Max TS)

- Any activity that can start first
- Minimum activity duration
- Maximum activity duration

The project analyst must carefully analyze the prevailing project situation and decide which of the several rules will be most suitable for the resource allocation requirements. Since there are numerous rules, it is often difficult to know which rule to apply. Experience and intuition are often the best guides for selecting and implementing scheduling heuristics. Some of the shortcomings of heuristics include subjectivity of the technique, arbitrariness of the procedures, lack of responsiveness to dynamic changes in the project scenario, and oversimplified assumptions.

There are advantages and disadvantages to using specific heuristics. For example, the shortest duration heuristic is useful for quickly reducing the number of pending activities. This may be important for control purposes. The smaller the number of activities to be tracked, the lower the burden of project control. The longest duration heuristic, by contrast, has the advantage of scheduling the biggest tasks in a project first. This permits the lumping of the smaller activities into convenient work packages later on in the project. Decomposition of large projects into subprojects can enhance the application of heuristics that are only effective for small project networks. Chapter 7 also discusses the use of a simulated annealing heuristic for the resource allocation problem.

5.4.12 Example of ACTIM

Brooks's algorithm uses ACTIM to determine which activities should receive limited resources first. The algorithm considers the single project, single resource case. Whitehouse and Brown (1979) used the following example to illustrate the use of the ACTIM scheduling heuristic. Figure 5–2 presents a project network based on the activity-on-arrow (AOA) convention. The arrows represent activities, while the nodes represent activity end points. Each activity is defined by its end points as $i - j$. The two numbers within parentheses represent activity duration and resource requirement (t, r) respectively. The network consists of seven actual activities and one dummy activity. The dummy activity (3–4) is required to show that activities (1–3) and (2–3) are predecessors for activity (4–5). Table 5–3 presents the tabular implementation of the steps in the algorithm for 3 units of resource.

STEP 1:   Develop the project network as in CPM. Identify activities, their estimated durations, and resource requirements.

STEP 2:   Determine for each activity the maximum time it controls through the network on any one path. This is equivalent to the critical path time minus the latest start time of the starting node of the activity. These times are then scaled from 0 to 100. This scaled network control time is designated as ACTIM.

STEP 3:   Rank the activities in decreasing order of ACTIM as shown in Table 5–3. The duration and resources required for each activity are those determined in the first step. The rows TEARL, TSCHED, TFIN, TNOW are explained as follows: TEARL is the earliest time of an

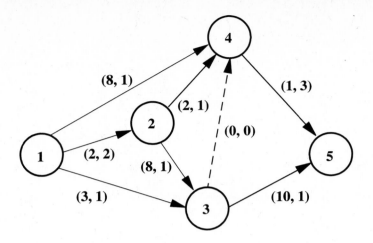

Figure 5-2. AOA Network for ACTIM Example

Table 5-3 Scheduling With ACTIM Heuristic when $Z_1 = 3$

	Activity						
	1–2	2–3	1–3	3–5	1–4	2–4	4–5
Duration	2	8	3	10	8	2	1
ACTIM	20	18	13	10	9	3	1
Scaled ACTIM	100	90	65	50	45	15	5
Resources required	2	1	1	1	1	1	3
TEARL	0	2	0	10	0	2	10
TSCHED	0	2	0	10	2	3	20
TFIN	2	10	3	20	10	5	21
TNOW	0	2	3	5	10	20	
Resources available	3,1,0	2,1,0	1,0	1	3,2	3,0	
ACT. ALLOW.	1–2	2–3	2–4	—	3–5	4–5	
	1–3	1–4			4–5		
	1–4	2–4					
Iteration number	1	2	3	4	5	6	

activity determined by traditional CPM calculations. TSCHED is the actual scheduled starting time of an activity as determined by Brooks's algorithm. TFIN is the completion time of each activity. TNOW is the time at which the resource allocation decision is being made.

STEP 4: Set TNOW to 0. The allowable activities (ACT. ALLOW.) to be considered for scheduling at TNOW of zero are those activities with TEARL of 0. These are 1–2, 1–3, 1–4. These are placed in the ACT. ALLOW. row in decreasing order of ACTIM. Ties are broken by scheduling the activity with longest duration first. The number of resources initially available (i.e., 3) is placed in the resources available column.

STEP 5: Determine if the first activity in ACT. ALLOW. (i.e., 1–2) can be scheduled. Activity 1–2 requires two resource units and three are available. So, 1–2 is scheduled. A line is drawn through 1–2 to indicate that it has been scheduled and the number of resources available is decreased by two. TSCHED and TFIN are then set for activity 1–2. This process is repeated for the remaining activities in ACT. ALLOW. until the resources available are depleted.

STEP 6: TNOW is raised to the next TFIN time of 2, which occurs at the completion of activity 1–2. The resources available are now 2. ACT. ALLOW. includes those activities not assigned at the previous TNOW (i.e., 1–4) and those new activities whose predecessors have been completed (i.e., 2–3 and 2–4).

STEP 7: Repeat this assignment process until all activities have been scheduled. The latest TFIN gives the duration of the project. For this example, the duration is 21 days. Figure 5–3 presents the Gantt chart for the final schedule.

5.4.13 Comparison of ACTIM, ACTRES, and TIMRES

Brooks's algorithm can be implemented with any other heuristic apart from ACTIM. Whitehouse and Brown (1979) used the example network in Figure 5–4 to compare the schedules generated by ACTIM, ACTRES, and TIMRES. Tables 5–4, 5–5, and 5–6 present the tabular of the implementation of Brooks's algorithm with scaled values of ACTIM, ACTRES, and TIMRES. Note that each heuristic yields a different project schedule. Even though the project durations obtained from ACTRES and TIMRES are the same (13 days), the specific scheduled times are different for the activities in each schedule. The larger the project network, the more differences that can be expected from the schedules generated by different heuristics. Figure 5–5 presents the Gantt charts comparing the schedules generated by ACTIM, ACTRES, and TIMRES.

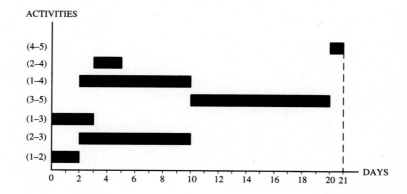

Figure 5-3. Gantt Chart ACTIM Schedule

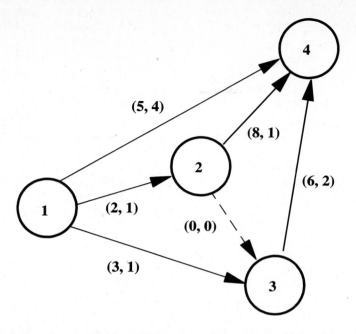

Figure 5-4. Network for ACTIM, ACTRES, and TIMRES Comparison

Table 5–4 ACTIM Schedule for Comparative Network when $Z_1 = 5$

	Activity				
	1–2	1–3	2–4	3–4	1–4
Duration	2	3	8	6	5
ACTIM	10	9	8	6	5
Scaled ACTIM	100	90	80	60	50
Resources required	1	1	1	2	4
TEARL	0	0	2	3	0
TSCHED	0	0	2	3	9
TFIN	2	3	10	9	14
TNOW	0	2	3	9	
Resources available	5, 4, 3	4, 3	4, 2	4	
ACT. ALLOW.	1–2	2–4	3–4	1–4	
	1–3	1–4	1–4		
	1–4				
Iteration Number	1	2	3	4	

174

Table 5-5 ACTRES Schedule for Comparative Network $Z_1 = 5$

	Activity				
	1–4	1–3	1–2	3–4	2–4
Duration	5	3	2	6	8
ACTRES	20	3	2	12	8
Scaled ACTRES	100	15	10	60	40
Resources required	4	1	1	2	1
TEARL	0	0	0	5	5
TSCHED	0	0	3	5	5
TFIN	5	3	5	11	13
TNOW	0	3	5		
Resources available	5, 1, 0	1, 0	5, 3, 2		
ACT. ALLOW.	1–4	1–2	3–4		
	1–3		2–4		
	1–2				
Iteration number	1	2	3		

Table 5-6 TIMRES Schedule for Comparative Network $Z_1 = 5$

	Activity				
	1–2	1–3	1–4	2–4	3–4
Duration	2	3	5	8	6
TIMRES	85	82.5	75	60	60
Scaled TIMRES	100	97	88	71	59
Resources required	1	1	4	1	2
TEARL	0	0	0	2	3
TSCHED	0	0	2	3	7
TFIN	2	3	7	11	13
TNOW	0	2	3	7	
Resources available	5, 4, 3	4, 0	1, 0	4, 2	
ACT. ALLOW.	1–2	1–4	2–4	3–4	
	1–3	2–4	3–4		
	1–4				
Iteration number	1	2	3	4	

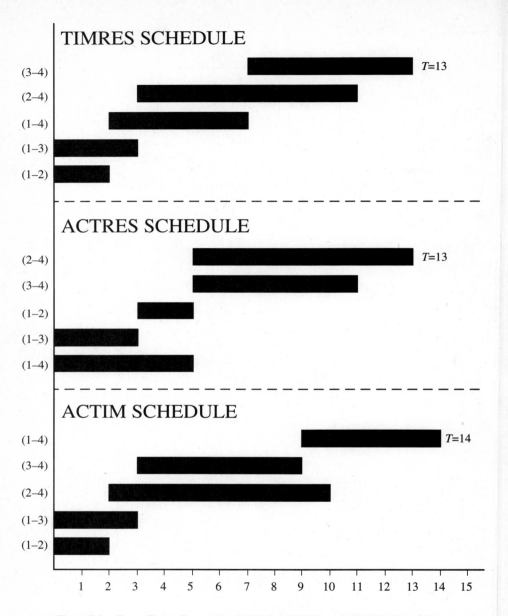

Figure 5-5. Gantt Charts Comparing ACTIM, ACTRES, and TIMRES Schedules

5 RESOURCE WORK RATE

Work rate and work time are essential components of estimating the cost of specific tasks in project management. Given a certain amount of work that must be done at a given work rate, the required time can be computed. Once the required time is known, the cost of the task can be computed on the basis of a specified cost per unit time. Work rate analysis is important for resource substitution decisions. The analysis can help identify where and when the same amount of work can be done with the same level of quality and within a reasonable time span by a less expensive resource. The results of learning curve analysis can yield valuable information about expected work rate. The general relationship among work, work rate, and time is given by

work done = (work rate)(time)

This is expressed mathematically as

$$w = rt,$$

where

w = the amount of actual work done expressed in appropriate units. Example of work units are miles of road completed, lines of computer code typed, gallons of oil spill cleaned, units of widgets produced, and surface area painted.
r = the rate at which the work is accomplished.
t = the total time required to perform the work excluding any embedded idle times.

It should be noted that work is defined as a physical measure of accomplishment with uniform density. That means, for example, that one line of computer code is as complex and desirable as any other line of computer code. Similarly, cleaning 1 gallon of oil spill is as good as cleaning any other gallon of oil spill within the same work environment. The production of one unit of a product is identical to the production of any other unit of the product. If uniform work density cannot be assumed for the particular work being analyzed, then the relationship presented above may lead to erroneous conclusions. Uniformity can be enhanced if the scope of the analysis is limited to a manageable size. The larger the scope of the analysis, the more the variability from one work unit to another, and the less uniform the overall work measurement will be. For example, in a project involving the construction of 50 miles of surface road, the work analysis may be done in increments of 10 miles at a time rather than the total 50 miles. If the total amount of work to be analyzed is defined as one whole unit, then the relationship below can be developed for the case of a single resource performing the work:

Resource	Work rate	Time	Work done
Machine A	r	t	1.0

where r is the amount of work accomplished per unit time. For a single resource to perform the whole unit of work, we must have the following:

$$rt = 1.0$$

For example, if machine A is to complete one work unit in 30 minutes, it must work at the rate of 1/30 of the work content per unit time. If the work rate is too low, then only a fraction of the required work will be performed. The information about the proportion of work completed may be useful for productivity measurement purposes. In the case of multiple resources performing the work simultaneously, the work relationship is as presented in the following table:

Resource, i	Work rate, r_i	Time, t_i	Work done, w
Machine A	r_1	t_1	$(r_1)(t_1)$
Machine B	r_2	t_2	$(r_2)(t_2)$
...	...	...	...
Machine n	r_n	t_n	$(r_n)(t_n)$
		Total	1.0

Even though the multiple resources may work at different rates, the sum of the work they all performed must equal the required whole unit. In general, for multiple resources, we have the following relationship:

$$\sum_{i=1}^{n} r_i t_i = 1.0$$

where

n = number of different resource types
r_i = work rate of resource type i
t_i = work time of resource type i

For partial completion of work, the relationship is

$$\sum_{i=1}^{n} r_i t_i = p$$

where p is the proportion of the required work actually completed.

Example

Machine A, working alone, can complete a given job in 50 minutes. After machine A has been working on the job for 10 minutes, machine B was brought in to work with machine A in completing the job. Both machines working together finished the remaining work in 15 minutes. What is the work rate for machine B?

    Solution

    The amount of work to be done is 1.0 whole unit.

    The work rate of machine A is 1/50.

    The amount of work completed by machine A in the 10 minutes it worked alone is (1/50)(10) = 1/5 of the required total work.

    Therefore, the remaining amount of work to be done is 4/5 of the required total work.

The two machines working together for 15 minutes yield the following results:

Resource, i	Work rate, r_i	Time, t_i	Work done, w
Machine A	1/50	15	15/50
Machine B	r_2	15	$15(r_2)$
		Total	4/5

$$\frac{15}{50} + 15(r_2) = \frac{4}{5}$$

which yields $r_2 = 1/30$. Thus, the work rate for machine B is 1/30. That means machine B, working alone, could perform the same job in 30 minutes. In this example, it is assumed that both machines produce an identical quality of work. If quality levels are not identical, then the project analyst must consider the potentials for quality/time trade-offs in performing the required work. The relative costs of the different resource types needed to perform the required work may be incorporated into the analysis as shown in the table:

Resource, i	Work rate, r_i	Time, t_i	Work done, w	Pay rate, p_i	Pay, P_i
Machine A	r_1	t_1	$(r_1)(t_1)$	p_1	P_1
Machine B	r_2	t_2	$(r_2)(t_2)$	p_2	P_2
...	...	...	...	...	...
Machine n	r_n	t_n	$(r_n)(t_n)$	p_n	P_n
		Total	1.0		Budget

    Using the above relationship for work rate and cost, the work crew can be analyzed to determine the best strategy for accomplishing the required work, within the required time, and within a specified budget.

6 RESOURCE-CONSTRAINED PDM NETWORK

The conventional precedence diagramming network with no resource limitations was presented in Chapter 4. In this section, we extend the previous example to a resource-constrained problem with probabilistic activity durations. Table 5–7 presents the modified project data after the activities are partitioned into segments (or subactivities)

Table 5-7   Resource-Constrained PDM Network Data

Activity number	Name	Predecessors	Duration	Duration variance	Resource 1 required	Resource 2 required
1	A_1	—	1.79	0.11	3	0
2	A_2	A_1	8.17	0.10	5	4
3	B_1	A_1	0.63	0.11	4	1
4	B_2	B_1	3.01	0.12	2	0
5	B_3	A_2, B_2	0.75	0.11	4	3
6	C_1	B_1	11.20	0.13	2	7
7	C_2	B_3, C_1	2.29	0.09	6	2
					$Z_1 = 8$	$Z_2 = 10$

based on the lead-lag PDM restrictions. There are three main activities. These are partitioned into seven subactivities. Two resource types are involved in this example. There are 8 units of resource type 1 available and 10 units of resource type 2 available. The resource requirements for the individual segments of each activity are shown in the two right-hand columns of the table. The seven activity segments are ranked by the CAF heuristic in the following priority order for resource allocation purposes:

Activity 6 (C_1): $CAF = 100$
Activity 2 (A_2): $CAF = 71.7$
Activity 5 (B_3): $CAF = 65.2$
Activity 3 (B_1): $CAF = 49.6$
Activity 7 (C_2): $CAF = 45.2$
Activity 4 (B_2): $CAF = 29.6$
Activity 1 (A_1): $CAF = 28.7$

     Figure 5–6 presents the resulting Gantt chart for the resource-constrained schedule. Note that the expected project duration is 24.08 time units. This may be compared to the duration of 18 time units obtained in Chapter 4 without resource limitations.

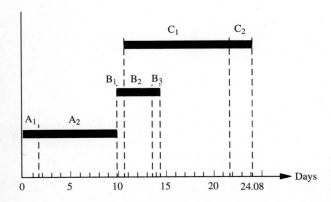

Figure 5-6.   Resource-Constrained PDM Schedule

5.7 CRITICAL RESOURCE DIAGRAM

Resource management is a major function in any organization. In a project management environment, project goals are achieved through the strategic allocation of resources to tasks. Several analytical and graphical tools are available for activity planning, scheduling, and control. Examples are the critical path method (CPM), the program evaluation and review technique (PERT), and the precedence diagramming method (PDM). Unfortunately, similar tools are not available for resource management. There is a need for simple tools for resource allocation planning, scheduling, tracking, and control. In this section, a simple extension of the CPM diagram is developed for resource management purposes. The extension, called the *critical resource diagram* (CRD), is a graphical tool that brings the well-known advantages of the CPM diagram to resource scheduling. The advantages of CRD include simplified resource tracking and control, better job distribution, better information to avoid resource conflicts, and better tools for resource leveling.

5.7.1 Resource Management Constraints

Resource management is a complex task that is subject to several limiting factors including:

- Resource interdependencies
- Conflicting resource priorities
- Mutual exclusivity of resources
- Limitations on resource availability
- Limitations on resource substitutions
- Variable levels of resource availability
- Limitations on partial resource allocation

The above factors invariably affect the criticality of certain resource types. It is logical to expect different resource types to exhibit different levels of criticality in a resource allocation problem. For example, some resources are very expensive; some resources possess special skills; some are in very limited supply. The relative importance of different resource types should be considered when carrying out resource allocation in activity scheduling. The critical resource diagram helps in representing resource criticality.

5.7.2 CRD Network Development

Figure 5–7 shows an example of a CRD for a small project requiring 6 different resource types. Each node identification, RESj, refers to a task responsibility for resource type j. In a CRD, a node is used to represent each resource unit. Thus, there are eight nodes in Figure 5–7 because there are 2 units of resource type 1 and 2 units of resource type 4. The interrelationships between resource units are indicated by arrows. The arrows are referred to as resource-relationship (R-R) arrows. For example, if the job of resource 1 must precede the job of resource 2, then an arrow is drawn from the node for resource 1 to the node for resource 2. Task durations are included in a CRD to provide further details about resource relationships. Unlike activity diagrams, a resource unit may appear at more than one location in a CRD provided there are no time or task conflicts. Such multiple locations indicate the number of different jobs for which the resource is responsible. This information may be useful for task distribution and resource leveling purposes. In Figure 5–7, resource 1 (RES 1) and resource 4 (RES 4) appear at two different nodes, indicating that each is responsible for two different jobs within the same work scenario.

5.7.3 CRD Computations

The same forward and backward computations used in CPM are applicable to a CRD diagram. However, the interpretation of the critical path may be different, since a single resource may appear at multiple nodes. Figure 5–8 presents an illustrative computational analysis of the CRD network in Figure 5–7. Task durations (days) are given below the resource identifications. Earliest and latest times are computed and appended to each resource node in the same manner as in CPM analysis. RES 1, RES 2, RES 5, and RES 6 form the critical resource path. These resources have no slack times with respect to the completion of the given project. Note that only one of the two tasks of RES 1 is on the critical resource path. Thus, RES 1 has slack time for performing one job, while it has no slack time for performing the other. Neither of the two tasks of RES 4 is on the critical resource path. For RES 3, the task duration is specified as zero. Despite this favorable task duration, RES 3 may turn out to be a bottleneck resource. RES 3 may be a senior manager whose task is signing a work order. But if he or she is not available to sign at the appropriate time, then the tasks of several other resources may be adversely affected. A major benefit of a CRD is that both senior-level and lower-level resources can be included in the resource planning network.

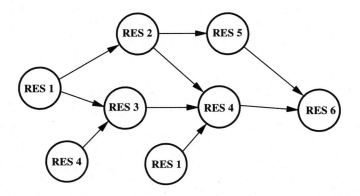

Figure 5-7.   CRD Network Example

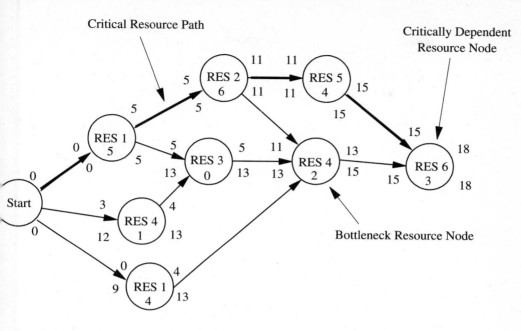

Figure 5-8.   CRD Network Analysis

5.7.4 CRD Node Classifications

A *bottleneck* resource node is defined as a node at which two or more arrows merge. In Figure 5–8, RES 3, RES 4, and RES 6 have bottleneck resource nodes. The tasks to which bottleneck resources are assigned should be expedited in order to avoid delaying dependent resources. A *dependent* resource node is a node whose job depends on the job of immediately preceding nodes. A *critically dependent* resource node is defined as a node on the critical resource path at which several arrows merge. In Figure 5–8, RES 6 is both a critically dependent resource node and a bottleneck resource node. As a scheduling heuristic, it is recommended that activities that require bottleneck resources be scheduled as early as possible. A *burst* resource node is defined as a resource node from which two or more arrows emanate. Like bottleneck resource nodes, burst resource nodes should be expedited, since their delay will affect several following resource nodes.

5.7.5 Resource Schedule Chart

The critical resource diagram has the advantage that it can be used to model partial assignment of resource units across multiple tasks in single or multiple projects. A companion chart for this purpose is the resource schedule (RS) chart. Figure 5–9 shows an example of an RS chart based on the earliest times computed in Figure 5–8. A horizontal bar is drawn for each resource unit or resource type. The starting point and the length of each resource bar indicate the interval of work for the resource. Note that the two jobs of RES 1 overlap over a four-day time period. By comparison, the two jobs of RES 4 are separated by a period of six days. If RES 4 is not to be idle over those six days, "fill-in" tasks must be assigned to it. For resource jobs that overlap, care must be taken to ensure that the resources do not need the same tools (e.g., equipment, computers, and lathe) at the same time. If a resource unit is found to have several jobs overlapping over an extensive period of time, then a task reassignment may be necessary to offer some relief for the resource. The RS chart is useful for a graphical representation of the utilization of resources. Although similar information can be obtained from a conventional resource loading graph, the RS chart gives a clearer picture of where and when resource commitments overlap. It also shows areas where multiple resources are working concurrently.

5.7.6 CRD and Work Rate Analysis

When resources work concurrently at different work rates, the amount of work accomplished by each may be computed by the procedure for work rate analysis. The critical resource diagram and the resource schedule chart provide information to identify when, where, and which resources work concurrently.

Example

Suppose the work rate of RES 1 is such that it can perform a certain task in 30 days. It is desired to add RES 2 to the task so that the completion time of the task can be reduced. The work rate of RES 2 is such that it can perform the same task alone in 22 days. If RES 1 has already worked 12 days on the task before RES 2 comes in, find the completion time of the task. Assume that RES 1 starts the task at time 0.

Solution

The amount of work to be done is 1.0 whole unit (i.e., the full task).

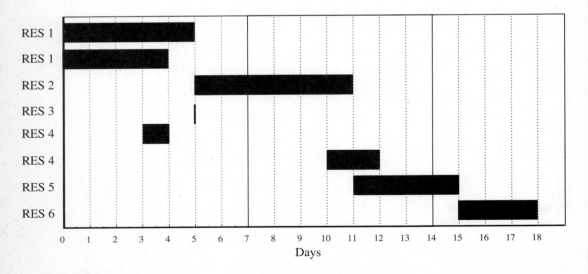

Figure 5-9.   Resource Schedule Chart

The work rate of RES 1 is 1/30 of the task per unit time.

The work rate of RES 2 is 1/22 of the task per unit time.

The amount of work completed by RES 1 in the 12 days it worked alone is $(1/30)(12) = 2/5$ (or 40%) of the required work.

Therefore, the remaining work to be done is 3/5 (or 60%) of the full task.

Let T be the time for which both resources work together.

The two resources working together to complete the task yield the following table:

Resource type i	Work rate, r_i	Time, t_i	Work done, w_i
RES 1	1/30	T	$T/30$
RES 2	1/22	T	$T/22$
		Total	3/5

Thus, we have

$$T/30 + T/22 = 3/5$$

which yields $T = 7.62$ days. Thus, the completion time of the task is $(12 + T) = 19.62$ days from time zero. The results of this example are summarized graphically in Figure 5–10. It is assumed that both resources produce identical quality of work and that the respective work rates remain consistent. The respective costs of the different resource types may be incorporated into the work rate analysis. The CRD and RS charts are simple extensions of very familiar tools. They are simple to use and they convey resource information quickly. They can be used to complement existing resource management tools. Users can find innovative ways to modify or implement them for specific resource planning, scheduling, and control purposes. For example, resource-dependent task durations and resource cost can be incorporated into the CRD and RS procedures to enhance their utility for resource management decisions.

8 RESOURCE LOADING AND LEVELING

Resource loading refers to the allocation of resources to work elements in a project network. A resource loading graph is a graphical representation of resource allocation over time. Figure 5–11 shows an example of a resource loading graph. A resource loading graph may be drawn for the different resource types involved in a project.

The graph provides information useful for resource planning and budgeting purposes. A resource loading graph gives an indication of the demand a project will place on an organization's resources. In addition to resource units committed to activities, the graph may also be drawn for other tangible and intangible resources of an organization. For example, a variation of the graph may be used to present information about the depletion rate of the budget available for a project. If drawn for multiple resources, it can help identify potential areas of resource conflicts. For situations where a single resource unit is assigned to multiple tasks, a variation of the resource loading graph can be developed to show the level of load (responsibilities) assigned to the resource over time.

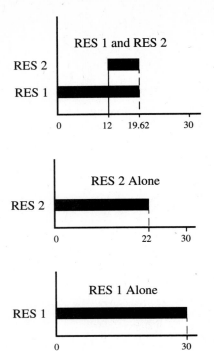

Figure 5-10. Resource Schedule Charts for RES 1 and RES 2

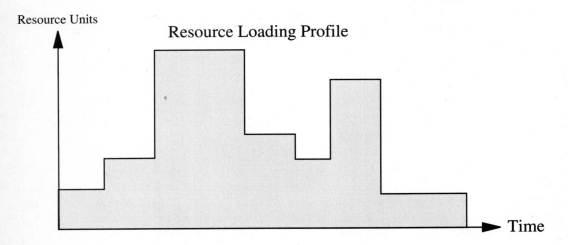

Figure 5-11. Example of Resource Loading Graph

5.8.1 Resource Leveling

Resource leveling refers to the process of reducing the period-to-period fluctuation in a resource loading graph. If resource fluctuations are beyond acceptable limits, actions are taken to move activities or resources around in order to level out the resource loading graph. For example, it is bad for employee morale and public relations when a company has to hire and lay people off indiscriminately. Proper resource planning will facilitate a reasonably stable level of the workforce. Other advantages of resource leveling include simplified resource tracking and control, lower cost of resource management, and improved opportunity for learning. Acceptable resource leveling is typically achieved at the expense of longer project duration or higher project cost. Figure 5–12 shows a leveled resource loading.

When attempting to level resources, note that

1. Not all of the resource fluctuations can be eliminated.
2. Resource leveling often leads to an increase in project duration.

Resource leveling attempts to minimize fluctuations in resource loading by shifting activities within their available slacks. For small networks, resource leveling can be attempted manually through trial-and-error procedures. For large networks, resource leveling is best handled by computer software techniques. Most of the available commercial project management software packages have internal resource leveling routines. One heuristic procedure for leveling resources, known as *Burgess's Method* (Woodworth and Willie 1975), is based on the technique of minimizing the sum of the squares for the resource requirements in each period.

5.8.2 Resource Idleness Graph

A resource idleness graph is similar to a resource loading graph except that it is drawn for the number of unallocated resource units over time. The area covered by the resource idleness graph may be used as a measure of the effectiveness of the scheduling strategy employed for a project. Suppose two scheduling strategies yield the same project duration, and suppose a measure of the resource utilization under each strategy

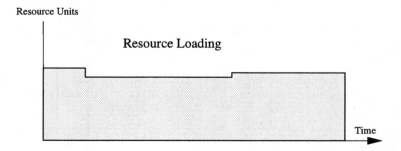

Figure 5-12.   Resource Leveling Graph

is desired as a means to compare the strategies. Figure 5–13 shows two hypothetical resource idleness graphs for the alternate strategies. The areas are computed as
follows:

$$\text{Area A} = 6(5) + 10(5) + 7(8) + 15(6) + 5(16) = 306 \text{ resource-units-time}$$
$$\text{Area B} = 5(6) + 10(9) + 3(5) + 6(5) + 3(3) + 12(12) = 318 \text{ resource-units-time}$$

Since area A is less than area B, it is concluded that strategy A is more effective for
resource utilization than strategy B. Similar measures can be developed for multiple
resources. However, for multiple resources, the different resource units must all be
scaled to dimensionless quantities before computing the areas bounded by the resource
idleness graphs.

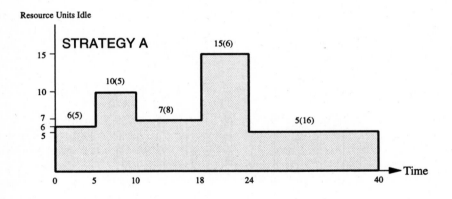

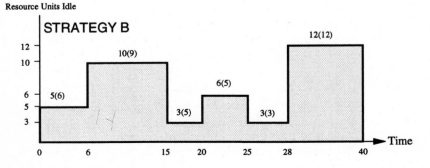

Figure 5-13.   Resource Idleness Graphs for Resource Allocation

.9 PROBABILISTIC RESOURCE UTILIZATION

In a nondeterministic project environment, probability information can be used to analyze resource utilization characteristics. Suppose the level of availability of a resource is probabilistic in nature. For simplicity, we will assume that the level of availability, X, is a continuous variable whose probability density function is defined by $f(x)$. This is true for many resource types such as funds, natural resources, and raw materials. If we are interested in the probability that resource availability will be within a certain range of x_1 and x_2, then the required probability can be computed as

$$P(x_1 \le X \le x_2) = \int_{x_1}^{x_2} f(x)\, dx$$

Similarly, a probability density function can be defined for the utilization level of a particular resource. If we denote the utilization level by U and its probability density function by $f(u)$, then we can calculate the probability that the utilization will exceed a certain level, u_0, by the following expression:

$$P(U \ge u_0) = \int_{u_0}^{\infty} f(u)\, du$$

Example

Suppose a critical resource is leased for a large project. There is a graduated cost associated with using the resource at a certain percentage level, U. The cost is specified as $10,000 per 10% increment in the utilization level above 40%. A flat cost of $5,000 is charged for utilization levels below 40%. The utilization intervals and the associated costs are presented as follows:

$$U < 40\%, 5{,}000$$
$$40\% <= U < 50\%, \$10{,}000$$
$$50\% <= U < 60\%, \$20{,}000$$
$$60\% <= U < 70\%, \$30{,}000$$
$$70\% <= U < 80\%, \$40{,}000$$
$$80\% <= U < 90\%, \$50{,}000$$
$$90\% <= U < 100\%, \$60{,}000$$

Thus, a utilization level of 50% will cost $20,000, while a level of 49.5% will cost $10,000. Suppose the utilization level is a normally distributed random variable with mean of 60% and standard deviation of 4%. Find the expected cost of using this resource.

Solution.   The solution procedure involves finding the probability that the utilization level will fall within each of the specified ranges. The expected value formula will then be used to compute the expected cost:

$$E[C] = \sum_k x_k P(x_k),$$

where x_k represents the kth interval of utilization. The standard deviation of utilization is 4%.

$$P(U < 40) = P\left(z \le \frac{40 - 60}{4}\right) = P(z \le -5) = 0.0$$

$$P(40 \le U < 50) = P\left(z < \frac{50-60}{4}\right) - P\left(z \le \frac{40-60}{4}\right)$$
$$= P(z \le -2.5) - P(z \le -5)$$
$$= 0.0062 - 0.0$$
$$= 0.0062$$

$$P(50 \le U < 60) = P\left(z < \frac{60-60}{4}\right) - P\left(z \le \frac{50-60}{4}\right)$$
$$= P(z \le 0) - P(z \le -2.5)$$
$$= 0.5000 - 0.0062$$
$$= 0.4938$$

$$P(60 \le U < 70) = P\left(z < \frac{70-60}{4}\right) - P\left(z \le \frac{60-60}{4}\right)$$
$$= P(z \le 2.5) - P(z \le 0)$$
$$= 0.9938 - 0.5000$$
$$= 0.4938$$

$$P(70 \le U < 80) = P\left(z < \frac{80-60}{4}\right) - P\left(z \le \frac{70-60}{4}\right)$$
$$= P(z \le 5) - P(z \le 2.5)$$
$$= 1.0 - 0.9938$$
$$= 0.0062$$

$$P(80 \le U < 90) = P\left(z < \frac{90-60}{4}\right) - P\left(z \le \frac{80-60}{4}\right)$$
$$= P(z \le 7.5) - P(z \le 5)$$
$$= 1.0 - 1.0$$
$$= 0.0$$

$$E(C) = \$5,000(0.0) + \$10,000(0.0062) + \$20,000(0.4938)$$
$$+ \$30,000(0.4938) + \$40,000(0.0062) + \$50,000(0.0)$$
$$= \$25,000.$$

Thus, it can be expected that leasing this critical resource will cost $25,000 in the long run. A decision can be made whether to lease or buy the resource. Resource substitution may also be considered on the basis of the expected cost of leasing.

Other analytical calculations using the normal distribution are presented in Chapter 6 under statistical project control.

5.10 LEARNING CURVE ANALYSIS

Learning curves are important for resource allocation decisions. Learning curves present the relationship between cost (or time) and level of activity on the basis of the effect of learning. An early study by Wright (1936) disclosed the "80 percent learning" effect, which indicates that a given operation is subject to a 20 percent productivity improvement each time the activity level or production volume doubles. A learning curve can serve as a predictive tool for obtaining time estimates for tasks in a project environment. Typical learning rates that have been encountered in practice range from

70 percent to 95 percent. A learning curves is also referred to as a *progress function,* a *cost-quantity relationship,* a *cost curve,* a *product acceleration curve,* an *improvement curve,* a *performance curve,* an *experience curve,* and an *efficiency curve.*

Several alternate models of learning curves have been presented in the literature (see Belkaoui 1986; J. Smith 1989; Teplitz 1991). Some of the most notable models are the *log-linear model,* the *S-curve,* the *Stanford-B model, DeJong's learning formula, Levy's adaptation function, Glover's learning formula, Pegels' exponential function, Knecht's upturn model,* and *Yelle's product model.* The univariate learning curve expresses a dependent variable (e.g., production cost) in terms of some independent variable (e.g., cumulative production). The log-linear model is by far the most popular and most used of all the learning curve models.

5.10.1 The Log-Linear Model

The log-linear model states that the improvement in productivity is constant (i.e., it has a constant slope) as output increases. There are two basic forms of the log-linear model: the average cost model and the unit cost model.

Average cost model.   The average cost model is used more than the unit cost model. It specifies the relationship between the cumulative average cost per unit and cumulative production. The relationship indicates that cumulative cost per unit will decrease by a constant percentage as the cumulative production volume doubles. The model is expressed as

$$A_x = C_1 x^b$$

where

A_x = cumulative average cost of producing x units
C_1 = cost of the first unit
x = cumulative production count
b = the learning curve exponent (i.e., constant slope of on log-log paper)

The relationship between the learning curve exponent, b, and the learning rate percentage, p, is given by

$$b = \frac{\log p}{\log 2} \text{ or } p = 2^b$$

The derivation of the above relationship can be seen by considering two production levels where one level is double the other, as shown next.

Let level I = x_1 and level II = $x_2 = 2x_1$. Then,

$$A_{x_1} = C_1(x_1)^b \qquad \text{and} \qquad A_{x_2} = C_1(2x_1)^b$$

The percent productivity gain is then computed as

$$p = \frac{C_1(2x_1)^b}{C_1(x_1)^b} = 2^b$$

When linear graph paper is used, the log-linear learning curve is a hyperbola of the form shown in Figure 5–14. On log-log paper, the model is represented by the following straight line equation:

$$\log A_x = \log C_1 + b \log x$$

where b is the constant slope of the line. It is from this straight line that the name *log-linear* was derived.

Example

Assume that 50 units of an item are produced at a cumulative average cost of $20 per unit. Suppose we want to compute the learning percentage when 100 units are produced at a cumulative average cost of $15 per unit. The learning curve analysis would proceed as follows:

Initial production level = 50 units; Average cost = $20

Double production level = 100 units; Cumulative average cost = $15

Using the log relationship, we obtain the following equations:

$$\log 20 = \log C_1 + b \log 50$$
$$\log 15 = \log C_1 + b \log 100$$

Solving the equations simultaneously yields

$$b = \frac{\log 20 - \log 15}{\log 50 - \log 100} = -0.415$$

Thus,

$$p = (2)^{-0.415} = 0.75$$

That is a 75% learning rate. In general, the learning curve exponent, b, may be calculated directly from actual data or computed analytically. That is,

$$b = \frac{\log A_{x_1} - \log A_{x_2}}{\log x_1 - \log x_2}$$

or

$$b = \frac{\ln(p)}{\ln(2)}$$

where

x_1 = first production level
x_2 = second production level
A_{x_1} = cumulative average cost per unit at the first production level
A_{x_2} = cumulative average cost per unit at the second production level
p = learning rate percentage

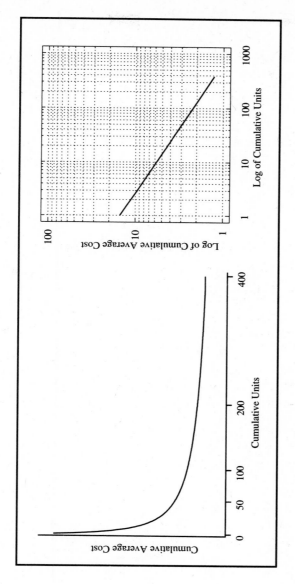

Figure 5-14. The Log-Linear Learning Curve

193

Using the basic cumulative average cost function, the total cost of producing x units is computed as

$$TC_x = (x)A_x = (x)C_1x^b = C_1x^{(b+1)}$$

The unit cost of producing the xth unit is given by

$$U_x = C_1x^{(b+1)} - C_1(x-1)^{(b+1)}$$
$$= C_1[x^{(b+1)} - (x-1)^{(b+1)}]$$

The marginal cost of producing the xth unit is given by

$$MC_x = \frac{d[TC_x]}{dx} = (b+1)C_1x^b$$

Example

Suppose in a production run of a certain product it is observed that the cumulative hours required to produce 100 units is 100,000 hours with a learning curve effect of 85%. For project planning purposes, an analyst needs to calculate the number of hours spent in producing the fiftieth unit. Following the notation used previously, we have the following information:

$$p = 0.85$$
$$X = 100 \text{ units}$$
$$A_x = 100,000 \text{ hours}/100 \text{ units} = 1,000 \text{ hours/unit}$$

Now,

$$0.85 = 2^b$$

Therefore, $b = -0.2345$
Also,

$$1,000 = C_1(100)^b$$

Therefore, $C_1 = 2,944.42$ hours. Thus,

$$C_{50} = C_1(50)^b$$
$$= 1,176.50 \text{ hours}$$

That is, the cumulative average hours for 50 units is 1,176.50 hours. Therefore, cumulative total hours for 50 units = 58,824.91 hours. Similarly,

$$C_{49} = C_1(49)^b$$
$$= 1,182.09 \text{ hours}$$

That is, the cumulative average hours for 49 units is 1,182.09 hours. Therefore, cumulative total hours for 49 units = 57,922.17 hours. Consequently, the number of hours for the fiftieth unit is given by

$$58,824.91 \text{ hours} - 57,922.17 \text{ hours} = 902.74 \text{ hours}.$$

Unit cost model.   The unit cost model is expressed in terms of the specific cost of producing the xth unit. The unit cost formula specifies that the individual cost per unit will decrease by a constant percentage as cumulative production doubles. The formulation of the unit cost model is presented below. Define the average cost as A_x.

$$A_x = C_1 x^b$$

The total cost is defined as

$$TC_x = (x)A_x = (x)C_1 x^b = C_1 x^{(b+1)}$$

and the marginal cost is given by

$$MC_x = \frac{d[TC_x]}{dx} = (b+1)C_1 x^b$$

This is the cost of one specific unit. Therefore, define the unit cost model as

$$U_x = (1+b)C_1 x^b$$

U_x is the cost of producing the xth unit. To derive the relationship between A_x and U_x,

$$U_x = (1+b)C_1 x^b$$
$$\therefore \frac{U_x}{(1+b)} = C_1 x^b = A_x$$
$$\therefore A_x = \frac{U_x}{(1+b)}$$
$$U_x = (1+b)A_x$$

To derive an expression for finding the cost of the first unit, C_1, we will proceed as follows. Since $A_x = C_1 x^b$, we have

$$C_1 x^b = \frac{U_x}{(1+b)}$$
$$\therefore C_1 = \frac{U_x x^{-b}}{(1+b)}$$

Figure 5–15 presents a plot comparing the unit cost model to the average cost model for the previous example, where $C_1 = \$2,944$ and $b = -0.2345$.

For the case of continuous product volume (e.g., chemical processes), we have the following corresponding expressions:

$$TC_x = \int_0^x U(z)\,dz \quad C_1 \int_0^x z^b\,dz = \frac{C_1 x^{(b+1)}}{b+1}$$

$$Y_x = \left(\frac{1}{x}\right) \frac{C_1 x^{(b+1)}}{b+1}$$
$$MC_x = \frac{d[TC_x]}{dx} = \frac{d\left[\frac{C_1 x^{(b+1)}}{b+1}\right]}{dx} = C_1 x^b$$

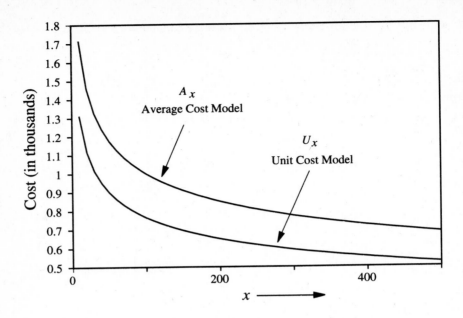

Figure 5-15.   Comparison of the Unit Cost and Average Cost Models

5.10.2 Graphical Analysis

Suppose an observation of the cumulative hours required to produce a unit of an item was recorded at irregular intervals during a production cycle. The recorded observations are presented in Table 5–8.

The project analyst would like to perform the following computational analyses:

Table 5-8   Learning Curve Observations

Cumulative units produced (X)	Cumulative average hours (A_x)
10	92.5
15	71.2
25	50.0
40	35.0
50	26.2
60	20.0
85	11.3
115	10.0
165	7.5
190	6.3

1. Calculate the learning curve percentage when cumulative production doubles from 10 to 20 units.
2. Calculate the learning curve percentage when cumulative production doubles from 20 units to 40 units.
3. Calculate the learning curve percentage when cumulative production doubles from 40 units to 80 units.
4. Calculate the learning curve percentage when cumulative production doubles from 80 units to 160 units.
5. Compute the average learning curve percentage for the given operation.
6. Estimate a standard time for performing the given operation if the steady production level per cycle is 200 units.

A plot of the recorded data is shown in Figure 5–16. A regression model fitted to the data is also shown in the figure. The fitted model is expressed mathematically as

$$A_x = 634.22x^{-0.8206}$$

with an R^2 value of 98.6%. Thus, we have a highly significant model fit. The fitted model can be used for estimation and planning purposes. Time requirements for the

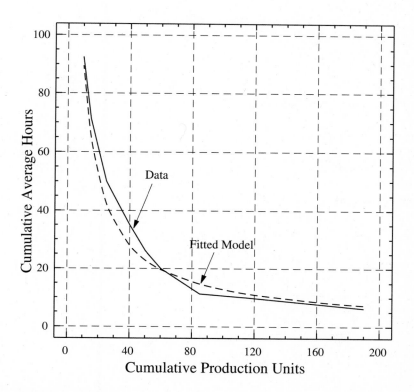

Figure 5-16.   Graphical Analysis of Empirical Learning Curve

operation at different production levels can be estimated from the model. From the model, we have an estimated cost of the first unit as

$$C_1 = \$634.22$$
$$b = -0.8206$$
$$p = 2^{(-0.8206)} = 56.62\% \text{ learning rate}$$

By using linear interpolation for the recorded data, we can estimate the percentage improvement from one production level to another. For example, when production doubles from 10 to 20 units, we obtain an estimated cumulative average hours of 60.6 hours by interpolating between 71.2 hours and 50 hours. Similarly, cumulative average hours of 13.04 hours were obtained for the production level of 80 units, and cumulative average hours of 7.75 hours were obtained for the production level of 160 units. Now, these average hours are used to compute the percent improvement over the various production levels. For example, the percentage improvement when production doubles from 10 units to 20 units is obtained as $p = 60.6/92.5 = 65.5\%$. The calculated percent improvement levels are presented in Table 5–9. The average percent is found to be 53.76%. This compares favorably with the 56.62% suggested by the fitted regression model.

Using the fitted model, the estimated cumulative average hours per unit when 200 units are produced is estimated as

$$A_x = 634.22(200)^{-0.8206}$$
$$= 8.20 \text{ hours}$$

Caution should be exercised in using the fitted learning curve for extrapolation beyond the range of the data used to fit the model.

5.10.3 Multivariate Learning Curves

Extensions of the single factor learning curve are important for realistic analysis of productivity gain. In project operations, several factors can intermingle to affect performance. Heuristic decision making, in particular, requires careful consideration of qualitative factors. There are numerous factors that can influence how fast, how far, and how well a worker learns within a given time span. Multivariate models are useful for performance analysis in project planning and control. One form of the multivariate learning curve is defined as

Table 5-9 Learning Curve Percentage Analysis

Initial production level	Doubled production level	Learning percentage
10	20	65.50%
20	40	57.76
40	80	37.26
80	160	59.43
	Average	54.99

$$A_x = K \prod_{i=1}^{n} c_i x_i^{b_i}$$

where

A_x = cumulative average cost per unit

K = cost of first unit of the product

x = vector of specific values of independent variables

x_i = specific value of the ith factor

n = number of factors in the model

c_i = coefficient for the ith factor

b_i = learning exponent for the ith factor

A bivariate form of the model is presented as

$$C = \beta_0 x_1^{\beta_1} x_2^{\beta_2}$$

where C is a measure of cost and x_1 and x_2 are independent variables of interest. Some of the multivariate models that have been reported in the literature are discussed briefly.

Alchian (1963) modeled learning curves that estimate direct labor per pound of airframe needed to manufacture the Nth airframe in a cumulative production of N planes based on World War II data. He studied the alternate functions presented below to describe the relationships between direct labor per pound of airframe (m), cumulative production (N), time (T), and rate of production per month (DN):

1. $\log m = a_2 + b_2 T$
2. $\log m = a_3 + b_3 T + b_4 DN$
3. $\log m = a_4 + b_5 (\log T) + b_6 (\log DN)$
4. $\log m = a_5 + b_7 T + b_8 (\log DN)$
5. $\log m = a_6 + b_9 T + b_{10} (\log N)$
6. $\log m = a_7 + b_{11} (\log N) + b_{12} (\log DN)$

The multiplicative power function, often referred to as the Cobb-Douglas function, was investigated by Goldberger (1968) as a model for the learning curve. The model is of the general form

$$C = b_0 x_1^{b_1} x_2^{b_2} \ldots x_n^{b_n} \epsilon$$

where

C = estimated cost

b_0 = model coefficient

x_i = ith independent variable ($i = 1, 2, \ldots, n$)

b_i = exponent of the ith variable

ϵ = error term

For parametric cost analysis, Waller and Dwyer (1981) studied an additive model of the form

$$C = c_1 x_1^{b_1} + c_2 x_2^{b_2} + \ldots + c_n x_n^{b_n} + \epsilon$$

where $c_i (i = 1, 2, \ldots, n)$ is the coefficient of the ith independent variable. The model was reported to have been fitted successfully for missile tooling and equipment cost. A variation of the power model was used by Bemis (1981) to study weapon system production. Cox and Gansler (1981) discuss the use of a bivariate model for the assessment of the costs and benefits of a single-source versus multiple-source production decision with variations in quantity and production rate in major Department of Defense (DOD) programs. A similar study by Camm, Evans, and Womer (1987) also uses the multiplicative power model to express program costs in terms of cumulative quantity and production rate in order to evaluate contractor behavior.

McIntyre (1977) introduced a nonlinear cost-volume-profit model for learning curve analysis. The nonlinearity in the model is effected by incorporating a nonlinear cost function that expresses the effects of employee learning. The profit equation for the initial period of production for a product subject to the usual learning function is expressed as

$$P = px - c(ax^{b+1}) - f$$

where

P = profit

p = price per unit

x = cumulative production

c = labor cost per unit time

f = fixed cost per period

b = index of learning

The profit function for the initial period of production with n production processes operating simultaneously is given as

$$P = px - nca \left(\frac{x}{n} \right)^{b+1} - f$$

where x is the number of units produced by n labor teams consisting of one or more employees each. Each team is assumed to produce x/n units. This model indicates that when additional production teams are included, more units are produced over a given time period. However, the average time for a given number of units increases because more employees are producing while they are still learning. That is, more employees with low (but improving) productivity are engaged in production at the same time. The preceding model is extended to the case where employees with different skill levels produce different learning parameters between production runs. This is modeled as

$$P = p \sum_{i=1}^{n} x_i - c \sum_{i=1}^{n} a_i x_i^{b_i+1} - f$$

where a_i and b_i denote the parameters applicable to the average skill level of the ith production run, and x_i represents the output of the ith run in a given time period. This model could be useful for manufacturing systems that call for concurrent engineering. Womer (1979) presents a multivariate model that incorporates cumulative production, production rate, and program cost. His approach involves a production function that relates output rate to a set of inputs with variable utilization rates.

Bivariate example.   A bivariate model is used here to illustrate the nature and modeling approach for general multivariate models. An experiment presented by Badiru (1992b) models a learning curve containing two independent variables: *cumulative production* (x_1) and *cumulative training time* (x_2). The following model was chosen for illustration:

$$A_{x_1 x_2} = K c_1 x_1^{b_1} c_2 x_2^{b_2}$$

where

> A_x = cumulative average cost per unit for a given set
>
> $\quad X$, of factor values
>
> K = intrinsic constant
>
> x_1 = specific value of first factor
>
> x_2 = specific value of second factor
>
> c_i = coefficient for the ith factor
>
> b_i = learning exponent for the ith factor

The set of test data used for the modeling is shown in Table 5–10.

Two data replicates are used for each of the 10 combinations of cost and time values. Observations are recorded for the number of units representing double production volumes. The model is transformed to the natural logarithmic form below:

$$\ln A_x = [\ln K + \ln(c_1 c_2)] + b_1 \ln x_1 + b_2 \ln x_2$$
$$= \ln a + b_1 \ln x_1 + b_2 \ln x_2$$

where a represents the combined constant in the model such that $a = (K)(c_1)(c_2)$. A regression approach yielded the fitted model

$$\ln A_x = 5.70 - 0.21(\ln x_1) - 0.13(\ln x_2)$$
$$A_x = 298.88 x_1^{-0.21} x_2^{-0.13}$$

with an R^2 value of 96.7%. The variables in the model are explained as follows:

> $\ln(a) = 5.70$(i.e., $a = 298.88$)
>
> x_1 = cumulative production units
>
> x_2 = cumulative training time in hours.

Table 5-10 Data for Modeling Bivariate Learning Curve
(From Badiru 1992b) reprinted with permission

Treatment number	Observation number	Cumulative average cost ($)	Cumulative production (units)	Cumulative training time (hours)
1	1	120	10	11
	2	140	10	8
2	3	95	20	54
	4	125	20	25
3	5	80	40	100
	6	75	40	80
4	7	65	80	220
	8	50	80	150
5	9	55	160	410
	10	40	160	500
6	11	40	320	660
	12	38	320	600
7	13	32	640	810
	14	36	640	750
8	15	25	1,280	890
	16	25	1,280	800
9	17	20	2,560	990
	18	24	2,560	900
10	19	19	5,120	1,155
	20	25	5,120	1,000

Figure 5–17 shows the response surface for the fitted model. As in the univariate case, the bivariate model indicates that the cumulative average cost decreases as cumulative production and training time increase. For a production level of 1,750 units and a cumulative training time of 600 hours, the fitted model indicates an estimated cumulative average cost per unit as

$$A_{(1750,600)} = (298.88)(1750^{-0.21})(600^{-0.13}) = 27.12$$

Similarly, a production level of 3,500 units and training time of 950 hours yield the following cumulative average cost per unit:

$$A_{(3500,950)} = (298.88)(3,500^{-0.21})(950^{-0.13}) = 22.08$$

To use the fitted model, consider the following problem. The standards department of a manufacturing plant has set a target average cost per unit of $12.75 to be achieved after 1,000 hours of training. We want to find the cumulative units that must be produced in order to achieve the target cost. From the fitted model, the following expression is obtained:

$$\$12.75 = (298.88)(X^{-0.21})(1,000^{-0.13})$$
$$\therefore X = 46,409.25$$

On the basis of the large number of cumulative production units required to achieve the expected standard cost, the standards department may want to review the cost standard. The standard of $12.75 may not be achievable if there is a limited market

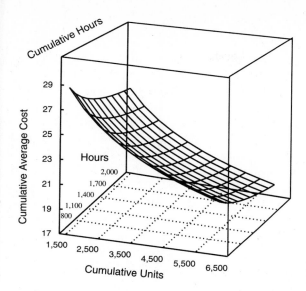

Figure 5-17.   Bivariate Learning Curve Model

demand (i.e., demand is much less than 46,409 units) for the particular product being considered. The relatively flat surface of the learning curve model as units and training time increase implies that more units will need to be produced in order to achieve any additional significant cost improvements. Thus, even though an average cost of $22.08 can be obtained at a cumulative production level of 3,500 units, it takes several thousand additional units to bring the average cost down to $12.75 per unit.

5.10.4 Interruption of Learning

Interruption of the learning process can adversely affect expected performance. J. Smith (1989) accounts for the effect of interruption in the learning process by developing what he calls a *manufacturing interruption ratio*. The ratio considers the learning decay that occurs when a learning process is interrupted. He suggests the following expression:

$$Z = (C_1 - A_x)\frac{(t-1)}{11}$$

where

      Z = per-product loss of learning costs due to manufacturing interruption

      $t = 1, 2, \ldots, 11$ (months of interruption from 1 month to 12 months)

      C_1 = cost of the first unit of the product

      A_x = cost of the last unit produced before a production interruption

    Thus, the unit cost of the first unit produced after production begins again is given by

$$
\begin{aligned}
A_{(x+1)} &= A_x + Z \\
&= A_x + (C_1 - A_x)\frac{(t-1)}{11}
\end{aligned}
$$

Badiru (1994) also presents models that consider interruptions in the learning process by incorporating forgetting functions into regular learning curves. In any practical situation, an allowance must be made for the potential impacts that forgetting may have on performance. Potential applications of the combined learning and forgetting models include design of training programs, manufacturing economic analysis, manpower scheduling, production planning, labor estimation, budgeting, and resource allocation.

5.11 EXERCISES

5.1. For the project data in Table 5–2, redefine the activity durations in terms of the PERT estimates presented below. If the other activity data are the same as presented in Table 5–2, compute the scaled CAF for each activity in the project and use the CAF criterion to schedule the project. Compare the CAF schedule to the ACTIM schedule.

Activity	a, m, b
1–2	1, 2, 3
2–3	7, 8, 9
1–3	2, 3, 4
3–5	9, 10, 11
1–4	7, 8, 9
2–4	1, 2, 3
4–5	0, 1, 2

5.2. Presented in the table is the activity data for a project that is subject to variable resource availability. It is assumed that the scaled CAF weights for the activities are already known as given in the last column of the tabulated data. There are two resource types. Resource availability varies from day to day on the basis of the resource schedule tabulated after the project data.

Activity	Predecessor	Duration (day)	Resource Requirement (type 1, type 2)	Scaled CAF
A	—	6	2, 5	50.1
B	—	5	1, 3	57.1
C	—	2	3, 1	56.8
D	A, B	4	2, 0	100
E	B	5	1, 4	53.2
F	C	6	3, 1	51.8
G	B, F	1	4, 3	44.9
H	D, E	7	2, 3	54.9

Variable resource availability for the project is as follows:

Time period (days)	Units available (type 1, type 2)
0 to 5	3, 5
5$^+$ to 10	2, 3
10$^+$ to 14	3, 4
14$^+$ to 20	4, 4
20$^+$ to 22	5, 3
22$^+$ to 30	4, 5
30$^+$ to 99	6, 6

Use the CAF weights to develop the Gantt chart for the project considering the daily resource availability. A task that is already in progress will be temporarily suspended whenever there are not enough units of resources to continue it. The task will resume (without increasing its duration) whenever enough resources become available.

5.3. Use the minimum total slack (Min TS) heuristic to schedule the resource-constrained project below. If a tie occurs, use minimum activity duration to break the tie.

Units of resource type 1 available = 10
Units of resource type 2 available = 15

Activity	Predecessor	a	m	b	Resource units Type 1	Resource units Type 2
A	—	1	2	4	3	0
B	—	5	6	7	5	4
C	—	2	4	5	4	1
D	A	1	3	4	2	0
E	C	4	5	7	4	3
F	A	3	4	5	2	7
G	B, D, E	1	2	3	6	2

5.4. For Exercise 5.3, draw the resource loading diagram for each resource type on the same graph. Which resource type exhibits more fluctuations in resource loading?

5.5. Using the computational approach presented in this chapter, compute the scaled CAF weight for each activity in Exercise 5.3.

5.6. Schedule the project in Exercise 5.3 using each of the following heuristics: CAF, LS, ES, LC, EC, TS, GRD, and ACTIM. What is the shortest project duration obtained? Which heuristics yield that shortest project duration?

5.7. Draw the resource loading graph for each schedule in Exercise 5.6.

5.8. Draw the resource idleness graph for each schedule in Exercise 5.6.

5.9. Use the computational measure presented in this chapter to compute the resource utilization level (area) for each schedule in Exercise 5.6.

5.10. Three project crews are awarded a contract to construct an automated plant. The crews can work simultaneously on the project if necessary. Crew 1, working alone, can complete the project in 82 days. Crew 2, working alone, can complete the project in 50 days. Crew 3, working alone, can complete the project in 64 days. Crew 2 started the project. Ten days after the project started, crew 1 joined the project. Fifteen days after crew 1 joined the project, crew 3 also joined the project. Crew 2 left the project 5 days before it was completed. Crew 1 is paid $500,000 per day on the project. Crew 2 is paid $750,000 per day. Crew 3 is paid $425,975 per day.
   (a) Find how many days it took to complete the project (assume that fractions of a day are permissible).
   (b) Based on the project participation described above, find the minimum budget needed to complete the project.

5.11. A project involves laying 100 miles of oil distribution pipe in the Alaskan wilderness. It is desired to determine the lowest-cost crew size and composition that can get the job done within 90 days. Three different types of crew are available. Each crew can work independently or work simultaneously with other crews. Crew 1 can lay pipes at the rate of 1 mile per day. If Crew 2 is to do the whole job alone, it can finish it in 120 days. Crew 3 can lay pipes at the rate of 1.7 miles per day, but this crew already has other imminent project commitments that will limit its participation in the piping project to only 35 contiguous days. All three crews are ready, willing, and available to accept their contracts at the beginning of the project. Crew 3 charges $50,000 per mile of pipe laid. Crew 2 charges $35,000 per mile and Crew 1 charges $40,000 per mile. Contracts can be awarded only in increments of 10 miles. That is, a crew can get a contract only for 0 mile, 10 miles, 20 miles, 30 miles, and so on. *Required*: Determine the best composition of the project crews and the minimum budget needed to finish the project in the shortest possible time.

5.12. Suppose the utilization level of a critical resource is defined by a normal distribution with a mean of 70 percent and a variance of 24 percent squared. Compute the probability that the utilization of the resource will exceed 85 percent.

5.13. Suppose rainfall is a critical resource for a farming project. The availability of rainfall in terms of inches during the project is known to be a random variable defined by a triangular distribution with a lower end point of 5.25 inches, a mode of 6 inches, and an upper end point of 7.5 inches. Compute the probability that there will be between 5.5 and 7 inches of rainfall during the project.

5.14. A scarce resource is to be leased for an engineering project. There is a graduated cost associated with using the resource at a certain percentage level, U. A step function has been defined for the cost rates for different levels of utilization. The step cost function is presented below:

$$\text{If } U < 50\%, \text{cost} = \$0$$
$$\text{If } 50\% \le U < 60\%, \$5,000$$
$$\text{If } 60\% \le U < 67\%, \$20,000$$
$$\text{If } 67\% \le U < 85\%, \$45,000$$
$$\text{If } U \ge 85\%, \$55,000$$

Suppose the utilization level is a random variable following a triangular distribution with a lower limit of 40%, a mode of 70%, and an upper limit of 95%.
   (a) Plot the step function for the cost of leasing the resource.
   (b) Plot the triangular probability density function for the utilization level.
   (c) Find the expected cost of using this resource.

5.15. Suppose three resource types (RES 1, RES 2, and RES 3) are to be assigned to a certain task. RES 1 working alone can complete the task in 35 days, RES 2 working alone can complete the task in 40 days, and RES 3 working alone can complete the task in 60 days. Suppose 1 unit of RES 1 and 1 unit of RES 2 start working on the task together at time zero. Two units of RES 3 joined the task after 45% of the task has been completed. RES 2 quits the task when there is 15% of the task remaining. Determine the completion time of the task and the total number of days that each resource type worked on the task. Assume that work rates are constant and units of the same resource type have equal work rates.

5.16. Three resource types (RES 1, RES 2, and RES 3) are to be assigned to a certain task. RES 1 working alone can complete the task in 35 days, RES 2 working alone can complete the task in 40 days, and RES 3 working alone can complete the task in 60 days. Suppose 1 unit of RES 1 and 1 unit of RES 2 start working on the task together at time zero. Two units of RES 3 are brought in to join the task some time after time zero. RES 2 quits the task when there is 20% of the task remaining. If the full task is desired to be completed 13 days from time zero, determine how many days from time zero the 2 units of RES 3 should be brought in to join the task. Assume that work rates are constant and units of the same resource type have equal work rates.

5.17. Develop a quantitative methodology for incorporating different levels of quality of work by different resource types into the procedure for work rate analysis presented in this chapter.

5.18. Use the approach presented in Chapter 4 to compute the project network complexity for the data in Exercise 5.3.

5.19. For the project data in the table, use Badiru's approach to compute the network complexity for the following alternate levels of resource availability: 2, 3, 4, 5, 6, 7, 8, 9, 10, 11, and 12. Plot the complexity measures against the respective resource availability levels. Discuss your findings.

Activity number	PERT estimates (a, m, b)	Preceding activities	Required resources
1	1, 3, 5	—	1
2	0.5, 1, 3	—	2
3	1, 1, 2	—	1
4	2, 3, 6	Activity 1	2
5	1, 3, 4	Activity 2	1
6	1.5, 2, 2	Activity 3	2

5.20. For the project data in the table, use Badiru's approach to compute the network complexity for all possible combinations of resource availability (Z_1, Z_2), where the possible values of Z_1 are 4, 6, 7, 8, and 9 and the possible values of Z_2 are 2, 3, 4, 5, and 6. Note that this will generate 25 network complexity values. Fit an appropriate multiple regression function to the data you generated, where network complexity is the dependent variable and Z_1 and Z_2 are the independent variables.

Activity number	PERT Estimates (a, m, b)	Preceding activities	Required resources (x_{i1}, x_{i2})
1	1, 3, 5	—	1, 0
2	0.5, 1, 3	—	1, 1
3	1, 1, 2	—	1, 1
4	2, 3, 6	Activity 1	2, 0
5	1, 3, 4	Activity 2	1, 0
6	1.5, 2, 2	Activity 3	4, 2

5.21. There are three resource types available for a certain project. One unit of each resource type is available. The project manager wants to evaluate the project cost on the basis of how resource teams are made up. She has the option of using a resource team, where the team can consist of only one resource type or a combination of resource types. If the team consists of more than one resource type, the resources will start and stop working at the same time. Fractions of a work day are permissive with prorated cost.

> Resource 1 can complete the project alone in 50 days at a cost of $2,500 per day.
> Resource 2 can complete the project alone in 35 days at a cost of $5,750 per day.
> Resource 3 can complete the project alone in 75 days at a cost of $1,500 per day.

The resources produce identical quality of work.
(a) Determine the team composition that will yield the minimum project cost.
(b) What is the project duration corresponding to that minimum cost? Plot all the project costs versus the respective project durations.

5.22. Three resource types (RES 1, RES 2, and RES 3) are to be assigned to a certain task. RES 1 working alone can complete the task in 30 days, RES 2 working alone can complete the task in 45 days, and RES 3 working alone can complete the task in 50 days. Suppose 2 units of RES 1 and 1 unit of RES 2 start working on the task together at time zero. One unit of RES 3 is brought in to join the task some time after time zero. RES 2 quits the task when there is 30% of the task remaining. If the full task is desired to be completed 20 days from time 0, determine when, if at all, RES 3 should be brought in to join the task. Assume that work rates are constant and units of the same resource type have equal work rates. Show all work completely and clearly.

5.23. Suppose snowfall is a critical resource for a skiing business. The availability of snowfall in inches during a season is known to be a random variable defined by a triangular distribution with a lower end point of 3 inches, a mode of 6 inches, and an upper end point of 10.5 inches. Compute the probability that there will be between 5 and 9 inches of snowfall during a season.

5.24. Compute scaled CAF priority weight for each of the activities in the project described below. Use the scaled values to schedule the activities. Compare the schedule obtained to the schedule in Exercise 5.3.

> Units of resource type 1 available = 10
> Units of resource type 2 available = 15

Activity	Predecessor	a	m	b	Resource units Type 1	Resource units Type 2
A	—	1	2	4	3	0
B	—	5	6	7	5	4
C	—	2	4	5	4	1
D	A	1	3	4	2	0
E	C	4	5	7	4	3
F	A	3	4	5	2	7
G	B, D, E	1	2	3	6	2

5.25. In a certain project operation, it was noted that a total assembly cost of $2,000 was incurred for the production of 50 units of a product. Suppose an additional 100 units were produced at a cost of $1,000. Determine the learning curve percentage for the operation.

5.26. Suppose an operation is known to exhibit a learning curve rate of 85%. Production was interrupted for three months after 100 units had been produced. If the cost of the very first unit is $50, find
(a) Unit cost of the first unit produced after production begins again.
(b) Cost of unit 150 of the product assuming no further production interruptions occur.

5.27. Suppose the first 20-unit batch of a product has an average cost of $72 per unit; the next 30-unit batch has an average cost of $60 per unit; and the next 50-unit batch has an average cost of $50 per unit. Based on this cost history, determine the appropriate learning curve percent to recommend for this operation.

5.28. The first performance of a task requires eight hours. The twentieth performance of the task requires only two hours. If this task is subject to a conventional learning process, determine
(a) Learning rate associated with the task.
(b) Number of hours it will take to perform the task the twelfth time.

5.29. Suppose unit 190 of a product requires 45 hours to produce under a learning curve rate of 80%. Determine the number of hours required by the first unit of the product and the number of hours required by unit 250.

5.30. The first 50-unit order of a job shop costs $1,500. It is believed that the shop experiences a 75% learning curve rate. Determine a reasonable price quote for the next 80-unit order of the same job.

5.31. Suppose the sixtieth unit of a product produced under a learning rate of 80% is $30. If the production standard is $20 per unit, determine how many more units must be produced before the standard can be reached.

6

Project Control

This chapter presents some approaches to project monitoring and control. The steps required to carry out project control are discussed. Schedule control through progress review is presented. Guidelines for performance control are developed. Project information systems needed for control are discussed. An approach to terminating projects as a managerial control is recommended. Our project management process has now gone through the following series of steps:

SET GOALS → **PLAN** → **ORGANIZE** → **SCHEDULE** →
ALLOCATE RESOURCES → **CONTROL**

6.1 ELEMENTS OF PROJECT CONTROL

The three factors (time, budget, and performance) that form the basis for the operating characteristics of a project also help determine the basis for project control. Project control is the process of reducing the deviation between actual performance and planned performance. To control a project, we must be able to measure performance. The ability to measure accurately is a crucial aspect of control. Measurements are taken on each of the three components of project constraints: time, performance, and cost. These constraints are often encountered in conflicting terms and they cannot be fully satisfied simultaneously. The conflicting nature of the three constraints on a project is represented in Figure 6–1.

The higher the desired performance, the more time and resources (cost) will be required. It will be necessary to compromise one constraint in favor of another. There are some projects where performance is the sole focus. Time and cost may be of secondary importance in such projects. In other projects, cost reduction may be the main goal. Performance and time may be sacrificed to some extent in such projects. And, there are projects where schedule compression (time) is the ultimate goal. The

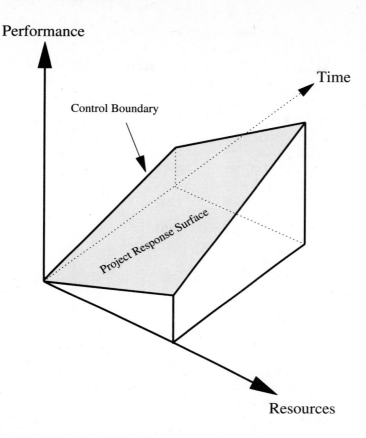

Figure 6-1.   Project Control Boundary

specific nature of a project will determine which constraints must be satisfied and which can be sacrificed. Some of the factors that can require control of a project are

1. FACTORS AFFECTING TIME

Supply delays
Missed milestones
Delay of key tasks
Change in customer specifications
Change of due dates
Unreliable time estimates
Increased use of expediting
Time-consuming technical problems
Impractical precedence relationships
New regulations that need time to implement

2. FACTORS AFFECTING PERFORMANCE

Poor design
Poor quality
Low reliability
Fragile components
Poor functionality
Maintenance problems
Complicated technology
Change in statement of work
Conflicting objectives
Restricted access to resources
Employee morale
Poor retention of experienced workforce
Sickness and injury

3. FACTORS AFFECTING COST

Inflation
New vendors
Incorrect bids
High labor cost
Budget revisions
High overhead costs
Inadequate budget
Increased scope of work
Poor timing of cash flows

Project control may be handled in a hierarchical manner starting with the global view of a project and ending with the elementary level of unit performance as shown in Figure 6–2.

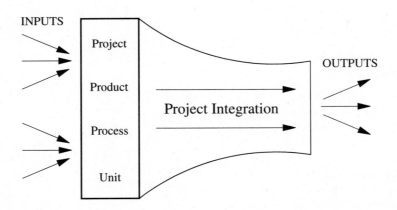

Figure 6-2.   Control Hierarchy from Inputs to Outputs

A product is project dependent. A process is product dependent. The performance of a unit depends on the process from which the unit is made. Such a control hierarchy makes the control process more adaptive, dynamic, and effective. The basic elements of adaptive project control are

- Continual tracking and reporting
- Modifying project implementation as objectives change
- Replanning to deal with new developments
- Evaluating achievement versus objectives
- Documenting project success and failures as guides for the future

.2 CONTROL STEPS

Parkinson's law states that a schedule will expand to fill available time and cost will increase to consume the available budget. Project control prevents a schedule from expanding out of control. Project control also assures that a project can be completed within budget. A recommended project control process is presented as follows:

STEP 1: Determine the criterion for control. This means that the specific aspects that will be measured should be specified.

STEP 2: Set performance standards. Standards may be based on industry practice, prevailing project agreements, work analysis, forecasting, and so on.

STEP 3: Measure actual performance. The measurement scale should be predetermined. The measurement approach must be calibrated and verified for accuracy. Quantitative and nonquantitative factors may require different measurement approaches. This step also requires reliable project tracking and reporting tools. Project status, no matter how unfavorable, must be reported.

STEP 4: Compare actual performance with the specified performance standard. The comparison should be done objectively and consistently based on the specified control criteria. Meet periodically to determine
- What has been achieved.
- What remains to be done.

STEP 5: Identify unacceptable variance from expectation.

STEP 6: Determine the expected impact of the variance on overall project performance.

STEP 7: Investigate the source of the poor performance.

STEP 8: Determine the appropriate control actions needed to overcome (nullify) the variance observed.

STEP 9: Implement the control actions with total dedication.

STEP 10: Ensure that the poor performance does not recur elsewhere in the project. The control steps can be carried out within the framework of the flowchart presented in Figure 6–3. The flow of capital, materials, and labor must be controlled throughout the project management process.

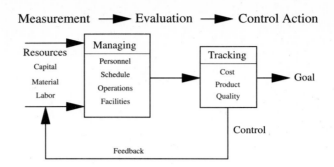

Figure 6-3.   Flowchart for Control Process

6.2.1 Formal and Informal Control

Informal control refers to the process of using unscheduled or unplanned approaches to assess project performance and using informal control actions. Informal control requires unscheduled visits and impromptu queries to track progress. The advantages of the informal control process are as follows:

- It allows the project manager to learn more about project progress.
- It creates a surprise element which keeps workers on their toes.
- It precludes the temptation for "doctored" progress reports.
- It allows peers and subordinates to assume control roles.
- It facilitates prompt appraisal of latest results.
- It gives the project manager more visibility.

A formal control process deals with the process of achieving project control through formal and scheduled reports, consultations, or meetings. Formal control is typically more time consuming.

- It can be used by only a limited (designated) group of people.
- It reduces the direct visibility of the project manager.
- It can impede the implementation of control actions.
- It encourages bureaucracy and "paper pushing."
- It requires a rigid structure.

Despite its disadvantages, formal control can be effective for all types of projects. With a formal control process, project responsibilities and accountability can be pursued in a structured manner. For example, standard audit questions may be posed in order to determine the current status of a project in order to establish the strategy for future performance. Examples of suitable questions are

- Where are we today?
- Where were we expected to be today?
- What are the prevailing problems?
- What problems are expected in the future?
- Where shall we be at the next reporting time?
- What are the major results since the last project review?
- What is the ratio of percent completion to budget depletion?
- Is the project plan still the same?
- What resources are needed for the next stage of the project?

A formal structured documentation of what questions to ask can guide the project auditor in carrying out project audits in a consistent manner. The availability of standard questions makes it unnecessary for the auditor to guess or ignore certain factors that may be crucial for project control.

.3 SCHEDULE CONTROL

The Gantt charts developed in the scheduling phase of a project can serve as the benchmark for measuring project progress. Project status should be monitored frequently. Symbols may be used to mark actual activity positions on the Gantt chart. A record should be maintained of the difference between the actual status of an activity and its expected status. This information should be conveyed to the appropriate personnel with a clear indication for the required control actions. The more milestones or control points there are in a project, the easier the control function. The larger number allows for more frequent and distinct avenues for monitoring the schedule. Problems can be identified and controlled before they accumulate into more serious problems. However, more control points mean higher cost of control.

Schedule variance magnitudes may be plotted on a time scale (e.g., on a daily basis). If the variance continues to get worse, drastic actions may be necessitated. Temporary deviations without a lasting effect on the project may not be a cause for concern. Some control actions that may be needed for project schedule delays are

- Job redesign
- Productivity improvement
- Revision of project scope
- Revision of project master plan
- Expediting or activity crashing
- Elimination of unnecessary activities
- Reevaluation of milestones or due dates
- Revision of time estimates for pending activities

6.3.1 Project Tracking and Reporting

Tracking and reporting provide the avenues for monitoring and evaluating project progress. This is an area where computerized tools are useful. Customized computer-based project tracking programs can be developed to monitor various aspects of project performance. An example is the program developed by B. Baker (1990). The program was developed with Lotus 1-2-3 Macro language. A modification of the Baker model is presented in Table 6–1. The model keeps track of several parameters for each activity including the following:

1. Days of work planned for an activity
2. Planned cost
3. Number of days worked
4. Number of days credited
5. Percent completion expected
6. Actual percent completion
7. Actual cost
8. Amount of budget shortfall or surplus

The activity parameters used in the model and shown in Table 6–1 are explained next.

- Work-days planned indicates the amount of work required for a task expressed in terms of the number of days required to perform the work. This is entered as an input by the user. In a more sophisticated model, the number of work-days required may be computed directly by the program based on a standard data base or some internal estimation formulas.
- Planned cost indicates the estimated cost of performing the full amount of work required by the activity. This can be entered as a user input or can be calculated internally by the program.

Table 6–1   Data for Project Performance Monitoring

Project Start Date: 1-15-93
Current Date: 7-15-93
Project Name: Installation of New Products Line

Task name	Work-days planned	Planned cost	Task weight	Work-days completed	Work-days credited
1. Define problem	16	$100,000	0.1524	16	16
2. Interviews	38	100,000	0.3619	38	38
3. Network analysis	7	100,000	0.0667	7	7
4. Improve flow	7	150,000	0.0667	7	7
5. New process	7	200,000	0.0667	14	7
6. Expert system	30	42,000	0.2857	0	0
	105		1.000		75

Table 6-1 (continued)

Task name	Expected % completion	Expected relative % completion	Actual % completion	Actual relative % completion	Actual cost	Cost deviation
1. Define problem	100	15.24	100	15.24	$59,000	($41,000)
2. Interviews	100	36.19	100	36.19	146,000	46,000
3. Network analysis	100	6.67	100	6.67	110,000	10,000
4. Improve flow	100	6.67	65	4.33	65,000	(85,000)
5. New process	100	6.67	10	0.67	216,000	16,000
6. Expert system	0	0	0	0		0
		71.43		0.6310		(54,000)

Planned Project % Completion = 71.43%
Project Tracking Index = -11.67%
Days Ahead/Behind Schedule = -8.75 days' worth of work
Cost Deviation = ($54,000) (Cost Saving!)

- Task weight represents the relative importance of the activity based on some stated criteria. In the model in Table 6–1, the task weight is computed on the basis of the work content of the activity relative to the overall project. The work content is expressed in terms of days. This is computed as

$$\text{task weight} = \frac{\text{(work-days for activity)}}{\text{(total work-days for project)}}$$

For example, in Table 6–1, the total work content for the project is 105 days. The work content for task 2 is 38 days. So, the weight for task 2 is 38/105 = 0.3619.

- Work-days completed indicates the amount of work actually completed out of the total planned work-days as of the current date. This value is based on actual measurement of work accomplishment. This may be done by monitoring the total number of cumulative hours worked and dividing that total by the number of hours per day to obtain number of work-days.

- Work-days credited indicates the number of productive work-days recognized out of the total work-days completed. For example, if no credit is given for setup times, then the work-days credited will be obtained by subtracting all setup times from work-days completed. For example, in Table 6–1, a total of 14 days were worked on task 5, but only 7 of the 14 days were credited to the task.

- Expected % completion indicates the percentage of the required work that is expected to be completed by the current date. This is computed as

$$\text{expected \% completion} = \frac{\text{(work-days completed)}}{\text{(work-days planned)}}$$

- Expected relative % completion indicates the percentage of work expected to be completed on the given activity relative to the total work on the project. This is computed as

$$\text{expected relative \% completion} = \text{(expected \% completion)} * \text{(task weight)}$$

For example, task 1 in Table 6–1 is expected to have completed 16 of the 105 days for the project by the current date. That translates to 16/105 = 0.1524 or 15.24%.

- Actual % completion indicates the actual percentage of work completed by the current date. This value will normally be obtained by a direct observation of the goal of the task. Note it is possible that even though the number of planned work-days have been completed, the actual amount of work completed may not be worth the number of days worked. This may be due to low employee productivity. For example, if a certain amount of work is expected to take 10 days, an employee may work diligently on the task for 10 days but he or she may accomplish only one-half of what is required to be done.

In Table 6–1, this occurs in the case of tasks 4 and 5. This is reflected in the actual % completion column. Even though the required 7 days for task 4 have been completed, only 65% of the required work has been done. The amount of work yet to be done on task 4 is equivalent to 35% of 7 days (i.e., 2.45 days' worth of work behind schedule). Similarly, the amount of work remaining to be done on task 5 is 90% of 7 days (i.e., 6.3 days' worth of work behind schedule). Consequently, the total project is behind schedule by 2.45 days + 6.3 days = 8.75 days. It should be noted that a project may be behind schedule either in terms of the physical schedule (i.e., number of days elapsed) or in terms of the amount of work accomplished. The 8.75 days is in terms of amount of work to be accomplished. That is, it will take 8.75 days to accomplish what is yet to be accomplished on tasks 4 and 5.

- Actual relative % completion indicates the relative percentage of work actually completed on the given activity as compared to other activities in the project. This is computed as

$$\text{actual relative \% completion} = (\text{actual \% completion}) * (\text{task weight})$$

- Actual cost indicates the actual total amount spent on the activity as of the current date.
- Cost deviation indicates the difference between the actual and planned costs. Cost overrun occurs when actual cost exceeds planned cost. Cost saving occurs when actual cost falls below planned cost. In Table 6–1, brackets are used to indicate cost savings. It is noted that tasks 2, 3, and 5 have cost overruns.

In addition to the activity parameters explained above, the following parameters are associated with the overall project:

- Planned project % completion indicates the percent of the project expected to be completed by the current date. This is computed as

$$\text{planned project \% completion} = \frac{(\text{work-days completed on project})}{\text{total work-days planned}}$$

This can also be obtained as the sum of the expected relative % completion for all activities. In Table 6–1, the planned % completion of the project is 71.43%.

- Project tracking index is a relative measure of the performance of the project based on the amount of work actually completed rather than the number of days elapsed. It is computed as

$$\text{index} = \frac{(\text{actual relative \% completion})}{(\text{expected relative \% completion})} - 1$$

If the project is performing better than expected (e.g., ahead of schedule), then the index will be positive. The index will be negative if the project performance is below expectation. The index, expressed as a percentage, may be viewed as a measure of the criticality of the control action needed on the project. For the example in Table 6–1, the project tracking index is computed as

$$(0.6310)/(0.7143) - 1 = -0.1167 \text{ or } -11.67\%$$

indicating that the project is behind schedule in work accomplishment. This implies that 11.67% of the work expected to be accomplished by the current date has not been accomplished.

- Days ahead/behind schedule refers to the performance of the project schedule. It indicates the amount of work by which the project deviates from expectation. The deviation is expressed in terms of days of work. This measure is computed as

> schedule performance = (project tracking index) ∗ (planned days credited)

If this number is negative, then the project is behind schedule in terms of work accomplishment. If the number is positive, then the project is ahead of schedule. It should be recalled that a project may be physically on schedule but still be behind schedule in terms of actual work accomplishment. The measure used in the Baker model provides a better basis for evaluating project performance than the traditional approach of merely looking at the physical schedule. In Table 6–1, the project is behind schedule by the following amount:

$$(-11.67\%) \ast (75 \text{ days}) = -8.75 \text{ days' worth of work}$$

It should be recalled that this same number, −8.75 days, was obtained earlier by adding up the work deficiencies of individual activities. Even though this project has been credited with 75 work-days, the actual amount of work accomplished is only 75 days − 8.75 days = 66.25 days.

6.4 PERFORMANCE CONTROL

Many project performance problems may not surface until after a project has been completed. This makes performance control very difficult. Effort should be made to measure all the interim factors that may influence final project performance. After-the-fact performance measurements are typically not effective for project control. Some of the performance problems may be indicated by time and cost deviations. So, when project time and cost have problems, an analysis of how the problems may affect performance should be made. Since project performance requirements usually relate to the performance of the end products, controlling performance problems may necessitate altering product specifications. Performance analysis will involve checking key elements of the product such as those discussed next.

1. Scope

    Is the scope reasonable based on the project environment?
    Can the required output be achieved with the available resources?

The current wave of *downsizing* and *rightsizing* in industry may be an attempt to define the proper scope of operations.

2. Documentation

Is the requirement specification accurate?

Are statements clearly understood?

3. Requirements

Is the technical basis for the specification sound?

Are requirements properly organized?

What are the interactions among specific requirements?

How does the prototype perform?

Is the raw material appropriate?

What is a reasonable level of reliability?

What is the durability of the product?

What are the maintainability characteristics?

Is the product compatible with other products?

Are the physical characteristics satisfactory?

4. Quality assurance

Who is responsible for inspection?

What are the inspection policies and methods?

What actions are needed for nonconforming units?

5. Function

Is the product usable as designed?

Can the expected use be achieved by other means?

Is there any potential for misusing the product?

Careful evaluation of performance on the basis of the above questions throughout the life cycle of a project should help identify problems early so that control actions may be initiated to forestall greater problems later.

6.4.1 Continuous Performance Improvement

Continuous performance improvement (CPI) is an approach to obtaining a steady flow of improvement in a project. The approach is based on the concept of continuous process improvement, which is used in quality management functions. The iterative decision processes in project management can benefit quite well from the concept of CPI. Continuous performance improvement is a practical method of improving performance in business, management, or technical processes. The approach is based on the following key points:

- Early detection of problems
- Incremental improvement steps
- Projectwide adoption of the CPI concept
- Comprehensive evaluation of procedures
- Prompt review of methods of improvement
- Prioritization of improvement opportunities

- Establishment of long-term improvement goals
- Continuous implementation of improvement actions

A steering committee is typically set up to guide the improvement efforts. The typical functions of the steering committee with respect to performance improvement include the following:

- Determination of organizational goals and objectives
- Communication with the personnel
- Team organization
- Administration of CPI procedures
- Education and guidance for companywide involvement
- Allocation or recommendation of resource requirements

Figure 6–4 represents the conventional fluctuating approach to performance improvement. In the figure, the process starts with a certain level of performance. A certain performance level is specified as the target to be achieved by time T. Without proper control, the performance will gradually degrade until it falls below the lower

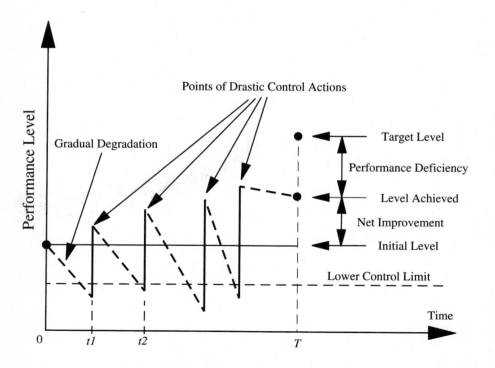

Figure 6-4. Conventional Approach to Control

control limit at time t_1. At that time, a drastic effort will be needed to raise the performance level. If neglected once again, the performance will go through another gradual decline until it again falls below the lower control limit at time t_2. Again, a costly drastic effort will be needed to improve the performance. This cycle of *degradation-innovation* may be repeated several times before time T is reached. At time T, a final attempt will be needed to suddenly raise the performance to the target level. But unfortunately, it may be too late to achieve the target performance.

There are many disadvantages of the conventional fluctuating approach to improvement. They are

1. High cost of implementation
2. Need for drastic control actions
3. Potential loss of project support
4. Adverse effect on personnel morale
5. Frequent disruption of the project
6. Too much focus on short-term benefits
7. Need for frequent and strict monitoring
8. Opportunity cost during the degradation phase

Figure 6–5 represents the approach of continuous improvement. In the figure, the process starts with the same initial quality level and it is continuously improved in a gradual pursuit of the target performance level. As opportunities to improve occur, they

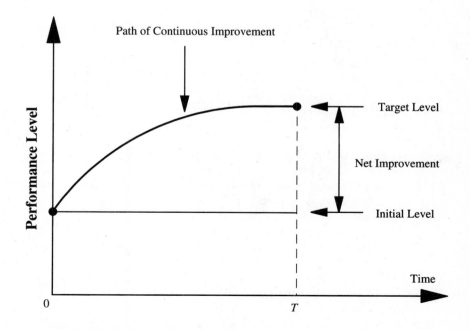

Figure 6-5.   Continuous Process of Control

are immediately implemented. The rate of improvement is not necessarily constant over the planning horizon. Hence, the path of improvement is shown as a curve rather than a straight line. The important aspect of the CPI is that each subsequent performance level is at least as good as the one preceding it.

The major advantages of continuous process of control include

1. Better client satisfaction
2. Clear expression of project expectations
3. Consistent pace with available technology
4. Lower cost of achieving project objectives
5. Conducive environment for personnel involvement
6. Dedication to higher quality products and services

6.5 COST CONTROL

Several aspects of a project can contribute to the overall cost of the project. These aspects must be carefully tracked during the project to determine when control actions may be needed. Some of the important cost aspects of a project are

- Cost estimation approach
- Cost accounting practices
- Project cash flow management
- Company cash flow status
- Direct labor costing
- Overhead rate costing
- Incentives, penalties, and bonuses
- Overtime payments

The process of controlling project cost covers several key issues that management must address. These include

1. Proper planning of the project to justify the basis for cost elements
2. Reliable estimation of time, resources, and cost
3. Clear communication of project requirements, constraints, and available resources
4. Sustained cooperation of project personnel
5. Good coordination of project functions
6. Consistent policy for project expenditures
7. Timely tracking and reporting of time, materials, and labor transactions
8. Periodic review of project progress
9. Revision of project schedule to adapt to prevailing project scenarios
10. Evaluation of budget depletion versus project progress

These items must be evaluated as an integrated control effort rather than as individual functions. The interactions among the various actions needed may be so unpredictable that the success achieved at one level may be nullified by failure at another level. Such uncoordinated analysis makes cost control very difficult. Project managers must be alert and persistent in the cost monitoring function.

Some government agencies have developed cost control techniques aimed at managing large projects that are typical in government contracts. The cost and schedule control system (C/SCS) is based on WBS (work breakdown structure) and it can quantitatively measure project performance at a particular point in a project. Another useful cost control technique is the accomplishment cost procedure (ACP). This is a simple approach for relating resources allocated to actual work accomplished. It presents costs based on scheduled accomplishments rather than as a function of time. In order to determine the progress of an individual effort with respect to cost, the cost/progress relationship in the project plan is compared to the cost/progress relationship actually achieved. The major aspect of the ACP technique is that it is not biased against high costs. It gives proper credit to high costs as long as comparable project progress is maintained. The cost analysis aspects of project management are covered in detail in a later chapter.

6 INFORMATION FOR PROJECT CONTROL

Complex projects require a well-coordinated communication system that can quickly reveal the status of each activity. Reports on individual project elements must be tied together in a logical manner to facilitate managerial control. The project manager must have prompt access to individual activity status as well as the status of the overall project. A critical aspect of this function is the prevailing level of communication, cooperation, and coordination in the project. The *project management information system (PMIS)* has evolved as the solution to the problem of monitoring, organizing, storing, and disseminating project information. Many commercial computer programs have been developed for the implementation of PMIS. The basic reporting elements in a PMIS may include the following:

- Financial reports
- Project deliverables
- Current project plan
- Project progress reports
- Material supply schedule
- Client delivery schedule
- Subcontract work and schedule
- Project conference schedule and records
- Graphical project schedule (Gantt Chart)
- Performance requirements evaluation plots
- Time performance plots (plan versus actual)
- Cost performance plots (expected versus actual)

Many standard forms have been developed to facilitate the reporting process. With the availability of computerized systems, manual project information systems are no longer used much in practice.

6.6.1 Measurement Scales

Project control requires data collection, measurement, and analysis. In project management, the manager will encounter different types of measurement scales depending on the particular items to be controlled. Data may need to be collected on project schedules, costs, performance levels, problems, and so on. The different types of data measurement scales that are applicable are discussed next.

Nominal scale of measurement. A *nominal scale* is the lowest level of measurement scales. It classifies items into categories. The categories are mutually exclusive and collectively exhaustive. That is, the categories do not overlap and they cover all possible categories of the characteristics being observed. For example, in the analysis of the critical path in a project network, each job is classified as either critical or not critical. Gender, type of industry, job classification, and color are some examples of measurements on a nominal scale.

Ordinal scale of measurement. An *ordinal scale* is distinguished from a nominal scale by the property of order among the categories. An example is the process of prioritizing project tasks for resource allocation. We know that first is above second, but we do not know how far above. Similarly, we know that better is preferred to good, but we do not know by how much. In quality control, the ABC classification of items based on the Pareto distribution is an example of a measurement on an ordinal scale.

Interval scale of measurement. An *interval scale* is distinguished from an ordinal scale by having equal intervals between the units of measure. The assignment of priority ratings to project objectives on a scale of 0 to 10 is an example of a measurement on an interval scale. Even though an objective may have a priority rating of 0, it does not mean that the objective has absolutely no significance to the project team. Similarly, the scoring of 0 on an examination does not imply that a student knows absolutely nothing about the materials covered by the examination. Temperature is a good example of an item that is measured on an interval scale. Even though there is a zero point on the temperature scale, it is an arbitrary relative measure. Other examples of interval scales are IQ measurements and aptitude ratings.

Ratio scale of measurement. A *ratio scale* has the same properties of an interval scale but with a true zero point. For example, an estimate of a zero time unit for the duration of a task is a ratio scale measurement. Other examples of items measured on a ratio scale are cost, time, volume, length, height, weight, and inventory level. Many of the items measured in a project management environment will be on a ratio scale.

Another important aspect of data analysis for project control involves the classification scheme used. Most projects will have both *quantitative* and *qualitative* data.

Quantitative data require that we describe the characteristics of the items being studied numerically. Qualitative data, on the other hand, are associated with object attributes that are not measured numerically. Most items measured on the nominal and ordinal scales will normally be classified into the qualitative data category while those measured on the interval and ratio scales will normally be classified into the quantitative data category.

The implication for project control is that qualitative data can lead to bias in the control mechanism because qualitative data are subject to the personal views and interpretations of the person using the data. Whenever possible, data for project control should be based on quantitative measurements.

Badiru and Whitehouse (1989) suggest a class of project data referred to as *transient data*. This is defined as a volatile set of data that is used for one-time decision making and is not then needed again. An example may be the number of operators that show up at a job site on a given day. Unless there is some correlation between the day-to-day attendance records of operators, this piece of information will have relevance only for that given day. The project manager can make his decision for that day on the basis of that day's attendance record. Transient data need not be stored in a permanent database unless it may be needed for future analysis or uses (e.g., forecasting, incentive programs, performance review).

Recurring data refers to data that is encountered frequently enough to necessitate storage on a permanent basis. An example is a file containing contract due dates. This file will need to be kept at least through the project life cycle. Recurring data may be further categorized into *static data* and *dynamic data*. A recurring data that is static will retain its original parameters and values each time that it is retrieved and used. A recurring data that is dynamic has the potential for taking on different parameters and values each time it is retrieved and used. Storage and retrieval considerations for project control should address the following questions:

1. What is the origin of the data?
2. How long will the data be maintained?
3. Who needs access to the data?
4. What will the data be used for?
5. How often will the data be needed?
6. Is the data for look-up purposes only (i.e., no printouts)?
7. Is the data for reporting purposes (i.e., generate reports)?
8. In what format is the data needed?
9. How fast will the data need to be retrieved?
10. What security measures are needed for the data?

6.6.2 Data Determination and Collection

It is essential to determine what data to collect for project control purposes. Data collection and analysis are basic components of generating information for project control. The requirements for data collection are discussed next.

Choosing the data.   This involves selecting data on the basis of their relevance and the level of likelihood that they will be needed for future decisions and whether or not they contribute to making the decision better. The intended users of the data should also be identified.

Collecting the data. This identifies a suitable method of collecting the data as well as the source from which the data will be collected. The collection method will depend on the particular operation being addressed. The common methods include manual tabulation, direct keyboard entry, optical character reader, magnetic coding, electronic scanner, and more recently, voice command. An input control may be used to confirm the accuracy of collected data. Examples of items to control when collecting data are the following:

Relevance Check. This checks if the data is relevant to the prevailing problem. For example, data collected on personnel productivity may not be relevant for a decision involving marketing strategies.

Limit Check. This checks to ensure that the data is within known or acceptable limits. For example, an employee overtime claim amounting to over 80 hours per week for several weeks in a row is an indication of a record well beyond ordinary limits.

Critical Value. This identifies a boundary point for data values. Values below or above a critical value fall in different data categories. For example, the lower specification limit for a given characteristic of a product is a critical value that determines whether or not the product meets quality requirements.

Coding the data. This refers to the technique used in representing data in a form useful for generating information. This should be done in a compact and yet meaningful format. The performance of information systems can be greatly improved if effective data formats and coding are designed into the system right from the beginning.

Processing the data. Data processing is the manipulation of data to generate useful information. Different types of information may be generated from a given data set depending on how it is processed. The processing method should consider how the information will be used, who will be using it, and what caliber of system response time is desired. If possible, processing controls should be used. This may involve

Control Total. Check for the completeness of the processing by comparing accumulated results to a known total. An example of this is the comparison of machine throughput to a standard production level or the comparison of cumulative project budget depletion to a cost accounting standard.

Consistency Check. Check if the processing is producing the same results for similar data. For example, an electronic inspection device that suddenly shows a measurement that is ten times higher than the norm warrants an investigation of both the input and the processing mechanisms.

Scales of Measurement. For numeric scales, specify units of measurement, increments, the zero point on the measurement scale, and the range of values.

Using the information. Using information involves people. Computers can collect data, manipulate data, and generate information, but the ultimate decision rests with people, and decision making starts when information becomes available. Intuition, experience, training, interest, and ethics are just a few of the factors that determine how people use information. The same piece of information that is positively used to further the progress of a project in one instance may also be used negatively in another instance. To assure that data and information are used appropriately, computer-based security measures can be built into the information system.

Project data may be obtained from several sources. Some potential sources are

- Formal reports
- Interviews and surveys
- Regular project meetings
- Personnel time cards or work schedules

The timing of data is also very important for project control purposes. The contents, level of detail, and frequency of data can affect the control process. An important aspect of project management is the determination of the data required to generate the information needed for project control. The function of keeping track of the vast quantity of rapidly changing and interrelated data about project attributes can be very complicated. The major steps involved in data analysis for project control are

- Data collection
- Data analysis and presentation
- Decision making
- Implementation of action

Data is processed to generate information. Information is analyzed by the decision maker to make the required decisions. Good decisions are based on timely and relevant information, which in turn is based on reliable data. Data analysis for project control may involve the following functions:

- Organizing and printing computer-generated information in a form usable by managers
- Integrating different hardware and software systems to communicate in the same project environment
- Incorporating new technologies such as expert systems into data analysis
- Using graphics and other presentation techniques to convey project information

Proper data management will prevent misuse, misinterpretation, or mishandling. Data is needed at every stage in the life cycle of a project from the problem identification stage through the project phaseout stage. The various items for which data may be needed are project specifications, feasibility study, resource availability, staff size, schedule, project status, performance data, and phaseout plan. The documentation of data requirements should cover the following:

- **Data summary.** A data summary is a general summary of the information and decision for which the data is required as well as the form in which the data should be prepared. The summary indicates the impact of the data requirements on the organizational goals.

- **Data processing environment.** The processing environment identifies the project for which the data is required, the user personnel, and the computer system to be used in processing the data. It refers to the project request or authorization and relationship to other projects and specifies the expected data communication needs and mode of transmission.

- **Data policies and procedures.** Data handling policies and procedures describe policies governing data handling, storage, and modification and the specific procedures for implementing changes to the data. Additionally, they provide instructions for data collection and organization.

- **Static data.** A static data description describes that portion of the data that is used mainly for reference purposes and it is rarely updated.

- **Dynamic data.** A dynamic data description describes that portion of the data that is frequently updated based on the prevailing circumstances in the organization.

- **Data frequency.** The frequency of data update specifies the expected frequency of data change for the dynamic portion of the data, for example, quarterly. This data change frequency should be described in relation to the frequency of processing.

- **Data constraints.** Data constraints refer to the limitations on the data requirements. Constraints may be procedural (e.g., based on corporate policy), technical (e.g., based on computer limitations), or imposed (e.g., based on project goals).

- **Data compatibility.** Data compatibility analysis involves ensuring that data collected for project control needs will be compatible with future needs.

- **Data contingency.** A data contingency plan concerns data security measures in case of accidental or deliberate damage or sabotage affecting hardware, software, or personnel.

6.6.3 Data Analysis and Presentation

Data analysis refers to the various mathematical and graphical operations that can be performed on data to elicit the inherent information contained in the data. The manner in which project data is analyzed and presented can affect how the information is perceived by the decision maker. The examples presented in this section illustrate how basic data analysis techniques can be used to convey important information for project control.

In many cases, data is represented as the answer to direct questions such as: When is the project deadline? Who are the people assigned to the first task? How many resource units are available? Are enough funds available for the project? What are the quarterly expenditures on the project for the past two years? Is personnel productivity low, average, or high? Who is the person in charge of the project? Answers to these types of questions constitute data of different forms or expressed on different scales. The resulting data may be qualitative or quantitative. Different techniques are available for analyzing the different types of data. This section discusses some of the basic techniques for data analysis. The data presented in Table 6–2 is used to illustrate the data analysis techniques.

Table 6-2 Quarterly Revenue from Four Projects (in $1,000s)

Project	Quarter 1	Quarter 2	Quarter 3	Quarter 4	Row total
A	3,000	3,200	3,400	2,800	12,400
B	1,200	1,900	2,500	2,400	8,000
C	4,500	3,400	4,600	4,200	16,700
D	2,000	2,500	3,200	2,600	10,300
Column total	10,700	11,000	13,700	12,000	47,400

Raw data.   Raw data consists of ordinary observations recorded for a decision variable or factor. Examples of factors for which data may be collected for decision making are revenue, cost, personnel productivity, task duration, project completion time, product quality, and resource availability. Raw data should be organized into a format suitable for visual review and computational analysis. The data in Table 6–2 represents the quarterly revenues from projects A, B, C, and D. For example, the data for quarter 1 indicates that project C yielded the highest revenue of $4,500,000 while project B yielded the lowest revenue of $1,200,000. Figure 6–6 presents the raw data of project revenue as a line graph. The same information is presented as a multiple bar chart in Figure 6–7.

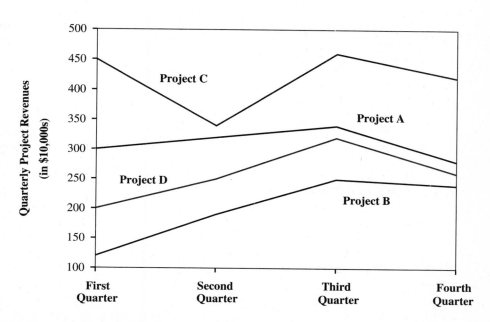

Figure 6-6. Line Graph of Quarterly Project Revenues

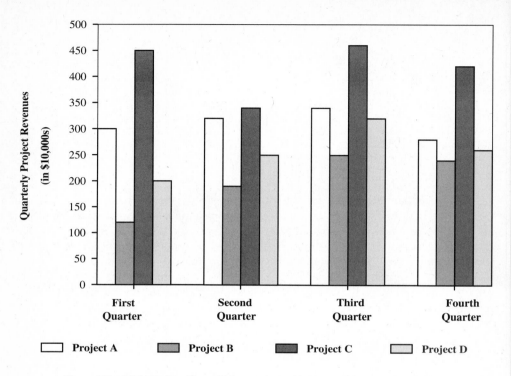

Figure 6-7.   Multiple Bar Chart of Quarterly Project Revenues

Total revenue.   A total or sum is a measure that indicates the overall effect of a particular variable. If $X_1, X_2, X_3, \ldots, X_n$ represent a set of n observations (e.g., revenues), then the total is computed as:

$$T = \sum_{i=1}^{n} X_i$$

For the data in Table 6–2, the total revenue for each project is shown in the last column. The totals indicate that project C brought in the largest total revenue over the four quarters under consideration while project B produced the lowest total revenue. The last row of the table shows the total revenue for each quarter. The totals reveal that the largest revenue occurred in the third quarter. The first quarter brought in the lowest total revenue. The grand total revenue for the four projects over the four quarters is shown as \$47,400,000 in the last cell in the table. Figure 6–8 presents the quarterly total revenues as stacked bar charts. Each segment in a stack of bars represents the revenue contribution from a particular project. The total revenues for the four projects over the four quarters are shown in a pie chart in Figure 6–9. The percentage of the overall revenue contributed by each project is also shown on the pie chart.

Average revenue.   Average is one of the most used measures in data analysis. Given n observations (e.g., revenues), $X_1, X_2, X_3, \ldots, X_n$, the average of the observations

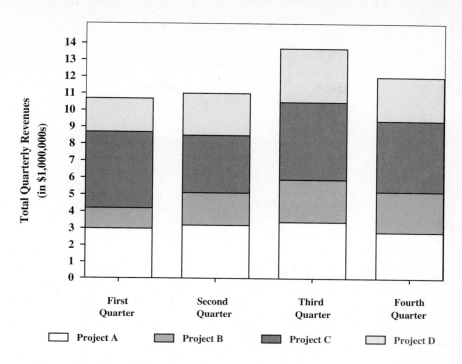

Figure 6-8. Stacked Bar Graph of Quarterly Total Revenues

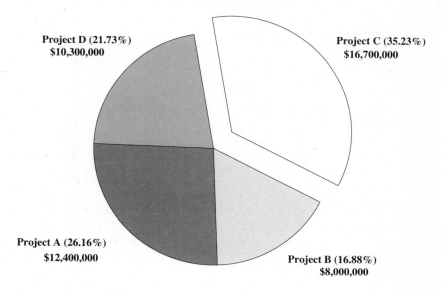

Figure 6-9. Pie Chart of Total Revenue per Project

is computed as

$$\overline{X} = \frac{\sum\limits_{i=1}^{n} X_i}{n}$$

$$= \frac{T_x}{n}$$

where T_x is the sum of n revenues. For our sample data, the average quarterly revenues for the four projects are

$$\overline{X}_A = \frac{(3{,}000 + 3{,}200 + 3{,}400 + 2{,}800)(\$1{,}000)}{4}$$

$$= \$3{,}100{,}000$$

$$\overline{X}_B = \frac{(1{,}200 + 1{,}900 + 2{,}500 + 2{,}400)(\$1{,}000)}{4}$$

$$= \$2{,}000{,}000$$

$$\overline{X}_C = \frac{(4{,}500 + 3{,}400 + 4{,}600 + 4{,}200)(\$1{,}000)}{4}$$

$$= \$4{,}175{,}000$$

$$\overline{X}_D = \frac{(2{,}000 + 2{,}500 + 3{,}200 + 2{,}600)(\$1{,}000)}{4}$$

$$= \$2{,}575{,}000$$

Similarly, the expected average revenues per project for the four quarters are

$$\overline{X}_1 = \frac{(3{,}000 + 1{,}200 + 4{,}500 + 2{,}000)(\$1{,}000)}{4}$$

$$= \$2{,}675{,}000$$

$$\overline{X}_2 = \frac{(3{,}200 + 1{,}900 + 3{,}400 + 2{,}500)(\$1{,}000)}{4}$$

$$= \$2{,}750{,}000$$

$$\overline{X}_3 = \frac{(3{,}400 + 2{,}500 + 4{,}600 + 3{,}200)(\$1{,}000)}{4}$$

$$= \$3{,}425{,}000$$

$$\overline{X}_4 = \frac{(2{,}800 + 2{,}400 + 4{,}200 + 2{,}600)(\$1{,}000)}{4}$$

$$= \$3{,}000{,}000$$

The above values are shown in a bar chart in Figure 6–10. The average revenue from any of the four projects in any given quarter is calculated as the sum of all the observations divided by the number of observations. That is,

$$\overline{\overline{X}} = \frac{\sum\limits_{i=1}^{N} \sum\limits_{j=1}^{M} X_{ij}}{K}$$

where

N = number of projects
M = number of quarters
K = total number of observations ($K = NM$)

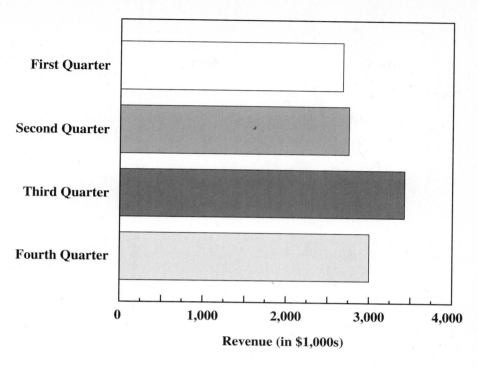

Figure 6-10.   Average Revenue per Project for Each Quarter

The overall average per project per quarter is

$$\overline{\overline{X}} = \frac{\$47,400,00}{16}$$
$$= \$2,962,500$$

As a cross-check, the sum of the quarterly averages should be equal to the sum of the project revenue averages, which is equal to the grand total divided by 4.

$$(2,675 + 2,750 + 3,425 + 3,000)(\$1,000) = (3,100 + 2,000 + 4,175 + 2,575)(\$1,000)$$
$$= \$11,800,000$$
$$= \$47,400,000/4$$

The cross-check procedure above works because we have a balanced table of observations. That is, we have four projects and four quarters. If there were only three projects, for example, the sum of the quarterly averages would not be equal to the sum of the project averages.

Median revenue.   The median is the value that falls in the middle of a group of observations arranged in order of magnitude. One-half of the observations are above the median and the other half are below the median. The method of determining the median depends on whether or not the observations are organized into a frequency distribution. For unorganized data, it is necessary to arrange the data in an increasing or decreasing order before finding the median. Given K observations (e.g., revenues), $X_1, X_2, X_3, \ldots, X_K$, arranged in increasing or decreasing order, the median is identified as the value in position $(K + 1)/2$ in the data arrangement if K is an odd number. If K is an even number, then the average of the two middle values is considered to be the median. If the sample data are arranged in increasing order, we would get the following:

1,200, 1,900, 2,000, 2,400, 2,500, 2,500, 2,600, 2,800, 3,000, 3,200, 3,200,

3,400, 3,400, 4,200, 4,500, 4,600

The median is then calculated as $(2,800+3,000)/2 = 2,900$. Half of the recorded revenues are expected to be above \$2,900,000 while half are expected to be below that amount. Figure 6–11 presents a bar chart of the revenue data arranged in increasing order. The median is anywhere between the eighth and ninth values in the ordered data.

      Quartiles and percentiles. The median is a position measure because its value is based on its position in a set of observations. Other measures of position are *quartiles* and *percentiles*. There are three quartiles which divide a set of data into four equal categories. The first quartile, denoted Q_1, is the value below which one-fourth of

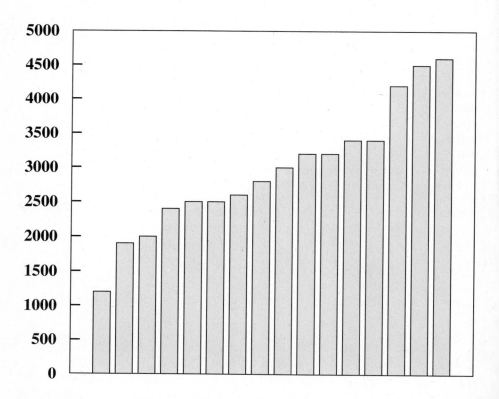

Figure 6-11. Bar Chart of Ordered Data

all the observations in the data set fall. The second quartile, denoted Q_2, is the value below which two-fourths or one-half of all the observations in the data set fall. The third quartile, denoted Q_3, is the value below which three-fourths of the observations fall. The second quartile is identical to the median. It is technically incorrect to talk of the fourth quartile because that will imply that there is a point within the data set below which all the data points fall: a contradiction! A data point cannot lie within the range of the observations and at the same time exceed all the observations, including itself.

The concept of percentiles is similar to the concept of quartiles except that reference is made to percentage points. There are 99 percentiles that divide a set of observations into 100 equal parts. The X percentile is the value below which X percent of the data fall. The 99 percentile refers to the point below which 99 percent of the observations fall. The three quartiles discussed previously are regarded as the 25th, 50th, and 75th percentiles. It would be technically incorrect to talk of the 100 percentile. In performance ratings, such as on an examination or product quality level, the higher the percentile of an individual or product, the better. In many cases, recorded data are classified into categories that are not indexed to numerical measures. In such cases, other measures of central tendency or position will be needed. An example of such a measure is the mode.

The Mode.   The mode is defined as the value that has the highest frequency in a set of observations. When the recorded observations can be classified only into categories, the mode can be particularly helpful in describing the data. Given a set of K observations (e.g., revenues), $X_1, X_2, X_3, \ldots, X_K$, the mode is identified as that value that occurs more than any other value in the set. Sometimes, the mode is not unique in a set of observations. For example, in Table 6–2, $2,500, $3,200, and $3,400 all have the same number of occurrences. Each of them is a mode of the set of revenue observations. If there is a unique mode in set of observations, then the data is said to be unimodal. The mode is very useful in expressing the central tendency for observations with qualitative characteristics such as color, marital status, or state of origin. The three modes in the raw data can be identified in Figure 6–11.

Range of revenue.   The range is determined by the two extreme values in a set of observations. Given K observations (e.g., revenues), $X_1, X_2, X_3, \ldots, X_K$, the range of the observations is simply the difference between the lowest and the highest observations. This measure is useful when the analyst wants to know the extent of extreme variations in a parameter. The range of the revenues in our sample data is ($4,600,000 - $1,200,000) = $3,400,000. Because of its dependence on only two values, the range tends to increase as the sample size increases. Furthermore, it does not provide a measurement of the variability of the observations relative to the center of the distribution. This is why the standard deviation is normally used as a more reliable measure of dispersion than the range.

The variability of a distribution is generally expressed in terms of the deviation of each observed value from the sample average. If the deviations are small, the set of data is said to have low variability. The deviations provide information about the degree of dispersion in a set of observations. A general formula to evaluate the variability of data cannot be based on the deviations. This is because some of the deviations are negative while some are positive and the sum of all the deviations is equal to zero. One possible solution to this is to compute the average deviation.

Average Deviation.   The average deviation is the average of the absolute values of the deviations from the sample average. Given K observations (e.g., revenues), $X_1, X_2, X_3, \ldots, X_K$, the average deviation of the data is computed as

$$\overline{D} = \frac{\sum_{i=1}^{K} |X_i - \overline{X}|}{K}$$

Table 6–3 shows how the average deviation is computed for our sample data. One aspect of the average deviation measure is that the procedure ignores the sign associated with each deviation. Despite this disadvantage, its simplicity and ease of computation make it useful. In addition, a knowledge of the average deviation helps in understanding the standard deviation, which is the most important measure of dispersion available.

Sample variance.   Sample variance is the average of the squared deviations computed from a set of observations. If the variance of a set of observations is large, the data is said to have a large variability. For example, a large variability in the levels of productivity of a project team may indicate a lack of consistency or improper methods

Table 6-3 Computation of Average Deviation, Standard Deviation, and Variance

Observation number (i)	Recorded observation X_i	Deviation from average $X_i - \overline{X}$	Absolute value $|X_i - \overline{X}|$	Square of deviation $\left(X_i - \overline{X}\right)^2$
1	3,000	37.5	37.5	1,406.25
2	1,200	−1762.5	1762.5	3,106,406.30
3	4,500	1537.5	1537.5	2,363,906.30
4	2,000	−962.5	962.5	926,406.25
5	3,200	237.5	237.5	56,406.25
6	1,900	−1062.5	1062.5	1,128,906.30
7	3,400	437.5	437.5	191,406.25
8	2,500	−462.5	462.5	213,906.25
9	3,400	437.5	437.5	191,406.25
10	2,500	−462.5	462.5	213,906.25
11	4,600	1637.5	1637.5	2,681,406.30
12	3,200	237.5	237.5	56,406.25
13	2,800	−162.5	162.5	26,406.25
14	2,400	−562.5	562.5	316,406.25
15	4,200	1237.5	1237.5	1,531,406.30
16	2,600	−362.5	362.5	131,406.25
Total	47,400.0	0.0	11,600.0	13,137,500.25
Average	2,962.5	0.0	725.0	821,093.77
Square root				906.14

in the project functions. Given K observations (e.g., revenues), $X_1, X_2, X_3, \ldots, X_K$, the sample variance of the data is computed as

$$s^2 = \frac{\sum_{i=1}^{K}(X_i - \overline{X})^2}{K-1}$$

The variance can also be computed by the following alternate formulas:

$$s^2 = \frac{\sum_{i=1}^{K}(X_i^2) - \left(\frac{1}{K}\right)\left[\sum_{i=1}^{K} X_i\right]^2}{K-1}$$

$$s^2 = \frac{\sum_{i=1}^{K} X_i^2 - K\left(\overline{X}^2\right)}{K-1}$$

Using the first formula, the sample variance of the data in Table 6–3 is calculated as

$$s^2 = \frac{13,137,500.25}{16-1}$$
$$= 875,833.33$$

The average calculated in the last column of Table 6–3 is obtained by dividing the total for that column by 16 instead of $16 - 1 = 15$. That average is not the correct value of the sample variance. However, as the number of observations gets very large, the average as computed in the table will become a close estimate for the correct sample variance. Analysts make a distinction between the two values by referring to the average calculated in the table as the population variance when K is very large and referring to the average calculated by the formulas above as the sample variance particularly when K is small. For our example, the population variance is given by

$$\sigma^2 = \frac{\sum_{i=1}^{K}(X_i - \overline{X})^2}{K}$$
$$= \frac{13,137,500.25}{16}$$
$$= 821,093.77$$

while the sample variance, as shown previously for the same data set, is given by

$$\sigma^2 = \frac{\sum_{i=1}^{K}(X_i - \overline{X})^2}{K-1}$$
$$= \frac{\$13,137,500.25}{(16-1)}$$
$$= \$875,833.33$$

Standard deviation. The sample standard deviation of a set of observations is the positive square root of the sample variance. The use of variance as a measure of variability has some drawbacks. For example, the knowledge of the variance is helpful only when two or more sets of observations are compared. Because of the squaring operation, the variance is expressed in square units rather than the original units of the raw data. To get a reliable feel for the variability in the data, it is necessary to restore the original units by performing the square root operation on the variance. This is why standard deviation is a widely recognized measure of variability. Given K observations (e.g., revenues), $X_1, X_2, X_3, \ldots, X_K$, the sample standard deviation of the data is computed as

$$s = \sqrt{\frac{\sum_{i=1}^{K}(X_i - \overline{X})^2}{K-1}}$$

As in the case of the sample variance, the sample standard deviation can also be computed by the following alternate formulas:

$$s = \sqrt{\frac{\sum_{i=1}^{K} X_i^2 - \left(\frac{1}{K}\right)\left[\sum_{i=1}^{K} X_i\right]^2}{K-1}}$$

$$s = \sqrt{\frac{\sum_{i=1}^{K} X_i^2 - K(\overline{X})^2}{K-1}}$$

Using the first formula, the sample standard deviation of the data in Table 6–3 is calculated as

$$s = \sqrt{\frac{13,137,500.25}{16-1}}$$
$$= \sqrt{875,833.33}$$
$$= 935.8597$$

We can say that the variability in the expected revenue per project per quarter is $935,859.70. The population sample standard deviation is given by

$$\sigma = \sqrt{\frac{\sum_{i=1}^{K}(X_i - \overline{X})^2}{K}}$$
$$= \sqrt{\frac{13,137,500.25}{16}}$$
$$= \sqrt{821,093.77}$$
$$= 906.1423$$

while the sample standard deviation is given by

$$
\begin{aligned}
s &= \sqrt{\frac{\sum_{i=1}^{K}(X_i - \overline{X})^2}{K - 1}} \\
&= \sqrt{\frac{13,137,500.25}{(16 - 1)}} \\
&= \sqrt{875,833.33} \\
&= 935.8597
\end{aligned}
$$

The results of data analysis can be reviewed directly to determine where and when project control actions may be needed. The results can also be used to generate control charts as discussed in the next section.

6.6.4 Control Charts

Control charts may be used to track project performance before deciding what control actions are needed. Control limits are incorporated into the charts to indicate when control actions should be taken. Multiple control limits may be used to determine various levels of control points. Control charts may be developed for various aspects such as cost, schedule, resource utilization, performance, and other criteria for project evaluation.

Figure 6–12 represents a case of periodic monitoring of project progress. Cost is monitored and recorded on a monthly basis. If the cost is monitored on a more frequent basis (e.g., days), then we may be able to have a more rigid control structure. Of course, one will need to decide whether the additional time needed for frequent monitoring is justified by the extra level of control provided. The control limits may be calculated with the same procedures used for X-bar and R charts in quality control or they may be based on custom project requirements. In addition to drawing control charts for cost, we can also draw control charts for other measures of performance such as task duration, quality, or resource utilization.

Figure 6–13 shows a control chart for cumulative cost. The control limits on the chart are indexed to the project percent complete. At each percent complete point, there is a control limit which the cumulative project cost is not expected to exceed. A review of the control chart shows that the cumulative cost is out of control at the 10%, 30%, 40%, 50%, 60%, and 80% completion points. The indication is that control actions should be instituted right from the 10% completion point. If no control action is taken, the cumulative cost may continue to be out of control and eventually exceed the budget limit by the time the project is finished.

The information obtained from the project monitoring capabilities of project management software can be transformed into meaningful charts that can quickly identify when control actions are needed. A control chart can provide information about resource overallocation as well as unusually slow progress of work. Figure 6–14 presents a control chart for budget tracking and control. Starting at an initial level of $B, the budget is depleted in one of three possible modes. In case 1, the budget is depleted in such a way that a surplus is available at the end of the project cycle. This may be viewed

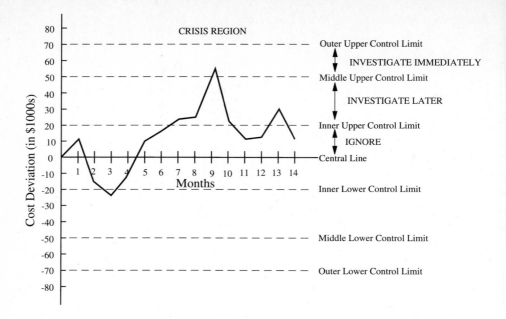

Figure 6-12. Control Chart for Project Monitoring

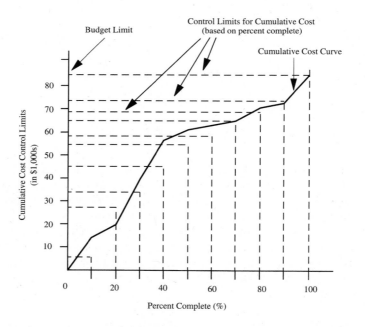

Figure 6-13. Control Chart for Cumulative Cost

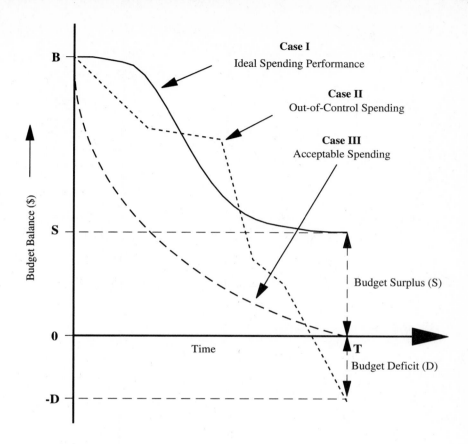

Figure 6-14.   Chart for Budget Tracking and Control

as the ideal spending pattern. In case 2, the budget is depleted in an out-of-control pattern that leaves a deficit at the end of the project. In case 3, the budget is depleted at a rate that is proportional to the amount of work completed. In this case, the budget is zeroed out just as the project finishes. Intermediate control lines may be included in the chart so that needed control actions can be initiated at the appropriate times.

Figure 6–15 presents a chart for monitoring revenues versus expenses in a project. As explained before, control limits may be added to the chart to determine when control actions should be taken.

Table 6–4 shows a report format for task progress analysis. The first three columns identify the task, its location, and its description. The activities column indicates the activities that make up the task. The next column would indicate the completed activities. Pending activities and past due activities are to be indicated in the next two columns. The last column is intended to display comments about problems encountered. This control table helps to focus control actions on specific activities in addition to the overall project control actions. Table 6–5 presents a format for task-based time analysis. The

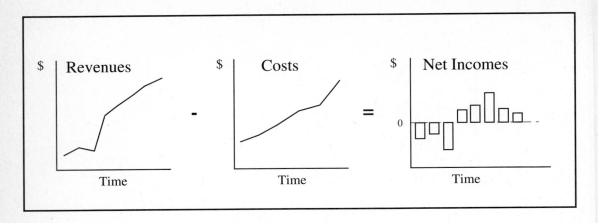

Figure 6-15.   Review Chart for Revenue versus Expenses

table has the additional feature of showing percent completion for each task. It also evaluates planned activities versus actual activities performed. Deviations from planned work should be explained.

Figure 6–16 shows a sketch of a graphical report on task progress. This bar chart analysis would generate reports showing expected completion and actual completion for task numbers, departments, or project segment. Figure 6–17 presents a chart that shows project progress versus resource loading. This is useful for identifying how the percent completion of a project is affected by the level of resource allocation.

Better performance can be achieved if more time and resources are available for a project. If lower costs and tighter schedules are desired, then performance may have to be compromised and vice versa. From the point of view of the project manager,

Table 6-4   Task Analysis Table

Column 1	Column 2	Column 3	Column 4	Column 5	Column 6	Column 7	Column 8
Task #	Department	Description	Activities	Completed	Pending	Past due	Comments

Table 6-5   Task-Based Time Analysis Table

Col. 1	Col. 2	Col. 3	Col. 4	Col. 5	Col. 6	Col. 7	Col. 8	Col. 9
Task #	Department	Description	Expected activity	Expected % completed	Actual activity	Actual % completed	Deviation	Explain

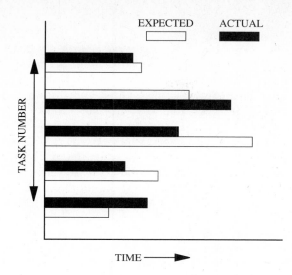

EXPECTED   ACTUAL

TASK NUMBER

TIME ➝

Figure 6-16. Graphical Report on Task Progress

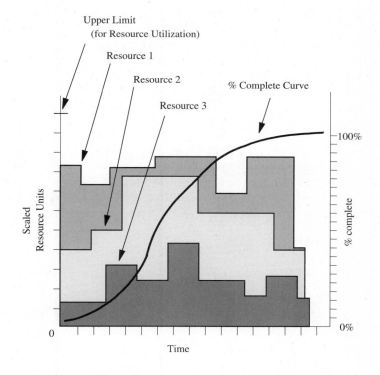

Upper Limit
(for Resource Utilization)

Resource 1

Resource 2

% Complete Curve

Resource 3

Scaled Resource Units

% complete

100%

0%

0

Time

Figure 6-17. Resource Loading versus Project Progress

the project should be at the highest point along the performance axis. Of course, this represents an extreme case of getting something for nothing. From the point of view of the project staff, the project should be at the point indicating highest performance, longest time, and most resources. This, of course, may be an unrealistic expectation since time and resources are typically in short supply. For project control, a feasible trade-off strategy must be developed.

Even though the control boundary is represented by a flat surface in Figure 6–1, it is obvious that the surface of the box will not be flat. If a multifactor mathematical model is developed for the three factors, the nature of the response surface will vary depending on the specific interactions of the factors for any given project. An example of a project performance response surface is presented in Figure 6–18. The desired trade-offs between the factors in the plot will help determine when and where control actions are required.

If we consider only two of the three constraints at a time, we can study their respective relationships better. Figure 6–19 shows some potential two-factor relationships. In the first plot, performance is modeled as the dependent variable, while cost is the independent variable. Performance increases as cost increases up to a point where performance levels off. If cost is allowed to continue to increase, performance eventually starts to drop. In the second plot, performance is modeled as being dependent on time. The more time that is allowed for a project, the higher the expected performance up to a point where performance levels off. In the third plot, cost depends on time. As project duration increases, cost increases. The increases in cost may be composed of labor cost, raw material cost, and/or cost associated with decreasing productivity. Note that there may be a fixed cost associated with a project even when a time schedule is not in effect. This is seen in the third plot.

Figure 6–20 shows an alternate time-cost trade-off relationship. In this case, the shorter the desired project duration, the higher the cost of the project. If more time is available for the project, then cost can be reduced. However, there is a limit to the possible reduction in cost. After some time, the cost function turns upward due to the increasing cost of keeping the workforce and resources tied up on the project for

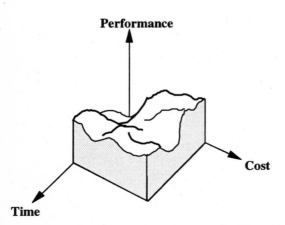

Figure 6-18.   Project Performance Response Surface

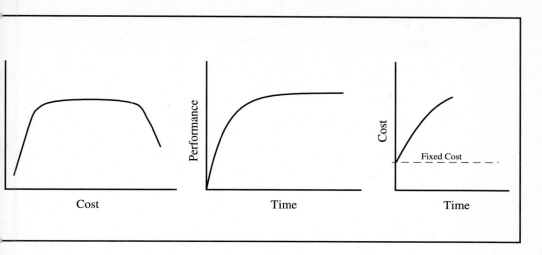

Figure 6-19.   Trade-off Relationships between Project Constraints

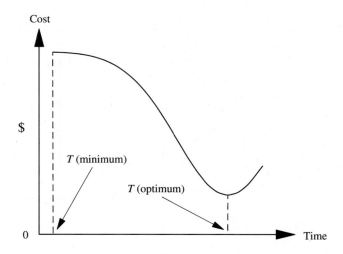

Figure 6-20.   Time Control versus Cost Control

a long period of time. The most cost-effective duration of the project corresponds to the point where the lowest cost is shown in the figure.

The basic data analysis presented in this section can play a significant role in conveying quick information about project requirements and performance so that prompt decisions can be made. The next section presents the use of statistical analysis for project control when project parameters are subject to variabilities.

6.7 STATISTICAL PROJECT CONTROL

Statistical control of a project requires that we recognize where risk and uncertainty exist in the project. Variability is a reality in project management. Uncertainty refers to the inability to predict the future accurately. Risk deals with the probability that something will happen and the loss that will result if the thing does happen. Risk and uncertainty may affect several project parameters including resource availability, activity durations, personnel productivity, budget, weather conditions, equipment failures, and cost. Statistical project control uses the techniques of probability and statistics to assess project performance and determine control actions. The different types of statistics relevant for project control are discussed next.

Descriptive statistics.   Descriptive statistics refers to analyses that are performed in order to describe the nature of a process or operation. The analyses presented in the previous section fall under the category of descriptive statistics because they are concerned with summary calculations and graphical displays of observations.

Inferential statistics.   Inferential statistics refers to the process of drawing inferences about a process based on a limited observation of the process. The techniques presented in this section fall under the category of inferential statistics. Inferential statistics is of interest because it is dynamic and provides generalizations about a population by investigating only a portion of the population. The portion of the population investigated is referred to as a sample. As an example, the expected duration of a proposed task can be inferred from several previous observations of the durations of identical tasks.

Deductive statistics.   Deductive statistics involves assigning properties to a specific item in a set based on the properties of a general class covering the set. For example, if it is known that 90 percent of projects in a given organization fail, then deduction can be used to assign a probability of 90 percent to the event that a specific project in the organization will fail.

Inductive statistics.   Inductive statistics involves drawing general conclusions from specific facts. Inferences about populations are drawn from samples. For example, if 95 percent of a sample of 100 people surveyed in a 5,000-person organization favor a particular project, then induction can be used to conclude that 95 percent of the personnel in the organization favor the project. The different types of statistics play important roles in project control. Sampling is an important part of drawing inferences.

6.7.1 Sampling Techniques

A *sample space* of an experiment is the set of all possible distinct outcomes of the experiment. An *experiment* is some process that generates distinct sets of observations. The simplest and most common example is the experiment of tossing a coin to observe whether heads or tails will show up. An *outcome* is a distinct observation resulting from a single trial of an experiment. In the experiment of tossing a coin, heads and tails are the two possible outcomes. Thus, the sample space consists of only two items.

There are several examples of statistical experiments suitable for project control. A simple experiment may involve checking to see whether it rains or not on a given day. Another experiment may involve counting how many tasks fall behind schedule during a project life cycle. An experiment may involve recording how long it takes to perform a given activity in each of several trials. The outcome of any experiment is frequently referred to as a *random outcome* because each outcome is independent and has the same chance of occurring. We cannot predict with certainty what the outcome of a particular trial of the experiment would be. An event can be a collection of outcomes.

Sample.   A sample is a subset of a population that is selected for observation and statistical analysis. Inferences are drawn about the population based on the results of the analysis of the sample. The reasons for using sampling rather than complete population enumeration are

1. It is more economical to work with a sample.
2. There is a time advantage to using a sample.
3. Populations are typically too large to work with.
4. A sample is more accessible than the whole population.
5. In some cases, the sample may have to be destroyed during the analysis.

There are three primary types of samples. They differ in the manner in which their elementary units are chosen.

Convenience Sample.   A convenience sample refers to a sample that is selected on the basis of how convenient certain elements of the population are for observation.

Judgment Sample.   A judgment sample is one that is obtained based on the discretion of someone familiar with the relevant characteristics of the population.

Random Sample.   A random sample refers to a sample whereby the elements of the sample are chosen at random. This is the most important type of sample for statistical analysis. In random sampling, all the items in the population have an equal chance of being selected for inclusion in the sample.

Since a sample is a collection of observations representing only a portion of the population, the way in which the sample is chosen can significantly affect the adequacy and reliability of the sample. Even after the sample is chosen, the manner in which specific observations are obtained may still affect the validity of the results. The possible bias and errors in the sampling process are discussed next.

Sampling Error.   A sampling error refers to the difference between a sample mean and the population mean that is due solely to the particular sample elements that are selected for observation.

Nonsampling Error.   A nonsampling error refers to an error that is due solely to the manner in which the observation is made.

Sampling Bias.   A sampling bias refers to the tendency to favor the selection of certain sample elements having specific characteristics. For example, a sampling bias may occur if a sample of the personnel is selected from only the engineering department in a survey addressing the implementation of high technology projects.

Stratified Sampling. Stratified sampling involves dividing the population into classes, or groups, called strata. The items contained in each stratum are expected to be homogeneous with respect to the characteristics to be studied. A random subsample is taken from each stratum. The subsamples from all the strata are then combined to form the desired overall sample. Stratified sampling is typically used for a heterogeneous population such as data on employee productivity in an organization. Under stratification, groups of employees are selected so that the individuals within each stratum are mostly homogeneous and the strata are different from one another. As another example, a survey of project managers on some important issue of personnel management may be conducted by forming strata on the basis of the types of projects they manage. There may be one stratum for technical projects, one for construction projects, and one for manufacturing projects.

A *proportionate stratified sampling* results if the units in the sample are allocated among the strata in proportion to the relative number of units in each stratum in the population. That is, an equal sampling ratio is assigned to all strata in a proportionate stratified sampling. In *disproportionate stratified sampling*, the sampling ratio for each stratum is inversely related to the level of homogeneity of the units in the stratum. The more homogeneous the stratum, the smaller its proportion included in the overall sample. The rationale for using disproportionate stratified sampling is that when the units in a stratum are more homogeneous, a smaller subsample is needed to ensure good representation. The smaller subsample helps reduce sampling cost.

Cluster Sampling. Cluster sampling involves the selection of random clusters, or groups, from the population. The desired overall sample is made up of the units in each cluster. Cluster sampling is different from stratified sampling in that differences between clusters are usually small. In addition, the units within each cluster are generally more heterogeneous. Each cluster, also known as *primary sampling unit*, is expected to be a scaled-down model that gives a good representation of the characteristics of the population.

All the units in each cluster may be included in the overall sample or a subsample of the units in each cluster may be used. If all the units of the selected clusters are included in the overall sample, the procedure is referred to as *single-stage sampling*. If a subsample is taken at random from each selected cluster and all units of each subsample are included in the overall sample, then the sampling procedure is called *two-stage sampling*. If the sampling procedure involves more than two stages of subsampling, then the procedure is referred to as *multistage sampling*. Cluster sampling is typically less expensive to implement than stratified sampling. For example, the cost of taking a random sample of 2,000 managers from different industry types may be reduced by first selecting a sample, or cluster, of 25 industries and then selecting 80 managers from each of the 25 industries. This represents a two-stage sampling that will be considerably cheaper than trying to survey 2,000 individuals in several industries in a single-stage procedure.

Once a sample has been drawn and observations of all the items in the sample are recorded, the task of data collection is completed. The next task involves organizing the raw data into a meaningful format. Frequency distribution is an effective tool for organizing data. Frequency distribution involves the arrangement of observations into classes so as to show the frequency of occurrences in each class. (Guidelines for constructing histograms are presented in the next section.)

Example

Suppose a set of data is collected about project costs in an organization. Twenty projects are selected for the study. The observations below are recorded in thousands of dollars:

$3,000	$1,100	$4,200	$800	$3,000
$1,800	$2,500	$2,500	$1,700	$3,000
$2,900	$2,100	$2,300	$2,500	$1,500
$3,500	$2,600	$1,300	$2,100	$3,600

Table 6–6 shows the tabulation of the cost data as a frequency distribution. Note how the end points of the class intervals are selected such that no recorded data point falls at an end point of a class. Note also that seven class intervals seem to be the most appropriate size for this particular set of observations. Each class interval has a spread of $500 which is an approximation obtained from the expression presented below.

$$W = \frac{X_{max} - X_{min}}{N}$$
$$= \frac{4200 - 800}{7}$$
$$= 485.71$$
$$\approx 500$$

Table 6–7 shows the relative frequency distribution. The relative frequency of any class is the proportion of the total observations that fall into that class. It is obtained by dividing the frequency of the class by the total number of observations. The relative frequency of all the classes should add up to 1. From the relative frequency table, it is seen that 25 percent of the observed project costs fall within the range of $2,250 and $2,750. It is also noted that only 15 percent (0.10 + 0.05) of the observed project costs fall in the upper two intervals of project costs.

Figure 6–21 shows the histogram of the frequency distribution for the project cost data. Figure 6–22 presents a plot of the relative frequency of the project cost data. The plot of the cumulative relative frequency is superimposed on the relative frequency plot.

Table 6-6 Frequency Distribution of Project Cost Data

Cost interval ($)	Midpoint ($)	Frequency	Cumulative frequency
750–1,250	1,000	2	2
1,250–1,750	1,500	3	5
1,750–2,250	2,000	3	8
2,250–2,750	2,500	5	13
2,750–3,250	3,000	4	17
3,250–3,750	3,500	2	19
3,750–4,250	4,000	1	20
Total		20	

Table 6-7   Relative Frequency Distribution of Project Cost Data

Cost interval ($)	Midpoint ($)	Frequency	Cumulative frequency
750–1,250	1,000	0.10	0.10
1,250–1,750	1,500	0.15	0.25
1,750–2,250	2,000	0.15	0.40
2,250–2,750	2,500	0.25	0.65
2,750–3,250	3,000	0.20	0.85
3,250–3,750	3,500	0.10	0.95
3,750–4,250	4,000	0.05	1.00
Total		1.00	

The relative frequency of the observations in each class represents the probability that a project cost will fall within that range of costs. The corresponding cumulative relative frequency gives the probability that project cost will fall below the midpoint of that class interval. For example, 85 percent of project costs in this example are expected to fall below or equal $3,000.

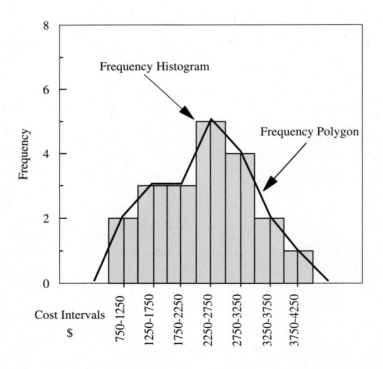

Figure 6-21.   Histogram of Project Cost Distribution Data

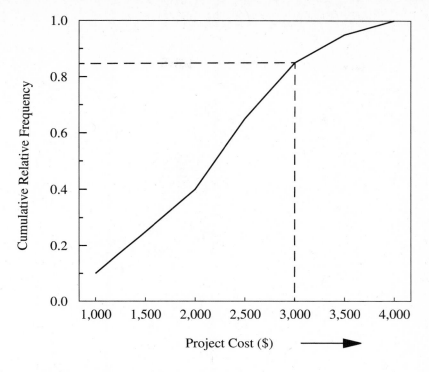

Figure 6-22. Plot of Cumulative Relative Frequency

6.7.2 Diagnostic Tools

To facilitate data analysis for diagnosing a project for control purposes, we recommend using available graphical tools. The tools include flowcharts, Pareto diagrams, cause-and-effect diagrams, check sheets, scatter plots, runs charts, and histograms. The tools are very effective for identifying problems that may need control actions.

Flowcharts. A flowchart is used to show the steps that a product or service follows from the beginning to the end of the process. It helps locate the value added parts of the process steps. It also helps in locating the unnecessary steps in the process where unnecessary cost and labor exist. These unnecessary steps can be reduced or permanently eliminated.

Pareto diagram. A Pareto diagram is used to display the relative importance or size of problems to determine the order of priority for projects. It can help identify the projects to concentrate on. For example, in Figure 6–23, analysts may tend to focus on project I since this is where the greatest dollar loss occurs. The criticality of a project may be determined by a combination of factors. The selection of project I as the most critical project to focus on should not be made solely on the largest dollar loss alone or any other single criterion. For example, if project I involves determining the number of accidents per year and project II involves determining the number of deaths per

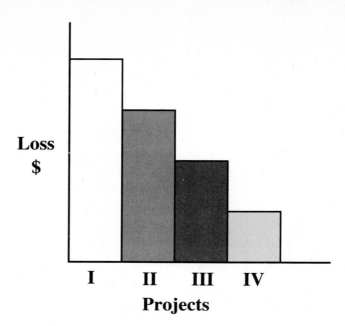

Figure 6-23.   Relative Dollar Losses of Quality Improvement Projects

year, then project II may have priority since focusing on the number of deaths may be more critical than focusing on the number of accidents, even though the frequency of accidents is more than the frequency of deaths.

Cause-and-effect (fishbone) diagram.   A fishbone diagram is used to develop a relationship between an effect and all the possible causes influencing it. It is also sometimes called a *tree* or *river* diagram. Figure 6–24 presents an example of a fishbone diagram. The diagram was originally developed for specifying the relationships between a quality characteristic and a set of factors. The diagram is now used for general applications in business and industry. The steps for developing a fishbone diagram are as follows:

Step 1:   Determine the characteristic or the response variable to be studied.

Step 2:   Write the characteristic on the right-hand side of a blank sheet of paper. Start with enough room on the paper because the diagram may expand considerably during the evaluation. Enclose the characteristic in a square. Now, write the primary causes which affect the quality characteristic as big branches (or bones). Enclose the primary causes in squares.

Step 3:   Write the secondary causes which affect the big branches as medium-sized branches. Write the tertiary (third-level) causes which affect the medium-sized bones as small bones.

CAUSES   EFFECT

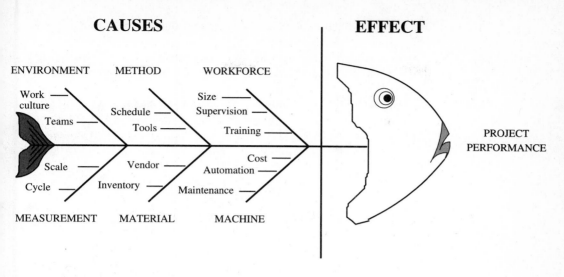

Figure 6-24. Fishbone Diagram

Step 4:   Assign relative importance ratings to the factors. Mark the particularly important factors that are believed to have a significant effect on the characteristic.

Step 5:   Append any necessary written explanation to the diagram.

Step 6:   Review the overall diagram for completeness. While it is important to expand the cause-and-effect relationships as much as possible, avoid cluttering the diagram. For a fishbone diagram to be presented to upper management, limit the contents to a few important details. At the operational level, more details will need to be provided.

Scatter plots.   A scatter plot is used to study the relationships between two variables. It is sometimes called an X-Y plot. The plot gives a visual assessment of the location tendencies of data points. The appearance of a scatter plot can help identify the type of statistical analyses that may be needed for the data. For example, in regression analysis, a scatter plot can help an analyst determine the type of models to be investigated.

Run charts and check sheets.   A run chart is a tool which can be used to monitor the trends in a process over time. A check sheet is a preprinted table layout that facilitates data collection. Items to be recorded are preprinted in the table. Observations are recorded by simply checking appropriate cells in the table. A check sheet helps to automatically organize data for subsequent analysis. If properly designed, a check sheet can eliminate the need for counting data points during data analysis.

Histogram.   A histogram is used to display the distribution of data by organizing the data points into evenly spaced numerical groupings that show the frequency of values in each group. Histograms can be used for quickly assessing the variation and distribution affecting a project. Important guidelines for drawing histograms are

Step 1:   Determine the minimum and maximum values to be covered by the histogram.

Step 2:   Select a number of histogram classes between 6 and 15. Having too few or too many classes will make it impossible to identify the underlying distribution.

Step 3:   Set the same interval length for the histogram classes such that every observation in the data set falls within some class. The difference between midpoints of adjacent classes should be constant and equal to the length of each interval. If N represents the number of histogram classes, determine the interval length as

$$W = \frac{X_{max} - X_{min}}{N}$$

where X_i represents an observation in the data set.

Step 4:   Count the number of observations that fall within each histogram class. This can be done by using a check sheet or any other counting technique.

Step 5:   Draw a bar for each histogram class such that the height of the bar represents the number of observations in the class. If desired, the heights can be converted to relative proportions in which the height of each bar represents the percentage of the data set that falls within the histogram class.

The number of classes should not be so small or so large that the true nature of the underlying distribution cannot be identified. Generally, the number of classes should be between 6 and 20. The interval length of each class should be the same. The interval length should be selected such that every observation falls within some class. The difference between midpoints of adjacent classes should be constant and equal to the length of each interval.

A frequency polygon may be obtained by drawing a line to connect the midpoints at the top of the histogram bars. The polygon will show the spread and shape of the distribution of the data set. Three possible patterns of distribution may be revealed by the polygon: *symmetrical, positively skewed*, and *negatively skewed*. In a symmetrical distribution, the two halves of the graph are identical. In a positively skewed distribution (skewed to the right), there is a long tail stretching to the right side of the distribution. In a negatively skewed distribution (skewed to the left), there is a long tail stretching to the left side of the distribution.

6.7.3 Probabilistic Decision Analysis

We deal with probability in most of our day-to-day activities. It is important to understand the basic principles of probability analysis for project control. The manager of an outside construction project may reschedule available personnel on the basis of weather forecasts, which are based on probability. Probability refers to the chances of occurrence of an event out of several possible events in a sample space. This is what people often refer to as the *law of averages*. If a coin is tossed a large number of times, say several million times, the proportion of heads tends to be one-half of the total number of tosses. In that case, the number, one-half, is referred to as the probability that heads will occur on one toss of the coin. If ten items with different colors are placed in a jar and one item is pulled out of the jar at random, the probability of pulling out one specific color is one-tenth. Some general facts about probability are

1. Probabilities are real numbers between 0 and 1 inclusive that reflect an individual's belief in the chances of the occurrence of events.
2. A probability value near 0 indicates that the event in question is not expected to occur. However, it does not mean that the event will not occur.
3. A probability near 1 indicates that the event in question is expected to occur. It does not mean that the event will definitely occur.
4. A probability of one-half indicates that the event in question has equal likelihood of occurring or not occurring.
5. The sum of the probabilities of all the mutually exclusive events in a sample space is 1. This is one of the most basic facts of probability. And yet, it is the most violated rule in probability analysis by practitioners.

Normal distribution.   A *probability density function* is a mathematical expression that describes the random behaviors of events in a sample space. Probability density functions are associated with continuous sample spaces where there is an infinite number of possible events. If the number of elements in a sample space is finite or countably finite, then the behavior of the events in the sample space would be described by a discrete probability distribution rather than a continuous probability density function. Countably infinite means that there is an unending sequence with as many elements as there are whole numbers. Probability distributions refer to discrete sample spaces while probability density functions refer to continuous sample spaces.

In most practical problems, continuous random variables represent measured data, such as all possible distances a car can travel, weights, temperatures, and task duration, while discrete random variables represent counted data, such as the number of absent employees on a given day, the number of late jobs in a project, and the amount of dollars available for a particular project. Examples of discrete probability distributions are the binomial distribution, the geometric distribution, and the Poisson distribution. This section presents some of the basic properties of the normal probability density function. Other examples of probability density functions are the exponential probability density function, gamma probability density function, chi-square probability density function, and Weibull probability density function.

The normal probability density function is the most important continuous probability density function in the entire field of statistics. It is often referred to as the normal distribution, normal curve, bell-shaped curve, or Gaussian distribution. This distribution fits many of the physical events in nature, hence, its popularity and wide appeal. The normal distribution is characterized by the following formula:

$$f(x) = \frac{1}{\sqrt{2\pi}\sigma} e^{-\frac{1}{2}\left(\frac{x-\mu}{\sigma}\right)^2}, \qquad -\infty < x < \infty$$

where

μ = mean of the distribution
σ = standard deviation of the distribution
e = 2.71828... (a natural constant)
π = 3.14159... (a natural constant)

The bell-shaped appearance of the normal distribution is shown in Figure 6–25. The values of μ and σ are the parameters that determine the specific appearance (fat, thin, long, short, narrow or wide) of the normal distribution. Theoretically, the tails of the curve trail on to infinity. When μ = 0 and σ = 1, the normal distribution is referred to as the *standard normal distribution*. A tabulation of the standard normal distribution is presented in appendix A.

Most of the analyses involving the normal curve are done in the standardized domain. This is done by using the transformation

$$Z = \frac{X - \mu}{\sigma}$$

where Z is the standard normal random variable and X is the general normal random variable with a mean of μ and standard deviation of σ. The variable, Z, is often referred to as the *normal deviate*. One important aspect of the normal distribution relates to the percent of observations within one, two, or three standard deviations. Approximately 68.27 percent of observations following a normal distribution lie within plus and minus one standard deviation from the mean. Approximately 95.45 percent of the observations lie within plus or minus two standard deviations from the mean, and approximately 99.73 percent of the observations lie within plus or minus three standard deviations from the mean. These are shown graphically in Figure 6–26.

To obtain probabilities for particular values of a random variable, it is necessary to know the probability distribution of the random variable. Because of the infinite possible combinations of means and standard deviation values, there is an infinite number of normal distributions. It is quite impractical to try and calculate probabilities directly from each one of them individually. The standard normal distribution can be applied to each and every possible normal random variable by using the transformation expression

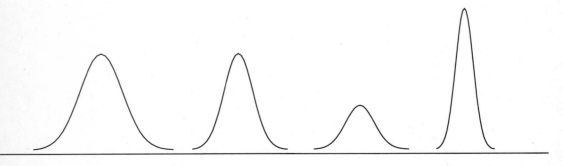

Figure 6-25.   Shapes of the Normal Curve

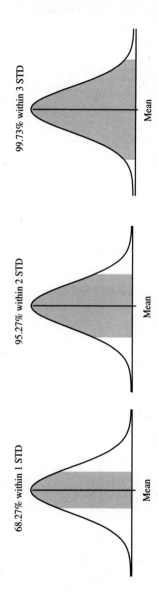

Figure 6-26. Areas under the Normal Curve

presented earlier. The standard normal distribution is of great importance in practice because it can be used to approximate many of the other discrete and continuous random variables.

Because the normal distribution represents a continuous random variable, it is impossible to calculate the probability of a single point on the curve. To determine probabilities, it is necessary to refer to intervals, such as the interval between point a and point b. In Figure 6–27, the area under the curve from a to b represents the probability that the random variable will lie between a and b. That probability is calculated as follows:

Given: Normal random variable X, representing task duration
Mean of X = 50 days
Standard deviation of X = 10

Required: The probability that the task duration will lie between 45 and 62 days.

Solution: Let X_1 = 45 and X_2 = 62
Then,

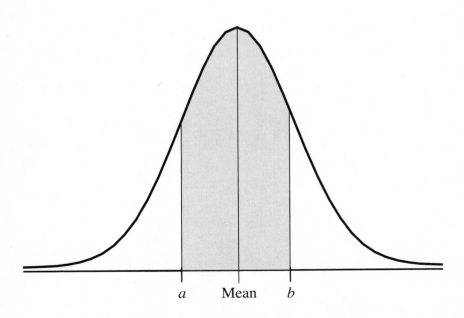

Figure 6-27. Probability of an Interval under the Normal Curve

$$z_1 = \frac{45 - 50}{10}$$
$$= -0.5$$
$$z_2 = \frac{62 - 50}{10}$$
$$= 1.2$$

Therefore, we have

$$P(45 < X < 62) = P(-0.5 < Z < 1.2)$$
$$= P(Z < 1.2) - P(Z < -0.5)$$
$$= P(Z < 1.2) - [1 - P(Z < 0.5)]$$
$$= 0.8849 - (1 - 0.6915)$$
$$= 0.8849 - 0.3085$$
$$= 0.5764$$

There is a 57.64 percent chance that this particular task will last between 45 and 62 days. Note that the area under the curve between 45 and 62 is calculated by first finding the total area to the left of 62 (i.e., 0.8849) and then subtracting the area to the left of 45 (i.e., 0.3085). Note also that $P(Z < -0.5)$ can be computed as $1 - P(Z < 0.5)$ for the case where the normal table does not contain negative values of z.

The respective probabilities are read off the normal probability table given in appendix A. To illustrate the use of the table, let us find the probability that Z will be less than 1.23. First, we locate the value of z equal to 1.2 in the left column of the table and then move across the row to the column under 0.03, where we read the value of 0.8907 inside the body of the table. Thus, $P(Z < 1.23) = 0.8907$. Using a similar process, the following additional examples are presented:

$$P(X < 55) = P\left(Z < \frac{55 - 50}{10}\right)$$
$$= P(Z < 0.5)$$
$$= 0.6915$$

$$P(X > 65) = P\left(Z > \frac{65 - 50}{10}\right)$$
$$= P(Z > 1.5)$$
$$= 1 - P(Z < 1.5)$$
$$= 1 - 0.9332$$
$$= 0.0668$$

Note that since the normal distribution table is constructed as cumulative probabilities from the left, $P(Z > 65)$ is calculated as $1 - P(Z < 65)$. Note that

$P(Z < k) = 1.0$, for any value k that is greater than 3.5
$P(Z < 0) = 0.5$
$P(Z < k) = 0.0$, for any value k that is less than -3.5

6.7.4 Decision Trees

Decision tree analysis is used to evaluate sequential decision problems. In project management, a decision tree may be useful in evaluating sequential project milestones. Schuyler (1993) discusses the importance of decision trees for decision analysis in the project environment. A decision problem under certainty has two elements: *action* and *consequence*. The decision maker's choices are the actions while the results of those actions are the consequences. For example, in CPM network scheduling, the choice of one task among three tasks at a specific time represents a potential action. The consequences of choosing one task over another may be characterized in terms of the slack time created in the network, the cost of performing the selected task, the resulting effect on the project completion time, or the degree to which a specified performance criterion is satisfied.

If the decision is made under uncertainty, as in PERT network analysis, a third element is introduced into the decision problem. This third element is defined as *event*. Extending the CPM task selection example to a PERT analysis, the actions may be defined as select task 1, select task 2, and select task 3. The durations associated with the three possible actions can be categorized as "long task duration," "medium task duration," and "short task duration." The actual duration of each task is uncertain. Each task has some probability of exhibiting long, medium, or short durations.

The events can be identified as weather incidents: rain or no rain. The incidents of rain or no rain are uncertain. The consequences may be defined as "increased project completion time," "decreased project completion time," and "unchanged project completion time." However, these consequences are uncertain due to the probabilistic durations of the tasks and the variable choices of the decision maker. That is, the consequences are determined partly by choice and partly by chance. The consequence is dependent on which event, rain or no rain, occurs.

To simplify the decision analysis, the decision elements may be summarized by using a decision table. A decision table indicates the relationship between pairs of decision elements. The decision table for the preceding example is presented in Table 6–8.

In the table, each row corresponds to an event and each column corresponds to an action. The consequences appear as entries in the body of the table. The consequences have been coded as I (increased), D (decreased), U (unchanged). Each event-action

Table 6-8   Decision Table for Task Selection

	Actions								
	Task 1			Task 2			Task 3		
Event	Long	Medium	Short	Long	Medium	Short	Long	Medium	Short
Rain	I	I	U	I	U	D	I	I	U
No rain	I	D	D	U	D	D	U	U	U

I = increased project duration; D = decreased project duration; U = unchanged project duration.

combination has a specific consequence associated with it. In some decision problems, the consequences may not be unique. A consequence that is associated with a particular event-action pair may also be associated with another event-action pair. The actions included in the decision table are the only ones that the decision maker wishes to consider. Subcontracting or task deletion could be other possible choices for the decision maker. The actions included in the decision problem are mutually exclusive and collectively exhaustive, so that exactly one will be selected. The events are also mutually exclusive and collectively exhaustive.

The decision problem can also be represented as a decision tree as shown in Figure 6–28. The tree representation is particularly convenient for decision problems with choices that must be made at different times over an extended period. For example, resource allocation decisions must be made several times during the life cycle of a project. The choice of actions is shown as a fork with a separate branch for each action. The events are also represented by branches in separate forks. To avoid confusion in large decision trees, the nodes for actions are represented by squares while the nodes for events are represented by circles.

The basic guideline for constructing a tree diagram is that the flow should be chronological from left to right. The actions are shown on the initial fork because the decision must be made before the actual event is known. The events are thus shown as branches in the third-stage forks. The consequence resulting from an event-action combination is shown as the end point of the corresponding path from the root of the tree.

The decision tree shows that there are six paths leading to an increase in the project duration, five paths leading to a decrease in project duration, and seven paths leading to an unchanged project duration. For a balanced tree, the total number of paths is given by

$$P = \prod_{i=1}^{N} n_i$$

where

P = total number of paths in the decision tree
N = number of decision stages in the tree
n_i = number of branches emanating from each node in stage i

The expression is not a general formula because the number of outcomes for each decision node may not be the same (i.e., unbalanced tree). In Figure 6–28, the number of paths is $P = (3)(3)(2) = 18$ paths. Some of the paths lead to identical consequences even though they are distinct paths. Probability values can be incorporated into the decision structure as shown in Figure 6–29. Note that the selection of a task at the decision node is based on choice rather than probability.

In this example, it is assumed that the probability of having a particular task duration is independent of whether or not it rains. In some cases, the weather sensitivity of a task may influence the duration of the task. Also, the probability of rain or no

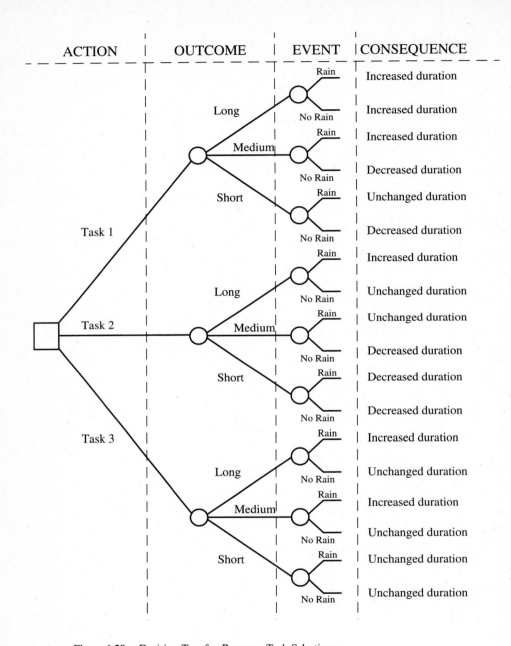

Figure 6-28. Decision Tree for Resource Task Selection

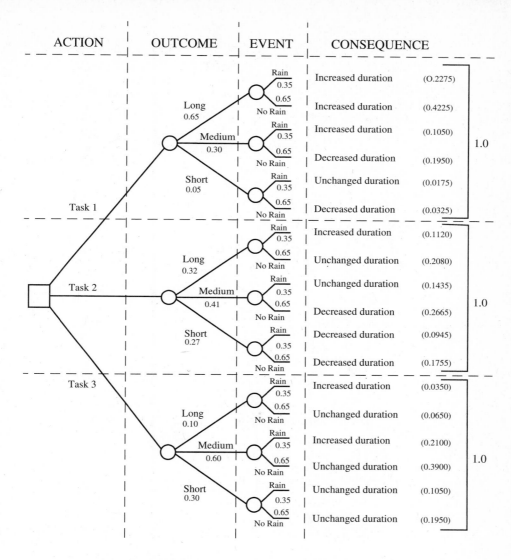

Figure 6-29. Probability Tree Diagram for Task Selection Example

rain is independent of any other element in the decision structure. If the items in the probability tree are interdependent, then the appropriate conditional probabilities would need to be computed. This will be the case if the duration of a task is influenced by whether or not it rains. In such a case, the probability tree should be redrawn as shown in Figure 6–30, which indicates that the weather event will need to be observed first before the task duration event can be determined. The conditional probability of each type of duration, given that it rains or it does not rain, will need to be calculated.

The respective probabilities of the three possible consequences are shown in Figure 6–29. The probability at the end of each path is computed by multiplying the

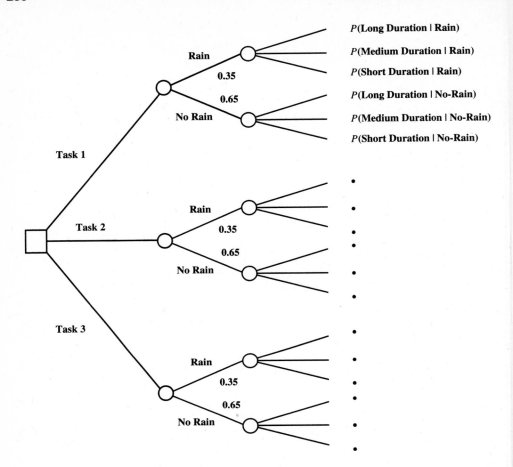

P(Long Duration | Rain)

P(Medium Duration | Rain)

P(Short Duration | Rain)

P(Long Duration | No-Rain)

P(Medium Duration | No-Rain)

P(Short Duration | No-Rain)

Figure 6-30.   Probability Tree for Weather-Dependent Task Durations

individual probabilities along the path. For example, the probability of having an increased project completion time along the first path (task 1, long duration, and rain) is calculated as

$$(0.65)(0.35) = 0.2275$$

Similarly, the probability for the second path (task 1, long duration, and no rain) is calculated as

$$(0.65)(0.65) = 0.4225$$

The sum of the probabilities at the end of the paths associated with each action (choice) is equal to 1 as expected. Table 6–9 presents a summary of the respective probabilities of the three consequences based on the selection of each task.

For example, the probability of having an increased project duration when task 1 is selected is calculated as

$$P(\text{increased project duration due to task 1}) = 0.2275 + 0.4225 + 0.105 = 0.755$$

Table 6-9 Probability Summary for Project Completion Time

Consequence	Selected task					
	Task 1		Task 2		Task 3	
Increased duration	0.2275 + 0.4225 + 0.105	0.755	0.112	0.112	0.035 + 0.21	0.245
Decreased duration	0.195 + 0.0325	0.2275	0.2665 + 0.0945 + 0.1755	0.5365	0.0	0.0
Unchanged duration	0.0175	0.0175	0.208 + 0.1435	0.3515	0.065 + 0.39 + 0.105 + 0.195	0.755
Sum of probabilities		1.0		1.0		1.0

Likewise, the probability of having an increased project duration when task 3 is selected is calculated as

$$P(\text{increased project duration due to task 3}) = 0.035 + 0.21 = 0.245$$

If the selection of tasks at the first node is probabilistic in nature, then the respective probabilities would be included in the calculation procedure. For example, Figure 6–31 shows a case where task 1 is selected 25 percent of the time, task 2 is selected 45 percent of the time, and task 3 is selected 30 percent of the time. The resulting end probabilities for the three possible consequences have been revised accordingly. Note that all the probabilities at the end of all the paths add up to 1 in this case. Table 6–10 presents the summary of the probabilities of the three consequences for the case of weather dependent task durations.

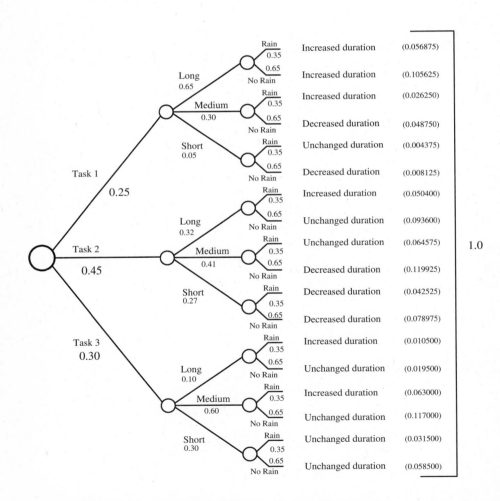

Figure 6-31. Modified Probability Tree for Task Selection Example

Table 6-10 Summary for the Case of Weather-Dependent Task Durations

Consequence	Path probabilities	Row total
Increased duration	0.056875 + 0.105625 + 0.02625 + 0.0504 + 0.0105 + 0.063	0.312650
Decreased duration	0.04875 + 0.119925 + 0.042525 + 0.078975	0.290175
Unchanged duration	0.004375 + 0.008125 + 0.0936 + 0.064575 + 0.0195 + 0.117 + 0.0315 + 0.0585	0.397175
	Column total	1.0

Example

As a project manager, Mr. Wizhead needs to decide which of three projects to postpone due to unplanned events that took place in his department. The worth of each project depends on the state of the company when the project is completed. He has come up with the benefit matrix shown in Table 6–11.

The probability that the company will be in any of the above states is

$$P(\text{above average}) = 0.60$$
$$P(\text{average}) = 0.30$$
$$P(\text{below average}) = 0.10$$

Mr. Wizhead can spend extra money and time and know more about the future state of the company. For this reason, he would like to know what the value of perfect information is so that he can make a decision as to whether to go along with extra analysis or make a decision now as to which project to postpone. The expected benefit for each project is calculated as

For Project A: $E(A) = 85(0.60) + 53(0.30) + 24(0.10) = 69.30$
For Project B: $E(B) = 90(0.60) + 50(0.30) + 32(0.10) = 72.20$
For Project C: $E(C) = 75(0.60) + 70(0.30) + 65(0.10) = 72.50$

Mr. Wizhead would have chosen to postpone project A with the above analysis. Therefore, the total expected benefits from projects B and C would be $72.20 + 72.50 = 144.70$. Had he known that the state of the company would be above average (perfect information), then

Table 6-11 Benefit Matrix

	State of the company		
	Above average	Average	Below average
Project A	85	53	24
Project B	90	50	32
Project C	75	70	65

he would have postponed project C and the total benefit from projects A and B would be 175. Similarly, if the state of the company were average, then the total benefit from projects A and C would be 123 and if the state of the company were below average, the total benefit from projects B and C would be 97. Hence, the maximum expected benefit under perfect information is

$$175(0.60) + 123(0.30) + 97(0.10) = 151.60$$

Therefore, the expected value of perfect information is $151.60 - 144.70 = 6.9$. At this point, Mr. Wizhead may decide not to undertake extra analysis since the added benefit may not justify it. Therefore, his final decision may be to postpone project A.

This example illustrates the use of simple probabilistic analysis to determine a decision for project control. Probabilistic and statistical analyses offer a robust approach to evaluating project performance. Measurement, evaluation, and control actions may be influenced by probabilistic events. For example, resource allocation decision problems under uncertainty can be handled by appropriate decision tree models. With the statistical approach, the overall function of project control can be improved.

6.8 PROJECT CONTROL THROUGH RESCHEDULING

This section presents project control through rescheduling based on project progress. CPM and PERT methods are generally used in the planning phase to enable managers to get an overall picture of how long the project will take, which set of tasks are critical to the completion of the project at the target date, and what the estimated costs will be to run the project. They can also be used to generate actual schedules and to monitor the progress of the project. Actual times and costs can be compared to the scheduled times and costs and actions that need to be taken if large deviations exist between the actual versus scheduled. In this section, we discuss how CPM and PERT can help managers in making decisions related to rescheduling.

As the project progresses, it may be that some activities take a longer time, or more resources than expected. If only a few activities are affected, then the effect on the total project completion time or the anticipated cost may be projected without having to recalculate early start times, early finish times, or new resource level requirements, and so on. However, if there have been several changes, then one needs to reevaluate the remainder of the project and maybe reallocate resources or reschedule activities in order to minimize deviations from the targeted realizations.

Consider the project in Figure 6–32. Suppose a week after the project has started an evaluation of the progress is made. It is found that activities a, b, c, d and e have been completed. Activity i started two days earlier and activities f and h have not yet started. Also the durations of activities h, m, and n have been modified to $d'(h)$, $d'(m)$, and $d'(n)$. Nodes 2, 3 and 4 were realized at times t_2, t_3 and t_4 respectively.

Figure 6–33 is the modified network which contains the start node, the set of nodes realized with unfinished immediate successors, and the set of nodes not yet realized. The start node is connected to each of the realized nodes with an arc whose duration is equal to the project elapsed time and does not require any resources. The durations of activities h, m and n are updated. The duration of activity i is decreased by two days

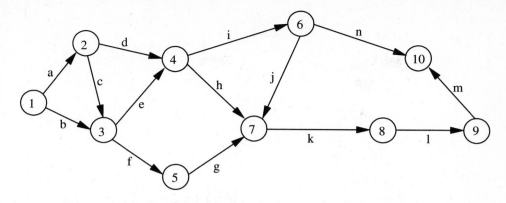

Figure 6-32.   Original Project Network for Rescheduling Example

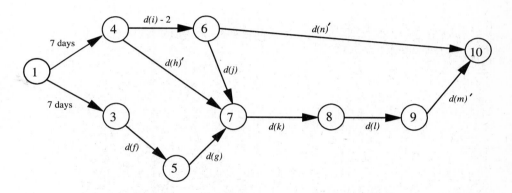

Figure 6-33.   Modified Network for Rescheduling

since two days have elapsed since the start of the activity. The modified network can then be used to reschedule the activities with the available methods described earlier.

.9 CONTROL THROUGH TERMINATION

Project termination is an important aspect of project control. Termination should be viewed as a control function since some projects can drag on unnecessarily if control is not instituted. There are several reasons for terminating projects. Some projects are terminated under cordial, arranged, and expected circumstances while others are terminated under unpleasant circumstances that call for managerial control. If necessary, a project audit should be conducted to ascertain the need to terminate a project. Some of the common reasons include the following:

- Cost overruns
- Alternate technology
- Missed deadline
- Product obsolescence
- Environmental concern
- Government requirement
- Excessive delay penalties
- Technically impossible goals
- Lack of project justification
- Poor performance beyond remedy
- Alternate objective to the initial plan
- Project objective accomplished
- Poor unachievable project plan
- Lack of required personnel or other resources

Even after the reasons for terminating the project have been identified, actual termination may not be easy to implement especially for long-range and large projects that have spread their tentacles throughout an organization. Problems of morale may develop. Some workers may have grown accustomed to the extra attention, recognition, or advancement opportunities associated with the project. They may not see the wisdom of terminating the project. The Triple C approach should be used in setting the stage for the termination of a project at the appropriate time. The termination process should cover the following items:

- Communicating with the personnel on the need for termination
- Retraining workers for new functions
- Reassigning workers to other functions
- Assuring the cooperation of those involved
- Returning workers to their previous functions
- Coordinating the required actions for termination
- Withdrawing funding from the project (pulling the plug)

If the termination is handled properly, workers will be less agonized by the loss and there will be a smooth transition to other projects.

6.10 EXERCISES

6.1. Give one definition of productivity that relates to managerial control.

6.2. List some common impediments to control.

6.3. Prepare a taxonomy of what should be included in a project management information system as measures of control.

6.4. How is the WBS (work breakdown structure) valuable in project monitoring and control?

6.5. How can schedule control be tied in with cost control?

6.6. For each of the causes of control problems listed in this chapter, discuss what corrective actions could be taken.

6.7. List some additional reasons for terminating a project as a measure of control.

6.8. Given the following data for three projects, perform the complete data analysis as was done for Table 6–1. How would you perform the average cross-check in the cell marked "XXX"?

Project	January	February	March	April	Row total	Row average
A	3000	3200	3400	2800		
B	1200	1900	2500	2400		
C	4500	3400	4600	4200		
Column total						
Column average						XXX

6.9. For each of the following measurement scales—nominal scale, ratio scale, ordinal scale, and interval scale—list at least five factors or data types associated with project management that can be measured on the scale.

6.10. The employees selected to work on a project are surveyed to select the type of organization structure suitable for the project. Which measure of central tendency would be the most appropriate to determine the preference by the greatest number of employees?

6.11. Use the raw data presented in Exercise 6.8 to verify that $\sum(X - \overline{X}) = 0$.

6.12. Using a software tool, such as a spreadsheet program, compute the average deviation, standard deviation and variance for the raw data presented in Exercise 6.8.

6.13. What types of data would you recommend to be collected for a project involving the construction of a new soccer playground in a small community?

6.14. The duration of a certain task is known to be normally distributed with a mean of 7 days and a standard deviation of 3 days. Find the following:
(a) The probability that the task can be completed in exactly 7 days
(b) The probability that the task can be completed in 7 days or less
(c) The probability that the task will be completed in more than 6 days

6.15. Alctrex Construction Company is bidding against Betatrex for a building project. Due to past performance of both companies, Alctrex knows that if the company bids a lower or an equal amount, it will win the bid. It will cost Alctrex \$9,500 to complete the project. Betatrex's bid is a random variable B with the following probabilities:

$$P(B = \$9,500) = 0.45$$
$$P(B = \$10,500) = 0.35$$
$$P(B = \$12,000) = 0.20.$$

Suppose that Alctrex is thinking of bidding between \$9,500 and \$12,000 in increments of \$500. Determine the profit matrix for Alctrex. What is the best decision for the company? What is the value of perfect information?

6.16. Consider the following payoff matrix for a decision-making problem:

	States	
	θ_1	θ_2
Alternative A	100,000	−40,000
Alternative B	50,000	−10,000
Alternative C	0	0

Determine the best alternative as a function of $P(\theta_1) = p$, where $0 \le p \le 1$. What is the best alternative for $p = 0.3$? What is the value of perfect information for $p = 0.3$?

6.17. Draw a fishbone diagram for evaluating the causes of deficiencies for each of the following project parameters:

Cost
Schedule
Performance

6.18. The *birthday problem* is a popular problem in probability. The problem involves finding the probability that at least two people have the same birthday in a group of N individuals. An activity scheduling formulation of the problem can be stated as follows: N activities must be scheduled at random during a given scheduling cycle consisting of 200 days. Each activity takes exactly one day to complete. The days all have equal likelihood of being selected for any of the activities. Find the probability of a schedule conflict. That is, the probability that two or more activities will be scheduled on the same day ($N \le 200$). Solve for $N = 32$.

6.19. Suppose N independent candidates are to be scheduled for interviews during one year (365 days). Each interview takes exactly one day. There is only one interviewer available and only one candidate can be interviewed at a time. Each candidate is requested to specify a preferred date for his or her interview. It is assumed that the candidates pick their interview days at random and independent of one another. Find the probability of having a conflict in scheduling the candidates.

6.20. Develop a computer simulation model to solve the problem in Question 6-18. Run the simulation for $N = 1, 2, 3, \ldots, 200$. Plot the probabilities versus the values of N. Discuss your findings.

Section 2

Optimization Models

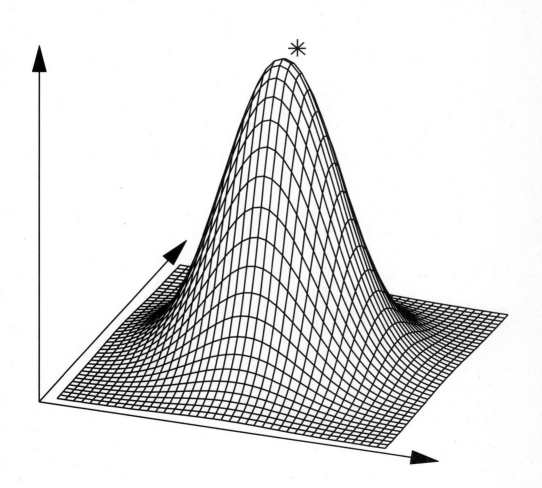

7

Project Modeling and Optimization

This chapter presents optimization models for project management. The models can be used at any stage of project planning, scheduling, and control. Optimization may focus on any of several performance measures of a project. Examples of performance measures are project duration, schedule composition, cost, resource allocation, and throughput. Several of these performance measures are addressed by the models discussed in this chapter. The current stage of our project management process can now be characterized as

SET GOALS → PLAN → ORGANIZE → SCHEDULE →
ALLOCATE RESOURCES → CONTROL → OPTIMIZE

7.1 PROJECT MODELING

Schedule optimization is often the major focus in project management. While heuristic scheduling is very simple to implement, it does have some limitations. The limitations of heuristic scheduling include subjectivity, arbitrariness, and simplistic assumptions. In addition, heuristic scheduling does not handle uncertainty very well. On the other hand, mathematical scheduling is difficult to apply to practical problems. However, the increasing access to low-cost high-speed computers has facilitated increased use of mathematical scheduling approaches that yield optimal project schedules. The advantages of mathematical scheduling include the following facts:

- It provides optimal solutions.
- It can be formulated to include realistic factors influencing a project.
- Its formulation can be validated.
- It has proven solution methodologies.

With the increasing availability of personal computers and software tools, there is very little need to solve optimization problems by hand nowadays. Computerized algorithms are now available to solve almost any kind of optimization problem. What is more important for the project analyst is to be aware of the optimization models available, the solution techniques available, and how to develop models for specific project optimization problems. It is crucial to know which model is appropriate for which problem and to know how to implement optimized solutions in practical settings. The presentations in this chapter concentrate on the processes for developing models for project optimization.

.2 GENERAL PROJECT SCHEDULING FORMULATION

Several mathematical models can be developed for project scheduling problems, depending on the specific objective of interest and the prevailing constraints. One general formulation is

$$
\begin{aligned}
\text{Minimize:} \quad & \{\max_{\forall i}\{s_i + d_i\}\} \\
\text{Subject to:} \quad & s_i \geq s_j + d_j, \text{for all } i; \qquad j \in P_i \\
& R_k \geq \sum_{i \in A_t} r_{ik}, \text{for all } t; \text{for all } k \\
& s_i \geq 0, \text{for all } i \\
& r_{ik} \geq 0, \text{for all } i; \text{for all } k
\end{aligned}
$$

where

s_i = start time of activity i

d_i = duration of activity i

P_i = set of activities which must precede activity i

R_k = availability level of resource type k over the project horizon

A_t = set of activities ongoing at time t

r_{ik} = number of units of resource type k required by activity i

The objective of the above model is to minimize the completion time of the last activity in the project. Since the completion time of the last activity determines the project duration, the project duration is indirectly minimized. The first constraint set ensures that all predecessors of activity i are completed before activity i may start. The second constraint set ensures that resource allocation does not exceed resource availability. The general model may be modified or extended to consider other project parameters. Examples of other factors that may be incorporated into the scheduling formulation include cost, project deadline, activity contingency, mutual exclusivity of activities, activity crashing requirements, and activity subdivision.

An *objective function* is a mathematical representation of the goal of an organization. It is stated in terms of maximizing or minimizing some quantity of interest. In a project environment, the objective function may involve any of the following:

- *Minimize project duration*
- *Minimize project cost*
- *Minimize number of late jobs*
- *Minimize idle resource time*
- *Maximize project revenue*
- *Maximize net present worth*

7.3 LINEAR PROGRAMMING FORMULATION

Linear programming is a mathematical technique for maximizing or minimizing some quantity, such as profit, cost, or time to complete a project. It is one of the most widely used quantitative techniques. It is a mathematical technique for finding the optimum solution to a linear objective function of two or more quantitative decision variables subject to a set of linear constraints. The technique is applicable to a wide range of decision-making problems. Its wide applicability is due to the fact that its formulation is not tied to any particular class of problems, as the CPM and PERT techniques are. Numerous research and application studies of linear programming are available in the literature.

The objective of a linear programming model is to optimize an objective function by finding values for a set of decision variables subject to a set of constraints. We can define the optimization problem mathematically as

$$
\begin{aligned}
\text{Optimize:} \quad & z = c_1 x_1 + c_2 x_2 + \ldots + c_n x_n \\
\text{Subject to:} \quad & a_{11} x_1 + a_{12} x_2 + \ldots + a_{1n} x_n \{\leq, =, \geq\} b_1 \\
& a_{21} x_1 + a_{22} x_2 + \ldots + a_{2n} x_n \{\leq, =, \geq\} b_2 \\
& \ldots \\
& \ldots \\
& a_{m1} x_1 + a_{m2} x_2 + \ldots + a_{mn} x_n \{\leq, =, \geq\} b_m \\
& x_1, x_2, \ldots, x_n \geq 0
\end{aligned}
$$

where

Optimize is replaced by *maximize* or *minimize* depending on the objective.
z is the value of the objective function for specified values of the decision variables.
$x_1, x_2, \ldots, x_n$ are the n decision variables.
$c_1, c_2, \ldots, c_n$ are the objective function coefficients.
$b_1, b_2, \ldots, b_n$ are the limiting values of the resources (*right-hand side*).
$a_{11}, a_{12}, \ldots, a_{mn}$ are the constraint coefficients (per-unit usage rates).

The word *programming* in LP does not refer to computer programming, as some people think. Rather, it refers to choosing a *program of action*. The word *linear* refers to the *linear relationships* among the variables in the model. The characteristics of an LP formulation are explained next.

Quantitative decision variables.   A decision variable is a factor that can be manipulated by the decision maker. Examples are number of resource units assigned to a task, number of product types in a product mix, and number of units of a product to produce. Each decision variable must be defined numerically in some unit of measurement.

Linear objective function.   The objective function relates to the measure of performance to be minimized or maximized. There is a linear relationship among the variables that make up the objective function. The coefficient of each variable in the objective function indicates its per-unit contribution (positive or negative) toward the value of the objective function.

Linear constraints.   Every decision problem is subject to some specific limitations or constraints. The constraints specify the restrictions on how the decision maker may manipulate the decision variables. Examples of decision constraints are capacity limitations, maximum number of resource units available, demand and supply requirements, and number of work hours per day. The relationships among the variables in a constraint must be expressed as linear functions represented as equations or inequalities.

Nonnegativity constraint.   The nonnegativity constraint is common to all linear programming problems. This requires that all decision variables are restricted to nonnegative values.

The general procedure for using a linear programming model to solve a decision problem involves an LP formulation of the problem and a selection of a solution approach. The procedure is summarized as follows:

1. Determine the decision variables in the problem.
2. Determine the objective of the problem.
3. Formulate the objective function as an algebraic expression.
4. Determine the real-world restrictions on the problem scenario.
5. Write each of the restrictions as an algebraic constraint. Make sure that units match throughout the constraints. Otherwise, the terms cannot be added.
6. Select a solution approach. The *graphical method* and the *simplex technique* are the two most popular approaches. The graphical method is easy to apply when the LP model contains just two decision variables. Several computer software packages are available for solving LP problems. Examples are LINDO, Linear Optimizer, LP88, MathPro, What-if Solver, and Turbo-Simplex. A comprehensive survey by Sharda (1992) lists several of the available LP packages.

An important aspect of using LP models is the interpretation of the results to make decisions. An LP solution that is optimal analytically may not be practical in a real-world decision scenario. The decision maker must incorporate his or her own subjective judgment when implementing LP solutions. Final decisions are often based on a combination of quantitative and qualitative factors. The examples presented in this chapter illustrate the application of optimization models to project planning and scheduling problems.

7.4 ACTIVITY PLANNING EXAMPLE

Activity planning is a major function in project management. Linear programming can be used to determine the optimal allocation of time and resources to the activities in a project. The example presented here is an adaptation of an example presented by Wu and Coppins (1981). Suppose a program planner is faced with the problem of planning a five-day development program for a group of managers in a manufacturing organization. The program includes some combination of four activities: a seminar, laboratory work, case studies, and management games. It is estimated that each day spent on an activity will result in productivity improvement for the organization. The productivity improvement will generate annual cost savings as shown in Table 7–1. The program will last five days and there is no time lost between activities. In order to balance the program, the planner must make sure that not more than three days are spent on active or passive elements of the program. The active and passive percentages of each activity are also shown in the table. The company wishes to spend at least half a day on each of the four activities. A total budget of $1,500 is available. The cost of each activity is shown in the tabulated data.

The program planner must determine how many days to spend on each of the four activities. The following variables are defined for the problem:

x_1 = number of days spent on seminar
x_2 = number of days of laboratory work
x_3 = number of days for case studies
x_4 = number of days with management games

The objective is to maximize the estimated annual cost savings. That is,

$$\text{Maximize} : f = 3200x_1 + 2000x_2 + 400x_3 + 2000x_4$$

subject to the following constraints:

1. The program lasts exactly 5 days:

$$x_1 + x_2 + x_3 + x_4 = 5$$

2. Not more than 3 days can be spent on active elements:

$$0.10x_1 + 0.40x_2 + x_3 + 0.60x_4 \leq 3$$

Table 7-1 Data for Activity Planning Problem (From Wu and Coppins, 1981 reprinted with permission.)

Activity	Cost savings ($/year)	% Active	% Passive	Cost ($/day)
Seminar	3,200,000	10	90	400
Laboratory work	2,000,000	40	60	200
Case studies	400,000	100	0	75
Management games	2,000,000	60	40	100

3. Not more than 3 days can be spent on passive elements:

$$0.90x_1 + 0.60x_2 + 0.40x_4 \leq 3$$

4. At least 0.5 day must be spent on each of the four activities:

$$x_1 \geq 0.50 \qquad x_3 \geq 0.50$$
$$x_2 \geq 0.50 \qquad x_4 \geq 0.50$$

5. The budget is limited to $1,500:

$$400x_1 + 200x_2 + 75x_3 + 100x_4 \leq 1,500.$$

The complete linear programming model for the example is presented below:

$$
\begin{array}{llll}
\text{Maximize:} & f = 3200x_1 + 2000x_2 + 400x_3 + 2000x_4 \\
\text{Subject to:} & x_1 & + x_2 & + x_3 & + x_4 & = 5 \\
& 0.1x_1 & + 0.4x_2 & + x_3 & + 0.6x_4 & \leq 3 \\
& 0.9x_1 & + 0.6x_2 & + 0x_3 & + 0.4x_4 & \leq 3 \\
& x_1 & & & & \geq 0.5 \\
& & x_2 & & & \geq 0.5 \\
& & & x_3 & & \geq 0.5 \\
& & & & x_4 & \geq 0.5 \\
& 400x_1 & + 200x_2 & + 75x_3 & + 100x_4 & \leq 1,500 \\
& & & x_1, x_2, x_3, x_4 & \geq 0
\end{array}
$$

The optimal solution to the problem is shown in Table 7–2. Most of the conference time must be allocated to the seminar (2.20 days). The expected annual cost savings due to this activity is $7,040,000. That is, 2.20 days × $3,200,000/year/day. Management games is the second most important activity. A total of 1.8 days for management games will yield annual cost savings of $3,600,000. Fifty percent of the remaining time (0.5 day) should be devoted to laboratory work, which will result in annual cost savings of $1,000,000. Case studies also require half a day with a resulting annual savings of $200,000. The total annual savings, if the LP solution is implemented, is $11,840,000. Thus, an investment of $1,500 in management training for the personnel can generate annual savings of $11,840,000: a huge rate of return on investment!

Table 7-2 LP Solution to the Activity Planning Example (From Wu and Coppins, 1981. Reprinted with permission)

Activity	Cost savings ($/year)	Number of days	Annual cost savings ($)
Seminar	3,200,000	2.20	7,040,000
Laboratory work	2,000,000	0.50	1,000,000
Case studies	400,000	0.50	200,000
Management games	2,000,000	1.80	3,600,000
Total		5	11,840,000

7.5 RESOURCE COMBINATION EXAMPLE

This example illustrates the use of LP for energy resource allocation (Badiru 1991a). Suppose an industrial establishment uses energy for heating, cooling, and lighting. The required amount of energy is presently being obtained from conventional electric power and natural gas. In recent years, there have been frequent shortages of gas, and there is a pressing need to reduce the consumption of conventional electric power. The director of the energy management department is considering a solar energy system as an alternate source of energy. The objective is to find an optimal mix of three different sources of energy to meet the plant's energy requirements. The three energy sources are

- Natural gas
- Conventional electric power
- Solar power

It is required that the energy mix yield the lowest possible total annual cost of energy for the plant. Suppose a forecasting analysis indicates that the minimum kwh (kilowatt-hour) needed per year for heating, cooling, and lighting, are 1,800,000 kwh, 1,200,000 kwh, and 900,000 kwh, respectively. The solar energy system is expected to supply at least 1,075,000 kwh annually. The annual use of conventional electric power must be at least 1,900,000 kwh due to a prevailing contractual agreement for energy supply. The annual consumption of the contracted supply of gas must be at least 950,000 kwh. The cubic foot unit for natural gas has been converted to kwh (1 cu. ft. of gas = 0.3024 kwh).

The respective rates of $6/kwh, $3/kwh, and $2/kwh are applicable to the three sources of energy. The minimum individual annual savings desired are $600,000 from solar power, $800,000 from conventional electric power, and $375,000 from natural gas. The savings are associated with the operating and maintenance costs. The energy cost per kwh is $0.30 for conventional electric power, $0.20 for natural gas, and $0.40 for solar power. The initial cost of the solar energy system has been spread over its useful life of ten years with appropriate cost adjustments to obtain the rate per kwh. The problem data is summarized in Table 7–3. If we let x_{ij} be the kwh used from source i for purpose j, then we would have the data organized as shown in Table 7–4.

The optimization problem involves the minimization of the total cost function, Z. The mathematical formulation of the problem is presented below.

Table 7-3 Energy Resource Combination Data

Energy source	Supply (1000s kwh)	Savings (1000s $)	Unit savings ($/kwh)	Unit cost ($/kwh)
Solar power	1,075	600	6	0.40
Electric power	1,900	800	3	0.30
Natural gas	950	375	2	0.20

Table 7–4 Tabulation of Data for LP Model

Energy source	Heating	Type of use Cooling	Lighting	Constraint
Solar power	x_{11}	x_{12}	x_{13}	$\geq 1{,}075\,\text{K}$
Electric power	x_{21}	x_{22}	x_{23}	$\geq 1{,}900\,\text{K}$
Natural gas	x_{31}	x_{32}	x_{33}	$\geq 950\,\text{K}$
Constraint	$\geq 1{,}800$	$\geq 1{,}200$	≥ 900	

$$
\text{Minimize:} \quad Z = 0.4 \sum_{j=1}^{3} x_{1j} + 0.3 \sum_{j=1}^{3} x_{2j} + 0.2 \sum_{j=1}^{3} x_{3j}
$$

$$
\begin{aligned}
\text{Subject to:} \quad & x_{11} + x_{21} + x_{31} \geq 1{,}800 \\
& x_{12} + x_{22} + x_{32} \geq 1{,}200 \\
& x_{13} + x_{23} + x_{33} \geq 900 \\
& 6(x_{11} + x_{12} + x_{13}) \geq 600 \\
& 3(x_{21} + x_{22} + x_{23}) \geq 800 \\
& 2(x_{31} + x_{32} + x_{33}) \geq 375 \\
& x_{11} + x_{12} + x_{13} \geq 1{,}075 \\
& x_{21} + x_{22} + x_{23} \geq 1{,}900 \\
& x_{31} + x_{32} + x_{33} \geq 950 \\
& x_{ij} \geq 0, \qquad i, j = 1, 2, 3
\end{aligned}
$$

Using the LINDO linear programming computer package (Schrage 1986), the solution presented in Table 7–5 was obtained. The table shows that solar power should not be used for cooling and lighting if the lowest cost is to be realized. The use of conventional electric power should be spread over the three categories of use. The solution indicates that natural gas should be used only for cooling purposes. In pragmatic terms, this LP solution may have to be modified before being implemented on the basis of the prevailing operating scenarios and the technical aspects of the units involved.

Table 7–5 LP Solution to the Resource Combination Example

Energy source	Heating	Type of use Cooling	Lighting
Solar power	1,075	0	0
Electric power	750	250	900
Natural gas	0	950	0

7.6 RESOURCE REQUIREMENTS ANALYSIS

Suppose a manufacturing project requires that a certain number of workers be assigned to a workstation. The workers produce identical units of the same product. The objective is to determine the number of workers to assign to the workstation in order to minimize the total production cost per shift. Each shift is eight hours long. Each worker can be assigned a variable number of hours and/or variable production rates to work during a shift. Four different production rates are possible: *slow rate, normal rate, fast rate,* and *high-pressure rate.* Each worker is capable of working at any of the production rates during a shift. The total number of work hours available per shift is determined by multiplying the number of workers assigned by the eight hours available in a shift.

There are variable costs and percent defective associated with each production rate. The variable cost and the percent defective increase as the production rate increases. At least 450 units of the product must be produced during each shift. It is assumed that the workers' performance levels are identical. The production rates (r_i), the respective costs (c_i), and percent defective (d_i), are presented as follows:

Operating Rate 1 (slow)

$r_1 = 10$ units per hour
$c_1 = \$5$ per hour
$d_1 = 5\%$

Operating Rate 2 (normal)

$r_2 = 18$ units per hour
$c_2 = \$10$ per hour
$d_2 = 8\%$

Operating Rate 3 (fast)

$r_3 = 30$ units per hour
$c_3 = \$15$ per hour
$d_3 = 12\%$

Operating Rate 4 (high pressure)

$r_4 = 40$ units per hour
$c_4 = \$25$ per hour
$d_4 = 15\%$

7.6.1 LP Formulation

Let x_i represent the number of hours worked at production rate i.
Let n represent the number of workers assigned.
Let u_i represent the number of good units produced at operation rate i.

$$u_1 = (10 \text{ units/hour}) \cdot (1 - 0.05) = 9.50 \text{ units/hour}$$
$$u_2 = (18 \text{ units/hour}) \cdot (1 - 0.08) = 16.56 \text{ units/hour}$$
$$u_3 = (30 \text{ units/hour}) \cdot (1 - 0.12) = 25.40 \text{ units/hour}$$
$$u_4 = (40 \text{ units/hour}) \cdot (1 - 0.15) = 34.00 \text{ units/hour}$$

Minimize: $z = 5x_1 + 10x_2 + 15x_3 + 25x_4$

Subject to: $x_1 + x_2 + x_3 + x_4 \le 8n$
$9.50x_1 + 16.56x_2 + 25.40x_3 + 34.00x_4 \ge 450$
$x_1, x_2, x_3, x_4 \ge 0$

The solution will be obtained by solving the LP model for different values of n. A plot of the minimum costs versus values of n can then be used to determine the optimum assignment policy. The complete solution is left as an exercise at the end of this chapter.

.7 SCHEDULING UNDER LIMITED RESOURCES: INTEGER PROGRAMMING APPROACH

Integer programming is a restricted model of linear programming that permits only solutions with integer values of the decision variables. Suppose we are interested in minimizing the project completion time while observing resource limitations and job precedences. The basic assumption is that once a job starts it has to be completed without interruption. We can construct several different integer programming models for the problem, all of which will give the same optimal solution but whose execution times will differ considerably. An efficient integer programming model for the problem should use as few integer variables as possible.

Define the variables as

x_{ij} :   1 if job i starts in period j
0 otherwise
t_p :   completion time of the project.

Only x_{ij}'s are restricted as integers. For each job i one can determine the early start and latest start times, ES_i and LS_i, respectively. Therefore, assuming that there are n jobs in the project, as $1 \le i \le n$ for each i, $ES_i \le j \le LS_i$. Let t_i denote the duration of job i. Resource availability constraints can be smartly handled by defining a vector $\mathbf{V}_{ij}$ which has 0's everywhere except positions $j, j + 1, \ldots, j + t_i - 1$, where it has 1's. It indicates the time period where job i uses the resource assuming that $x_{ij} = 1$. Let r_i and R_j be the resource required by job i and the resource available on day j, respectively. Let R be a row vector containing R_j .

Then, the integer programming model for the scheduling problem with limited resource can be defined as

Minimize: t_p

Subject to:

$$\sum_{j=ES_i}^{LS_i} x_{ij} = 1 \qquad\qquad \forall i = 1, \ldots, n \qquad\qquad (7.1)$$

$$-\sum_{j=ES_i}^{LS_i} jx_{ij} + \sum_{j=ES_k}^{LS_k} jx_{kj} \leq t_i \qquad\qquad \forall k \in S(i) \qquad\qquad (7.2)$$

$$\forall i = 1, \ldots, n$$

$$t_p - \sum_{j=ES_i}^{LS_i} jx_{ij} \geq t_i - 1 \qquad\qquad \forall i \text{ with } S(i) = \varnothing \qquad\qquad (7.3)$$

$$\sum_{i=1}^{n} \sum_{j=ES_i}^{LS_i} x_{ij} r_i V_{ij} \leq R \qquad\qquad\qquad\qquad\qquad (7.4)$$

$$x_{ij} = 0, 1 \qquad\qquad \forall j = ES_i, \ldots, LS_i$$

$$\forall i = 1, \ldots, n$$

where $S(i)$ = set of immediate successor jobs of job i.

Equation (7.1) indicates that each job must start on the same day. Equation (7.2) collectively makes sure that a job cannot start until all of the predecessor jobs are completed. Equation (7.3) determines the project completion time t_p. The project is completed after all the jobs without any successors are completed. The last set of equations makes sure that daily resource requirements are met. The indicator variable x_{ij} is restricted to the values 0 and 1.

The above integer programming model can be solved using LINDO (Schrage 1991) computer code and declaring x_{ij}'s as binary variables. The code uses the branch and bound method of integer programming (Winston 1987) to solve the problem. As an example, consider the project shown in Figure 7–1. Assume that the daily resource availability is 10 units. Figure 7–2 contains a V_{ij} vector chart for the example. The lines indicate the positions of 1's in the vector. Appendix B contains the LINDO program and the solution. It took a 386/33 MHz machine about 3 minutes to reach the optimal solution using the LINDO package. An interesting observation is that the optimal solution was found after examining 116 branches of the branch and bound tree. However, it was only after examining 1,609 branches that the method assured optimality of the solution. The optimal solution is illustrated in Figure 7–3.

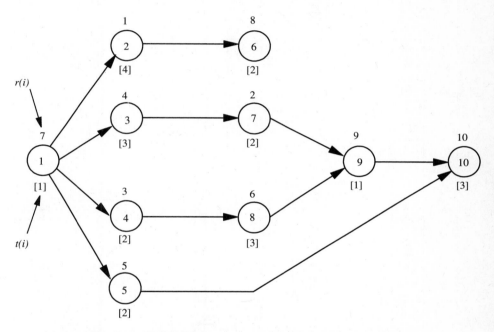

Figure 7–1. Example Network for Integer Programming Model (Adapted from Wiest and Levy, 1977 with permission)

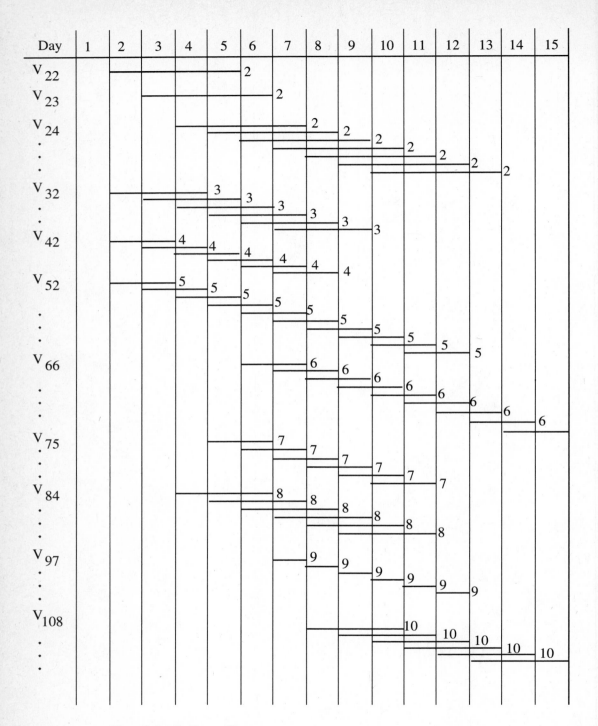

Figure 7–2. V_{ij} Vector Chart

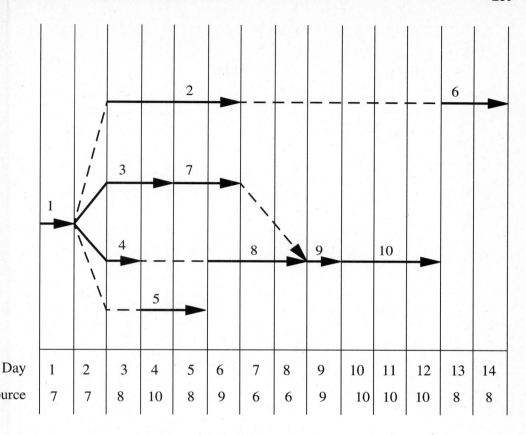

Day	1	2	3	4	5	6	7	8	9	10	11	12	13	14
ource	7	7	8	10	8	9	6	6	9	10	10	10	8	8

Figure 7–3.   Optimal Solution for Integer Programming

.8 TIME-COST TRADE-OFF MODEL

For projects involving several activities and complex precedence structure, optimization techniques provide efficient solutions for the time cost trade-off problem. In fact, one can generate the complete time-cost trade off curve with such techniques. We will provide the linear programming (LP) model for the time-cost trade-off problem and solve it using a network flow approach.

Let a_{ij} be the cost of crashing activity (i,j) by one time unit. The duration of activity (i,j) is denoted by y_{ij} which is bounded by the crash duration, d_{ij}, and the normal duration, u_{ij} as shown in Figure 7–4. If t_i represents realization time of node i, then the precedence relationship implies that

$$t_i + y_{ij} \leq t_j \qquad \text{for all } (i,j) \in \mathbf{A}$$

where $\mathbf{A}$ is the arc(activity) set. The above inequality states that node j cannot be realized before y_{ij} time units have elapsed after the realization time of node i.

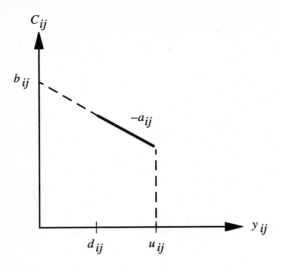

Figure 7–4. Linear Cost Function for y_{ij}

Given a project completion time T, the optimal activity durations can be found by solving the following LP model.

$$\text{Minimize:} \quad z = \sum_{(i,j)\in A} (b_{ij} - a_{ij} y_{ij}) \tag{7.5}$$

$$\text{Subject to:} \quad t_i + y_{ij} - t_j \leq 0, \qquad \forall (i,j) \in A \tag{7.6}$$

$$t_n = T \tag{7.7}$$

$$d_{ij} \leq y_{ij} \leq u_{ij}, \qquad \forall (i,j) \in A \tag{7.8}$$

where t_n denotes the realization time of the end node. By simply decreasing T and solving the above LP problem, the complete time-cost trade off curve can be generated. Clearly, $T_{min} \leq T \leq T_{max}$ where T_{min} and T_{max} are the minimum and maximum achievable project completion times. For a complete and rigorous derivation of the approach, the reader is referred to Elmaghraby (1977), Lawler (1976), and for a review of linear programming models to Bazaraa et al. (1990). We will explain a network flow procedure to solve the time-cost trade-off problem without getting into the theoretical details.

Minimizing the objective function of equation (7.5) is equivalent to maximizing $a_{ij}\, y_{ij}$, since b_{ij} is a constant. Given a project network $\mathbf{G} = (\mathbf{N}, \mathbf{A})$ where $\mathbf{N}$ is the node set and $\mathbf{A}$ is the arc set, the procedure starts with $y_{ij} = u_{ij}$ for all $(i,j) \in \mathbf{A}$. Node realization (early start) times are calculated using $t_1 = 0$ and the precedence constraint (7.6). At this point $t_n = T_{max}$. The activities which satisfy (7.6) as equality are called the critical activities and form the critical subgraph, $\mathbf{G}' = (\mathbf{N}, \mathbf{A}')$. The cheapest set of critical arcs are determined using a network flow approach and crashed until either an arc of the set reaches its crash duration or a noncritical activity becomes critical. The crashing decreases t_n and requires an update of $\mathbf{G}'$. The new set of cheapest critical activities is next determined and t_n is crashed further. The procedure continues until there exists a critical path from node 1 to node n with all arcs at their crash durations, in which case $t_n = T_{min}$.

We will now explain the maximum flow procedure which can be used to determine the cheapest set of critical arcs to crash.

7.8.1 Maximum Flow Procedure

For $t_n = T_{max}$ set $f_{ij} = 0$, $c_{ij} = a_{ij}$ for all $(i,j) \in A$. Otherwise, take the current f_{ij} and y_{ij} values.

Step 1:   Label node 1 with $(\infty, 0)$. In general, label node j from a labeled node i with (q_j, i) where

$$q_j = \begin{cases} (c_{ij} - f_{ij}) & \text{if } c_{ij} > 0 \text{ and } (i,j) \in A \\ f_{ij} & \text{if } f_{ji} > 0 \text{ and } (j,i) \in A \end{cases}$$

   Continue labeling until either node n is labeled or labeling terminates with node n unlabeled (nonbreakthrough). Go to step 2 with $k = n$ if $q_n > 0$ and is finite. Otherwise, stop. The maximum flow has been reached.

Step 2:   Node k has label $(q_k, \underline{i})$. If $(i,k) \in A$, then $f_{ik} = f_{ik} + q_n$. Otherwise, $(k,i) \in A$, set $f_{ki} = f_{ki} - q_n$. Set $k = i$ and repeat this step until $k = 1$.

The above procedure locates flow augmenting paths and augments flow of q_n units along the path. The maximum flow is determined when no more flow augmenting paths exist in the network. This is indicated by a nonbreakthrough. Then, the set of labeled nodes form the set $\mathbf{X}$ and the set of unlabeled nodes for $\overline{\mathbf{X}}$. The cut-set $\mathbf{C}$ is defined by

$$\mathbf{C} = \{(i,j) \in \mathbf{A} | i \in \mathbf{X}, j \in \overline{\mathbf{X}} \quad \text{or} \quad i \in \overline{\mathbf{X}}, j \in \mathbf{X}\}$$

The set of arcs in $\mathbf{C}$ define the minimum cutset separating nodes 1 and n. If u is the value of the maximum flow, then the maximum flow-minimum cut theorem (Lawler 1976) indicates that

$$v = \sum_{i \in \mathbf{X}, j \in \overline{\mathbf{X}}} c_{ij}$$

In the time-cost trade-off procedure the arc capacities are determined using a_{ij} values. The maximum flow value corresponds to the increase in cost per unit decrease in T value. The set of activities to be crashed is given by the set of critical arcs in $\mathbf{C}$ oriented from $\mathbf{X}$ to $\overline{\mathbf{X}}$. As it will be pointed out later, if there exists a critical arc$(i,j) \in \mathbf{C}$ with $i \in \overline{\mathbf{X}}$ and $j \in \mathbf{X}$, the duration of this arc will be lengthened when t_i is reduced.

7.8.2 Time-cost Trade-off Procedure

Step 1:   Determination of the critical subnetwork
Set $y_{ij} = u_{ij}$ for all $(i,j) \in A$
Set $t_1 = 0$
Find $t_j = \text{Max}_{(i,j) \in A}(t_i + y_{ij})$ for all $j \in N$
Define $A' = \{(i,j) | t_j = t_i + y_{ij}\}$

Step 2:   Preparation of the flow network
Set flow values, $f_{ij} = 0$ for all $(i,j) \in A'$
Set $c_{ij} = 0$ for all $(i,j) \in A - A'$
For all $(i,j) \in A'$, define

$$c_{ij} = \begin{cases} a_{ij}, & \text{if } y_{ij} = u_{ij} \\ 0, & \text{if } d_{ij} < y_{ij} < u_{ij} \\ \infty & \text{if } y_{ij} = d_{ij} \end{cases}$$

Step 3:   Use the maximum flow procedure (discussed earlier) to determine the maximum flow. Define,

$$C = C_{11} \cup C_{12} \cup C_{21} \cup C_{22}$$

where

$$C_{11} = \{(i,j) \in C \cap A' | i \in X, j \in \overline{X}\}$$
$$C_{12} = \{(i,j) \in C \cap A' | i \in \overline{X}, j \in X\}$$
$$C_{21} = \{(i,j) \in C \cap (A - A') | i \in X, j \in \overline{X}\}$$
$$C_{22} = \{(i,j) \in C \cap (A - A') | i \in X, j \in \overline{X}\}$$

Calculate

$$\Delta_1 = \min(\min_{(i,j) \in C_{11}}\{y_{ij} - d_{ij}\}, \min_{(i,j) \in C_{12}}\{u_{ij} - y_{ij}\})$$

and

$$\Delta_2 = \min_{(i,j) \in C_{21}}\{s_{ij}\} \qquad \text{where } s_{ij} = t_j - (t_i + y_{ij})$$

where s_{ij} represents the amount of slack for activity (i,j). Let $\Delta = \min(\Delta_1, \Delta_2)$.

Step 4:   Let $X, \overline{X}$ denote the set of labeled and unlabeled nodes leading to the nonbreakthrough condition, respectively. Update node realization times and activity durations as follows:

Set $t_j = t_j - \Delta$ for all $j \in \overline{X}$.
Set $y_{ij} = y_{ij} - \Delta$ for all $(i,j)C_{11}$
Set $y_{ij} = y_{ij} + \Delta$ for all $(i,j)C_{12}$

Update A' and c_{ij} for all $(i,j) \in A'$ according to the c_{ij} scale equation presented earlier. Return to step 3 with the updated flow network.

The procedure terminates when step 3 labels node n with $q_n = \infty$. At this point $T = T_{\min}$.

An application of the procedure to the sample network from Chapter 4 is as follows. Figure 7–5 is an AOA representation of Figure 4-2. The critical arcs are shown in Figure 7–6 by heavy arrows.

From Figure 7–6, we have $\mathbf{A'} = \{(1,2),(1,3),(3,4),(4,5)\}$. Figure 7–7 shows the flow network where $c_{ij} = a_{ij}$ for $(i,j) \in \mathbf{A'}$ with $y_{ij} = u_{ij}$. For arc$(1,2)$, $c_{12} = \infty$ since $y_{12} = u_{12} = d_{12}$. For $(i,j) \in \mathbf{A} - \mathbf{A'}$, $c_{ij} = 0$. The only path through which the flow can be augmented is 1-3-4-5 with the maximum flow of 25 units. Figure 7–8 performs labeling to detect this path. Figure 7–9 shows the updated flow values. The labeling in Figure 7–10 results in a nonbreakthrough, with $\mathbf{X} = \{$labeled nodes$\} = \{1,2,3,4\}, = \overline{\mathbf{X}}\{5\}$ and $\mathbf{C} = \{(2,5),(4,5)\}$.

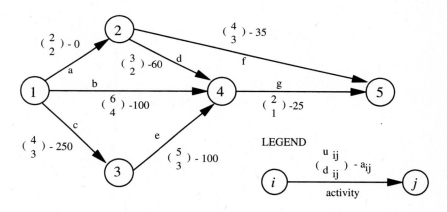

Figure 7–5. AOA Representation of Figure 4-2

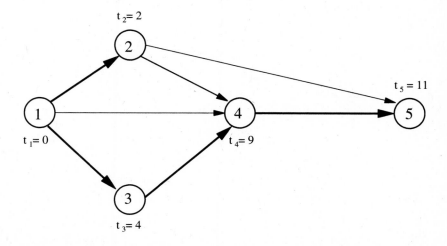

Figure 7–6. The Critical Subnetwork for $T = 11$

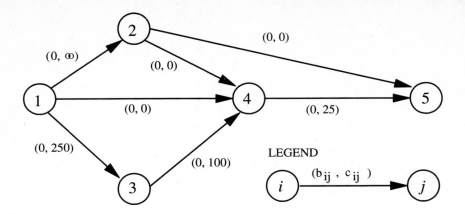

Figure 7–7. The Flow Network

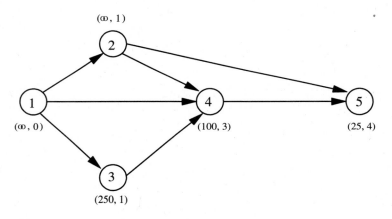

Figure 7–8. Maximum Flow Iteration 1

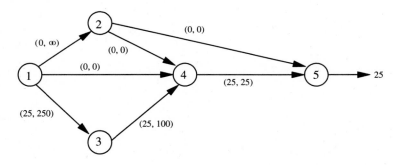

Figure 7–9. Maximum Flow Iteration 2

294

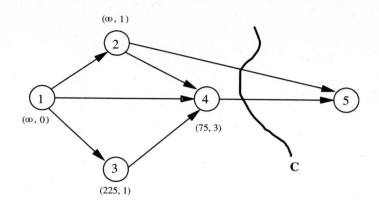

Figure 7–10.   Maximum Flow Iteration 3

Next, Δ is calculated as follows:
$$\Delta_2 = t_5 - t_2 - y_{25} = 11 - 2 - 4 = 5$$
$$\Delta_1 = y_{45} - d_{45} = 2 - 1 = 1$$

Therefore, $\Delta = min(\Delta_1, \Delta_2) = \Delta_1 = 1$. Reducing t_j by one unit for all $j \in \overline{\mathbf{X}}$, one gets $t_5 = 10$. The duration of $(4, 5)$ reaches its crash limit. The cost of crashing the project duration from 11 to 10 is $(a_{45}) \cdot (11 - 10) = \$25$. The critical path remains the same as shown in Figure 7–11. The flow network of Figure 7–12 reflects the current flow values and the updated c_{ij} values. $c_{45} = \infty$ since $y_{45} = d_{45}$. The only existing augmenting path is 1-3-4-5 through which 75 units of additional flow are augmented. Further labeling results in nonbreakthrough with $\mathbf{X} = \{1, 2, 3\}, \overline{\mathbf{X}} = \{4, 5\}$ and $\mathbf{C} = \{(1, 4), (2, 5), (2, 4), (3, 4)\}$. Figure 7–13 summarizes the steps of the maximum flow procedure. The Δ values are calculated as shown below:

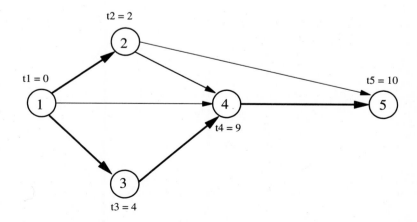

Figure 7–11.   The Critical Path Subnetwork for $T = 10$

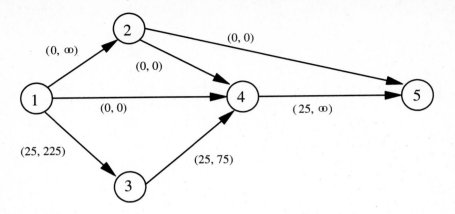

Figure 7–12.   The Flow Network for Computational Example

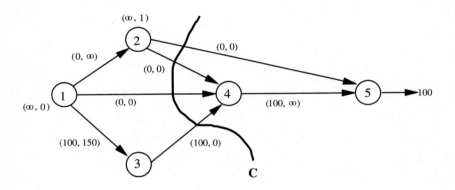

Figure 7–13.   The Maximum Flow Solution

$$\Delta_2 = \min\{(t_4 - t_1 - y_{14}), (t_5 - t_2 - y_{25}), (t_4 - t_2 - y_{24})\}$$
$$= \min\{(9 - 0 - 6), (10 - 2 - 4), (9 - 2 - 3)\} = 3$$
$$\Delta_1 = y_{34} - d_{34} = 5 - 3 = 2$$
$$\Delta = \min(\Delta_1, \Delta_2) = 2$$

Therefore, the project duration can be reduced another 2 units by crashing the duration of activity $(3, 4)$ to its crash duration. The slacks on arcs $(1, 4)$, $(2, 5)$ and $(2, 4)$ will be reduced by 2 units. Figure 7–14 contains the critical subnetwork and the new node realization times.

T is reduced from 10 to 8 at a cost of $a_{34}(\Delta) = \$200$. The updated flow network shown in Figure 7–15 indicates that a flow of 150 units can be augmented through the path 1-3-4-5. Further labeling results in a nonbreakthrough as shown in Figure 7–16. $\Delta = 1$, which results in $y_{13} = d_{13}$ and activity $(1, 4)$ joining the critical

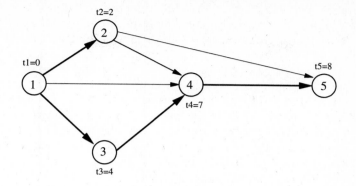

Figure 7–14. The Maximum Flow Solution

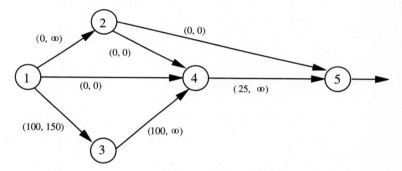

Figure 7–15. Network Flow Graph for Time-Cost Trade-Off Example

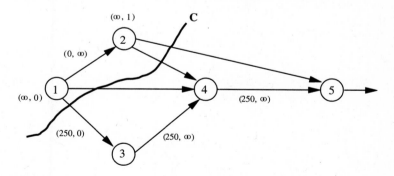

Figure 7–16. Maximum Flow Solution for Time-Cost Trade-Off Example

subgraph. Figures 7-17 and 7-18 give the expanded critical subgraph and the corresponding flow network, respectively. Labeling results in node 5 labeled with, which indicates that a critical path has reached its crash limit. Therefore, the procedure stops with $T = T_{min} = 7$.

The complete time-cost trade-off curve is given by Figure 7–19. The figure only illustrates the crashing cost as a function of project duration. In general, one needs to also consider the indirect cost which decreases as the project duration is decreased. The optimal project duration, T^* is the one which minimizes the total crashing cost and the indirect cost. This is illustrated in Figure 7–20.

If the crash duration of activity c was 2 time units, then we would have $c_{13} = 0$ in Figure 7–18, which would result in flow augmentation along 1–4–5 of 100 units. The maximum flow value of 350 units will indicate that further decrease in project duration will cost \$350 per time unit with activities b and c being the cheapest set of activities to crash.

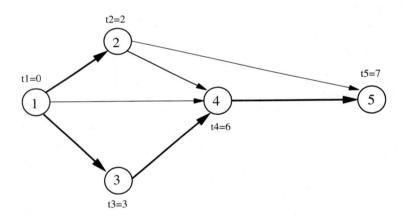

Figure 7–17.   Expanded Critical Subnetwork for Time-Cost Trade-Off Example

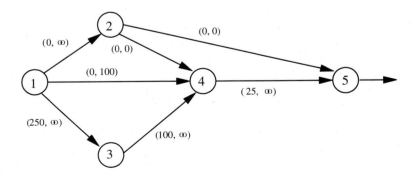

Figure 7–18.   Network Flow Graph for Time-Cost Trade-Off Example

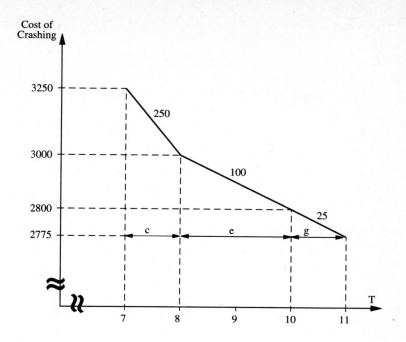

Figure 7–19. Time-Cost Trade-Off Curve

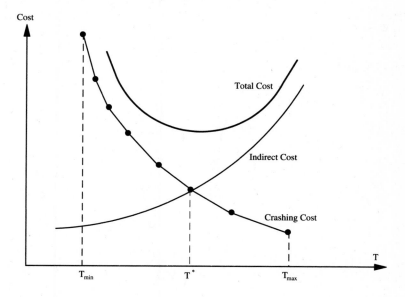

Figure 7–20. The Total Project Cost Curve

299

7.9 SENSITIVITY ANALYSIS FOR TIME-COST TRADE-OFF

In this section, we discuss the sensitivity analysis on the cost parameters for the time-cost trade-off problem. The cost of crashing an activity time, say for activity k by one unit, a_k, is generally estimated and is hence subject to error. Therefore, the project manager would like to know how sensitive the solution is to the estimated cost parameters. Consider the critical path subnetwork in Figure 7–21. Four cut-sets separate node 1 from node 4. Suppose the original costs were $a_a = 12$, $a_b = 6$, $a_c = 12$, and $a_d = 16$. The cheapest cutset is C_1 with the total cost of crashing being equal to $15. Suppose the manager has found out that the crashing costs for activities a and b were underestimated by 20 percent. Therefore, $a'_a = 12$, $a'_b = 6$, $a_c = 12$, and $a_d = 16$. C_1 is no longer the cheapest cutset. Cutset C_4 has crashing cost of $16 per time unit and hence is the cheapest cutset.

Clearly, if an activity has not been critical for $T_{min} \leq T \leq T_{max}$, then its cost will not be relevant for the time-cost trade-off problem. However, if an activity (i,j) has been added to the critical subnetwork at project completion time T_m, then we need to reevaluate the crashing decisions concerning $T_m \leq T \leq T_{max}$. Suppose the cost of activity (i,j) was overestimated, that is, the modified cost $a'_{ij} < a_{ij}$. This may affect the decisions over the period $[T_m, T_{max}]$, even though activity (i,j) was actually crashed when $T = T'$ where $T_m < T' < T_{max}$. However, if $a'_{ij} > a_{ij}$, then one needs to only reevaluate the decisions over the period $[T', T_{max}]$ even though activity (i,j) became critical at T_m. For the example given in Figure 7–5, if $a_c = 275$, then the reevaluations should start at Figure 7–15 with the modification C_{13}(residual capacity) = 175 instead of 150.

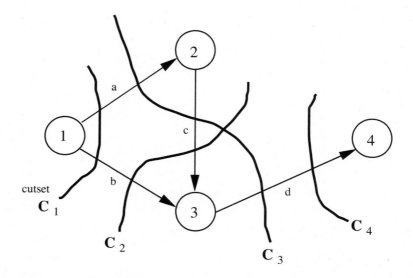

Figure 7–21.   Critical Path Subnetwork for Sensitivity Analysis

.10 THE KNAPSACK PROBLEM

The *knapsack problem* is a famous operations research problem dealing with resource allocation. The general nature of the problem is as follows: Suppose that n items are being considered for inclusion in a travel supply bag (knapsack). Each item has a certain per-unit value to the traveler. Each item also has a per-unit weight that contributes to the overall weight of the knapsack. There is a limitation on the total weight that the traveler can carry. Figure 7–22 shows a representation of the problem. The objective is to maximize the total value of the knapsack subject to the total weight limitation.

The mathematical formulation is for $j = 1, \ldots, n$, let $c_j > 0$ be the value per unit for item j. Let $w_j > 0$ be the weight per unit of item j. If the total weight limitation is W, then the problem of maximizing the total value of the knapsack is

$$
\begin{aligned}
\text{Maximize:} \quad & z = \sum_{j=1}^{n} c_j x_j \\
\text{Subject to:} \quad & \sum_{j=1}^{n} w_j x_j \leq W \\
& x_j \geq 0 \text{ integer}, \qquad j = 1, \ldots, n,
\end{aligned}
$$

where x_j is the number of units of item j included in the knapsack. All data in the knapsack formulation are assumed to be integers. The knapsack problem can be solved by any of the available general integer linear programming (ILP) algorithms. However,

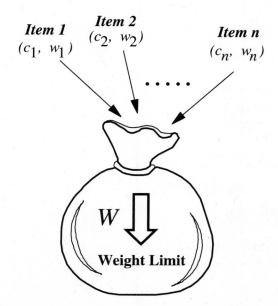

Figure 7–22. The Knapsack Problem

because it has only one constraint, more efficient solution algorithms are possible. For very simple models, the problem can be solved by inspection. For some special problems, such as project scheduling, it may be necessary to solve a large number of knapsack problems sequentially. Even though each knapsack formulation and solution may be simple, the overall problem structure may still be very complex.

7.11 KNAPSACK FORMULATION FOR SCHEDULING

The knapsack problem formulation can be applied to a project scheduling problem that is subject to resource constraints. In this case, each activity to be scheduled at a specific instant is modeled as an item to be included in the knapsack. The composition of the activities in a scheduling window is viewed as the knapsack. Figure 7–23 presents a representation of a scheduling window with its composition of scheduled activities at a given instant.

One important aspect of the Knapsack formulation for activity scheduling is that only one unit of each activity (item) can be included in the schedule at any given scheduling time. This is because the same activity cannot be scheduled more than once at the same time. However, in certain applications, such as concurrent engineering, it may be desired to have multiple executions of the same activity concurrently. For such cases, each execution of the activity may be modeled as a separate entity with one additional constraint to force concurrent execution. The formulation of the knapsack problem for activity scheduling is done at each and every scheduling time t. The objective

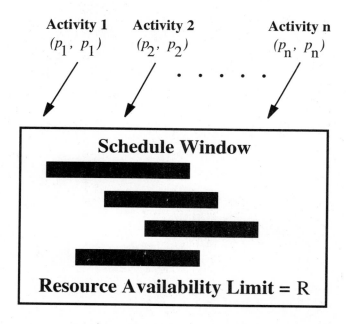

Figure 7–23. The Knapsack Formulation of Project Scheduling

is to schedule as many activities of high priority as possible while satisfying activity precedence relationships without exceeding the resource availability limitations. The activity precedence requirements are used to determine the activities that are eligible for scheduling at a given scheduling time.

For $j = 1, \ldots, N$ activities, let $p_j > 0$ be the priority value for activity j. The priority value is used to prioritize activities for resource allocation when activities compete for limited resources. It is assumed that the priority values for the activities do not change during the scheduling process. This is referred to as a *fixed prioritization* of activities. If the activity priority values can change depending on the state and time of the scheduling problem, then the activity prioritization is referred to as a *variable prioritization* of activities. Let $r_{ij} \geq 0$ be the number of units of resource type i (weight) required by activity j. It is also assumed that the resource requirements do not change during the scheduling process. Let R_{it} be the limit on the units of resource type i available at time t. The formulation of the scheduling problem for time t is as follows:

$$
\begin{aligned}
\text{Maximize:} \quad & z_t = \sum_{j \in S_t} p_j x_{jt} \\
\text{Subject to:} \quad & \sum_{j \in S_t} r_{it} x_{jt} \leq R_{it}, \qquad i = 1, \ldots, k \\
& x_{jt} = 0 \quad \text{or} \quad 1, \qquad j \in S_t
\end{aligned}
$$

where

z_t = overall performance measure of the schedule generated at time t

p_j = the priority measure (value) for activity j

t = current time of scheduling

x_{jt} = indicator variable specifying whether or not activity j is scheduled at time t

S_t = set of activities eligible for scheduling at time t

k = number of different resource types

r_{ij} = units of resource type i required by activity j

R_{it} = units of resource type i available at time t

The next scheduling time, t, for the knapsack problem is determined as the minimum of the finishing times of the scheduled and unfinished activities.

7.12 KNAPSACK SCHEDULING EXAMPLE

Suppose we have a project consisting of seven activities: activities A, B, C, D, E, F, and G, which are labeled 1 through 7.

The activity durations are specified as d_j :

$$d_1 = 2.17, d_2 = 6, d_3 = 3.83, d_4 = 2.83, d_5 = 5.17, d_6 = 4, d_7 = 2$$

The scaled priority values of the activities are specified as p_j :

$$p_1 = 55.4, p_2 = 100, p_3 = 72.6, p_4 = 54, p_5 = 88, p_6 = 66.6, p_7 = 75.3$$

There are two resource types with the following availability levels at time t, R_{jt}:

$$R_{1,0} = 10, R_{2,0} = 15$$

Units of *resource type* 1 required by the activities are specified as r_{1j} :

$$r_{11} = 3, r_{12} = 5, r_{13} = 4, r_{14} = 2, r_{15} = 4, r_{16} = 2, r_{17} = 6$$

Units of *resource type* 2 required by the activities are specified as r_{2j}:

$$r_{21} = 0, r_{22} = 4, r_{23} = 1, r_{24} = 0, r_{25} = 3, r_{26} = 7, r_{27} = 2$$

The precedence relationships between the activities are represented as $P(.) = $ {set of predecessors}:

$$P(A) = \emptyset, P(B) = \emptyset, P(C) = \emptyset, P(D) = \{A\}, P(E) = \{C\}, P(F) = \{A\},$$
$$P(G) = \{B, D, E\}$$

The successive Knapsack formulations and solutions for the problem are presented below.

$$t = 0:$$
$$\mathbf{S}_0 = \{A, B, C\}$$
Maximize: $z_0 = 55.4x_A + 100x_B + 72.6x_C$
Subject to: $3x_A + 5x_B + 4x_C \leq 10$
                 $0x_A + 4x_B + 1x_C \leq 15$
                 $x_A, x_B, x_C = 0 \text{ or } 1$

whose solution yields $x_A = 0, x_B = 1, x_C = 1$. Thus, activity B is scheduled at $t = 0$ and it will finish at $t = 6$; C is scheduled at $t = 0$ and it will finish at $t = 3.83$. The next scheduling time is $t = \text{Min}\{3.83, 6\} = 3.83$.

$$t = 3.83:$$
$$\mathbf{S}_{3.83} = \{A, E\}$$
Maximize: $z_{3.83} = 55.4x_A + 88.0x_E$
Subject to: $3x_A + 4x_E \leq 5$
                 $0x_A + 3x_E \leq 11$
                 $x_A, x_E = 0 \text{ or } 1$

whose solution yields $x_A = 0, x_E = 1$. Thus, activity E is scheduled at $t = 3.83$ and it will finish at $t = 9$. The next scheduling time is $t = \text{Min}\{6, 9\} = 6$.

$$t = 6:$$
$$\mathbf{S}_6 = \{A\}$$
Maximize: $z_6 = 55.4x_A$
Subject to: $3x_A \leq 6$
                 $x_A = 0 \text{ or } 1$

whose solution yields $x_A = 1$. Thus, activity A is scheduled at $t = 6$ and it will finish at $t = 8.17$. The next scheduling time is $t = \text{Min}\{8.17, 9\} = 8.17$.

$t = 8.17$:
$$\mathbf{S}_{8.17} = \{D, F\}$$
Maximize: $z_{8.17} = 54x_D + 66.6x_F$
Subject to: $2x_D + 2x_F \leq 6$
$\quad\quad\quad\quad\quad 0x_D + 7x_F \leq 8$
$\quad\quad\quad\quad\quad x_D, x_F = 0 \text{ or } 1$

whose solution yields $x_D = 1, x_F = 1$. Thus, activities D and F are scheduled at $t = 8.17$. D will finish at $t = 11$. F will finish at $t = 12.17$. The next scheduling time is $t = \text{Min}\{9, 11, 12.17\} = 9$.

$$t = 9:$$
$$\mathbf{S}_9 = \varnothing$$

No activity is eligible for scheduling at $t = 9$, since G cannot be scheduled until D finishes. The next scheduling time is $t = \text{Min}\{11, 12.17\} = 11$.

$t = 11$:
$$\mathbf{S}_{11} = \{G\}$$
Maximize: $z_9 = 75.3x_G$
Subject to: $6x_G \leq 8$
$\quad\quad\quad\quad\quad 2x_G \leq 8$
$\quad\quad\quad\quad\quad x_G = 0 \text{ or } 1$

whose solution yields $x_G = 1$. Thus, activity G is scheduled at $t = 11$ and it will finish at $t = 13$. All activities have been scheduled. The project completion time is $t = \text{Max}\{12.17, 13\} = 13$. Figure 7–24 shows the Gantt chart of the completed schedule.

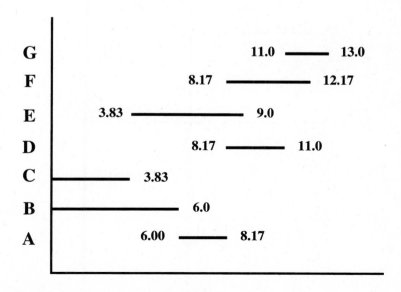

Figure 7–24. Gantt Chart of Knapsack Schedule

7.13 TRANSPORTATION PROBLEM

The *transportation problem* is a special class of optimization problem dealing with the distribution of items from sources of supply to locations of demand. This type of problem can occur in large or multiple project scheduling environments where supplies must be delivered to various project sites in a coordinated fashion to meet scheduling requirements. A specific algorithm, known as the *transportation method*, is available for solving transportation problems that satisfy the following assumptions:

1. The problem must concern a single product type to be transported.
2. There must be several *sources* from which the product is to be transported.
3. The amount of *supply* available at each source must be known.
4. There must be several *destinations* to which the product is to be transported.
5. The *demand* at each destination must be known.
6. The per-unit cost of transporting the product from any source to any destination must be known.

The general format of the transportation problem, referred to as the *transportation tableau*, is presented in Figure 7–25.

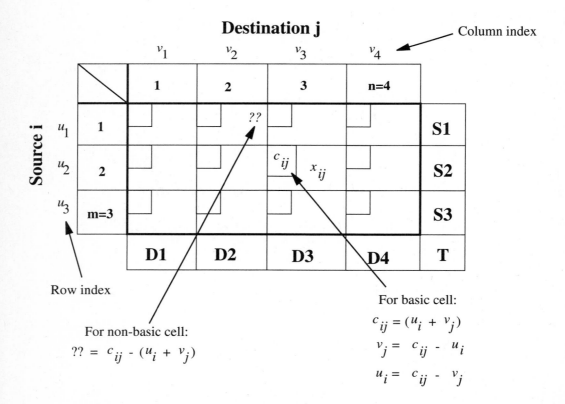

Figure 7–25. General Layout of the Transportation Problem

The objective of the transportation method is to determine the minimum cost plan to transport units from origins to destinations while satisfying all supply limitations and demand requirements. The following notation is used:

$Srci$ = source $i, i = 1, 2, \ldots, m$
$Desj$ = destination $j, j = 1, 2, \ldots, n$
S_i = available supply from source $i, i = 1, 2, \ldots, m$
D_j = total demand for destination $j, j = 1, 2, \ldots, n$
c_{ij} = cost of transporting one unit of the product from source i to destination j
x_{ij} = number of units of the product transported from source i to destination j.

A linear programming formulation of the transportation problem is presented next.

$$\text{Minimize:} \quad z = \sum_{i=1}^{m} \sum_{j=1}^{m} c_{ij} x_{ij}$$

$$\text{Subject to:} \quad \sum_{j=1}^{n} x_{ij} \leq S_i, \qquad i = 1, 2, \ldots, m$$

$$\sum_{i=1}^{n} x_{ij} \geq D_j, \qquad j = 1, 2, \ldots, n$$

$$x_{ij} \geq 0$$

The first set of constraints indicates that the sum of the shipments from a source cannot exceed its supply. The second set of constraints indicates that the sum of shipments to a destination must satisfy its demand. The above LP model implies that the total supply must at least equal the total demand. Although it is possible to model and solve a transportation problem as a linear programming problem, the special structure of the problem makes it possible to use the *transportation method* as a more efficient solution method. Like the LP simplex method, the transportation method is an iterative solution procedure, moving from one trial solution to another less costly trial solution. The iterative solution continues until the optimal solution is reached. Unlike the simplex method, however, the transportation method is limited to only certain types of problems.

7.13.1 Balanced versus Unbalanced Transportation Problems

A transportation problem is said to be *balanced* if the total available supply is equal to the total demand. For a balanced problem, the inequality constraints in the model will be modified to equality constraints. That is,

$$\sum_{j=1}^{n} x_{ij} = S_i, \qquad i = 1, 2, \ldots, m$$

$$\sum_{j=1}^{m} x_{ij} = D_j, \qquad i = 1, 2, \ldots, n$$

A transportation problem is *unbalanced* when total supply is not equal to total demand. Unbalanced transportation problems are more realistic in practical situations. A modification of the initial transportation tableau is required for unbalanced transportation problems. If total supply is greater than total demand, we will add one extra *dummy destination* to the initial tableau. The dummy destination is assigned a demand, D_{n+1}, that is equal to total available supply minus the total demand: That is,

$$D_{n+1} = \sum_{i=1}^{m} S_i - \sum_{j=1}^{n} D_j$$

The dummy destination serves to balance the problem. Allocations to cells associated with the dummy destination are simply not shipped from the sources. These unshipped units represent excess supply. The magnitude of the excess supply can be used for making capacity adjustment decisions at the sources if no "real" destinations can be found to absorb the excess units.

If total supply is less than total demand, then some of the demand cannot be fulfilled. In such a case, a *dummy source* is included in the formulation. Other options are to reduce demand levels or increase supply capacities. The supply from the dummy source may be viewed as subcontracted supply. If supply and demand levels are fixed, then the supply to be subcontracted out, S_{m+1}, will be

$$S_{m+1} = \sum_{j=1}^{n} D_j - \sum_{i=1}^{m} S_i$$

The solution to the transportation problem will always be integer. Hence, no integer restrictions need to be specified.

7.13.2 Initial Solution to the Transportation Problem

To use the transportation method, we must first identify an *initial feasible solution*. The *northwest-corner technique* is one way to obtain an initial solution to a transportation problem. The technique makes the first allocation to the northwest-corner cell in the transportation tableau. As much allocation as possible is made to that northwest-corner cell. This technique is a simple method to determine a *feasible solution* to the transportation problem. The feasible solution satisfies all supply and demand restrictions. The initial solution may not be optimal, but it serves as a starting point for procedures that are designed to find the optimal solution. A summary of the northwest-corner technique is presented below.

NORTHWEST-CORNER TECHNIQUE

Step 1: Set up tableau and enter the S_i, D_j, and c_{ij} values.
        a. Allocate $u_{11} = \text{Min}\{S_1, D_1\}$.
        b. Update supply and demand: $S_1 = S_1 - u_{11}$ and $D_1 = D_1 - u_{11}$. Set $i = 1, j = 1$.

Step 2: Move to a new cell.
        a. If current $S_i = 0$, move down one cell.
        b. If current $D_j = 0$, move one cell to the right. If both S_i and D_j are zero, perform step 2a or 2b only.

Step 3: Make allocation to the new cell ij
        a. Allocate $x_{ij} = \text{Min}\{S_i, D_j\}$.
        b. Update supply and demand: $S_i = S_i - x_{ij}$ and $D_j = D_j - x_{ij}$.

Step 4: Repeat steps 2 and 3 until all available supply and all demand have been allocated to cells within the initial tableau, that is, until $S_i = 0$ and $D_j = 0 (i = 1, 2, \ldots, m; j = 1, 2, \ldots, n)$.

The northwest-corner method, although very simple and fast, usually does not yield a good initial solution because it totally ignores costs. Better initial approximation methods are available. These other methods include the *least-cost method, Vogel's approximation method (VAM)*, and *Russell's approximation method* (Taha 1982; Hillier and Lieberman 1974).

Least-cost method.   The least-cost method uses the cell costs as the basis for making allocations to the cells in the transportation tableau.

Step 1: Assign as much as possible to the variable with the smallest unit cost in the tableau. Break ties arbitrarily. Cross out the satisfied row or column.

Step 2: If both a row and a column are satisfied simultaneously, only one may be crossed out. Adjust the supply and demand for all uncrossed rows and columns.

Step 3: Repeat the process by assigning as much as possible to the cell with the smallest uncrossed unit cost.

Step 4: The procedure is complete when all the rows and columns are crossed out

Vogel's approximation.   Vogel's approximation (Reinfeld and Vogel 1958) is better than the northwest-corner method. In fact, the method usually yields an optimal or close to optimal starting solution to the transportation problem. The steps of the method are as follows:

Step 1: Set up the transportation tableau and evaluate the row or column penalty.
        a. For each row, subtract the smallest cost element from the next smallest cost element in the same row.

b. For each column subtract the smallest cost element from the next smallest cost element in the same column.

Step 2: Identify the row or column with the largest penalty (break ties arbitrarily).

    a. Allocate as much as possible to the cell with the least cost in the selected row or column. That is, assign the smaller of row supply and column demand.

    b. If there is only one cell remaining in a row or column, choose the cell and allocate as many units as possible to it.

    c. Adjust the supply and demand and cross out the satisfied row or column.

Step 3: Determine satisfied rows or columns.

    a. If a row supply becomes zero, cross out the row and calculate the new column penalties. The other row penalties are not affected.

    b. If a column demand becomes zero, cross out the column and calculate new row penalties.

    c. If both a row supply and a column demand become zero at the same time, cross out only one of them. The one remaining will have a supply or demand of zero, which means an assignment of zero units in a subsequent step.

Step 4: Repeat steps 2 and 3 until an initial basic feasible solution has been obtained.

Russell's approximation method. Russell's approximation method (Russell 1969) is another approach to obtaining an initial solution for the transportation problem. The steps of the method are as follows:

Step 1: For each source row i remaining under consideration, determine u_i as $u_i = \text{Max}\{c_{ij}\}$ still remaining in that row. For each destination column j remaining under consideration, determine v_j as $v_j = \text{Max}\{c_{ij}\}$ still remaining in that column.

Step 2: For each cell not previously selected in the rows and columns, calculate $(c_{ij} - u_i - v_j)$.

Step 3: Select the cell having the largest negative value of $(c_{ij} - u_i - v_j)$. Break ties arbitrarily. Assign as many units as possible to the cell.

Step 4: Repeat steps 1 to 3 until an initial solution is obtained.

7.13.3 The Transportation Algorithm

The transportation algorithm generates an optimal solution to the transportation problem. The algorithm gets to the optimal solution by improving on an initial feasible solution. The steps of the algorithm are summarized below.

Step 1: Find an initial basic feasible solution.

Step 2: Define a row index, u_i, and a column index, v_j. This is called the MODI (modified distribution) or the uv method. There are m of the $u_i (i = 1, 2, \ldots, m)$ and n of the $v_j (j = 1, 2, \ldots, n)$. The indices are composed by solving the equations $c_{ij} = u_i + v_j$ for the basic cells. A basic cell is the cell where x_{ij} values were allocated. A basic cell may have $x_{ij} = 0$ in the case of degeneracy. This happens when an allocation satisfies the supply and demand simultaneously. There must be $m + n - 1$ basic cells. Thus, there are $m + n - 1$ equations in $m + n$ unknowns. This means that we can arbitrarily assign any value to one of the row or column indices (e.g., $u_1 = 0$), and solve the equations simultaneously for u_i and v_j values as explained in step 3.

Step 3:   After setting $u_1 = 0$, solve for the remaining indices by using the relationship $u_i + v_j = c_{ij}$ iteratively for each basic cell. Each time one index is computed, there will be another one that is immediately defined. Place the value of each index in the appropriate cell in the tableau.

Step 4:   Calculate the reduced cost of each nonbasic cell as follows:

$$c_{ij} - z_{ij} = c_{ij} - (u_i + v_j) = c_{ij} - u_i - v_j$$

Circle the $c_{ij} - (u_i + v_j)$ values in the nonbasic cells. The $c_{ij} - (u_i + v_j)$ value represents the reduced cost for the nonbasic cell. Because this is a cost minimization problem, the current basis is optimal if

$$c_{ij} - z_{ij} = c_{ij} - u_i - v_j \geq 0$$

for all nonbasic variables. Stop if all $c_{ij} - z_{ij} \leq 0$ which indicates the optimal solution.

Step 5:   Select the nonbasic variable with the most negative reduced cost to enter the basis. That is, select x_{ij} to enter the basis so that

$$c_{ij} - z_{ij} = \text{Min}_{pq}\{c_{pq} - z_{pq}\}$$

Step 6:   Use the *stepping-stone method* to determine the loop formed by the nonbasic cell (ij) and the basic cells. The stepping-stone method steps through cells in a closed loop path in the tableau based on the following rules:

a. Only one nonbasic variable (i.e., the entering variable) is included in the stepping-stone path.

b. The path is unique.

c. Start by placing a plus sign in the cell of the entering variable. The plus sign indicates that this variable will increase.

d. Place a minus sign in the cell of a basic variable so that the column or row of the incoming variable is balanced with one plus sign and one minus sign. The minus sign indicates that the cell will decrease by one unit for every unit increase in the entering variable.

e. Continue the assignment of plus and minus signs in the basic cells in such a way that each corner is a basic cell except one corner, and each vertical step is followed by a horizontal step with the signs alternated until the loop is closed. An example is illustrated later (see Figure 7–29).

    f. The stepping-stone path ends when both the row and the column of the entering variable are balanced. Basic variables that are not in the path contain no sign and will not change as the entering variable increases.

Step 7: Determine the exiting variable. This is the basic variable containing a minus sign and having the smallest x_{ij} value. Set $K = \text{Min}_{ij}\{x_{ij} \mid ij$ th cell sign is minus$\}$.

Step 8: Update the transportation tableau as follows:
    a. Add K units to each cell that has a plus sign in the closed path.
    b. Subtract K units from each that has a minus sign in the closed path.

Step 9: Return to step 2.

7.13.4 Example of Transportation Problem.

Suppose four project sites are to be supplied with units of a certain product from three different sources. It is desired to minimize the total cost of supplying the sites with the required units. The supply and demand data and the shipment costs are summarized as follows:

$$S_1 = 30, \quad S_2 = 50, \quad S_3 = 40$$
$$D_1 = 30, \quad D_2 = 20, \quad D_3 = 40, \quad D_4 = 30$$
$$c_{11} = \$5, \quad c_{12} = \$8, \quad c_{13} = \$3, \quad c_{14} = \$6$$
$$c_{21} = \$4, \quad c_{22} = \$5, \quad c_{23} = \$7, \quad c_{24} = \$4$$
$$c_{31} = \$6, \quad c_{32} = \$2, \quad c_{33} = \$4, \quad c_{34} = \$5$$

    Figure 7–26 shows the layout of the problem. This is a balanced transportation problem. The initial feasible solution obtained through the northwest-corner method is shown in Figure 7–27.

    There are $6 = m + n - 1$ basic cells, one of which $x_{12} = 0$. This is because when 30 units were allocated to x_{11}, it satisfied both S_1 and D_1. Crossing row 1, the northwest corner became cell(2,1), hence, 0 units were allocated to x_{12} and column 1 crossed out. The procedure continued with the next northwest corner cell(2,2). Let $u_2 = 0$. Determine

Destinations

Sources	1	2	3	4	
1	5	8	3	6	30
2	4	5	7	4	50
3	6	2	4	5	40
	30	20	40	30	120

Figure 7–26. Cost Matrix for Transportation Problem Example

Destinations

		1	2	3	4	
Sources	1	5 30	8	3	6	30
	2	4	5 20	7 30	4	50
	3	6	2	4 10	5 30	40
		30	20	40	30	120

Figure 7–27. Tableau 1 for Transportation Example

the remaining u_i and v_j values for the basic cells using the relationship $c_{ij} = (u_i + v_j)$. Using the u_i and v_j already available, compute $c_{ij} - (u_i + v_j)$ for each of the nonbasic cells. The resulting tableau is shown in Figure 7–28.

If all the $c_{ij} - (u_i + v_j)$ values for the nonbasic cells were positive, the solution would be optimal. Cell (1,3) has the most negative value, so it enters the solution. This is shown in Figure 7–29. Note that cell (2, 1) is treated as a basic cell since an *allocation* of 0 has been made to it. Allocate as much as possible to the entering cell, cell (1, 3), without violating the supply and demand constraints. The new solution is shown in Figure 7–30. The tableau generated by a return to step 2 of the algorithm is shown in Figure 7–31.

The optimal solution has not been reached. Therefore, continue with the algorithm steps. Figures 7–32 through 7–35 show the subsequent tableaus of the solution. Figure 7–36 shows the final tableau and the optimal solution.

		$v1 = 4$	$v2 = 5$	$v3 = 7$	$v4 = 8$	
		1	2	3	4	
$u1 = 1$	1	5 30	8 2	3 -5	6 -3	30
$u2 = 0$	2	4 0	5 20	7 30	4 -4	50
$u3 = -3$	3	6 5	2 0	4 10	5 30	40
		30	20	40	30	120

Figure 7–28. Tableau 2 for Transportation Example

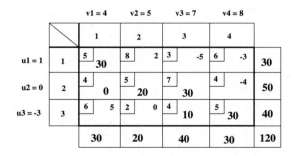

		1	2	3	4	
1		5 30 -K	8 2	3 -5 +K	6 -3	30
2		4 +K 0	5 20	7 -K 30	4 -4	50
3		6 5	2 0	4 10	5 30	40
		30	20	40	30	120

Figure 7–29. Tableau 3 for Transportation Example

Figure 7–30. Solution from Tableau 3 for Transportation Example

	1	2	3	4	
1	5 0	8	3 30	6	30
2	4 30	5 20	7	4	50
3	6	2	4 10	5 30	40
	30	20	40	30	120

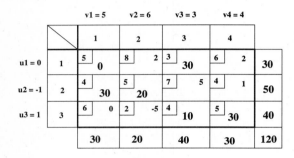

Figure 7–31. Tableau 4 for Transportation Example

	$v1 = 5$	$v2 = 6$	$v3 = 3$	$v4 = 4$	
	1	2	3	4	
$u1 = 0$ 1	5 0	8 2	3 30	6 2	30
$u2 = -1$ 2	4 30	5 20	7 5	4 1	50
$u3 = 1$ 3	6 0	2 -5	4 10	5 30	40
	30	20	40	30	120

Figure 7–32. Tableau 5 for Transportation Example

	1	2	3	4	
1	5 0 -K	8	3 30 +K	6	30
2	4 +K 30	5 -K 20	7	4	50
3	6	2 +K	4 10 -K	5 30	40
	30	20	40	30	120

Figure 7–33. Tableau 6 for Transportation Example

	1	2	3	4	
1	5	8	3 30	6	30
2	4 30	5 20	7	4	50
3	6	2 0	4 10	5 30	40
	30	20	40	30	120

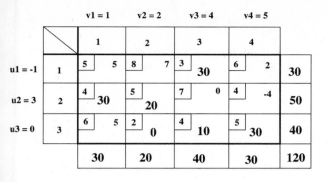

Figure 7–34. Tableau 7 for Transportation Example

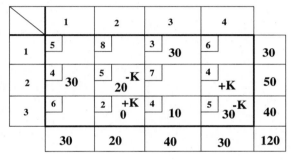

Figure 7–35. Tableau 8 for Transportation Example

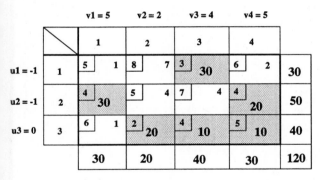

Figure 7–36. Final Tableau for Transportation Example

The solution indicates that we should ship 30 units from source 1 to destination 3; 30 units from source 2 to destination 1; 20 units from source 2 to destination 4; 20 units from source 3 to destination 2; 10 units from source 3 to destination 3; and 10 units from source 3 to destination 4. The minimum cost is

$$z = 3(30) + 4(30) + 4(20) + 2(20) + 4(10) + 5(10) = 420$$

7.13.5 Transshipment Formulation

The transshipment problem is a general model of the transportation problem. In this model, there can be *pure sources, pure destinations,* and *transshipment points* which can serve as both sources and destinations. It is possible for any source or destination to ship to any other source or destination. Thus, there may be many different ways of shipping from point i to point j in addition to the direct route. In the transportation problem, the way in which units are distributed from source i to destination j must be known in advance so that the corresponding cost per unit, c_{ij}, can be determined ahead of time. In the transshipment problem, units may go through intermediate points that offer lower total shipment cost. For example, instead of shipping units directly from source 2 to destination 3, it may be cheaper to include the units going to destination 3 with the units going to destination 4 and then ship those units from destination 4, which now serves as a source, to destination 3. A mathematical formulation of the transshipment problem is as follows. Let

x_{ij} = amount shipped from point i to point j, $i, j = 1, 2, \ldots, n$; $i \neq j$
c_{ij} = cost of shipping from point i to point j, $c_{ij} \geq 0$
r_i = net requirement at point i (negative for demand point, positive for supply point)

Each point must satisfy a balance equation stating that the amount shipped minus the amount received equals the net requirement at the point. It is also required that total demand equal total supply. Thus, we have

$$
\begin{aligned}
\text{Minimize:} \quad & z = \sum_{i=1}^{n} \sum_{j=1}^{n} c_{ij} x_{ij}, \qquad i \neq j \\
\text{Subject to:} \quad & \sum_{j=1, j \neq i}^{n} x_{ij} - \sum_{j=1, j \neq i}^{n} x_{ji} = r_i, \qquad i = 1, 2, \ldots, n \\
& \sum_{i=1}^{n} r_i = 0 \\
& x_{ij} \geq 0, \qquad j = 1, 2, \ldots, n; i \neq j
\end{aligned}
$$

7.14 ASSIGNMENT PROBLEM IN PROJECT OPTIMIZATION

Suppose there are n tasks which must be performed by n workers. The cost of worker i performing task j is c_{ij}. It is desired to assign workers to the tasks in a fashion that minimizes the cost of completing the tasks. This problem scenario is referred to as the *assignment problem*. The technique for finding the optimal solution to the problem is called the *assignment method*. Like the transportation method, the assignment method is an iterative procedure that arrives at the optimal solution by improving on a trial solution at each stage of the procedure.

CPM and PERT can be used in controlling projects to ensure that the project will be completed on time. As mentioned in Chapters 2 and 3, these two techniques do not consider the assignment of resources to the tasks that make up a project. The *assignment method* can be used to achieve an optimal assignment of resources to specific tasks in a project. Although the assignment method is cost-based, task duration can be incorporated into the modeling in terms of time-cost relationships. Of course, task precedence requirements and other scheduling restrictions will have to be accounted for in the final scheduling of the tasks. The objective is to minimize the total cost. Thus, the formulation of the assignment problem is as follows. Let

x_{ij} = 1 if worker i is assigned to task $j, i, j = 1, 2, \ldots, n$
x_{ij} = 0 if worker i is not assigned to task j
c_{ij} = cost of worker i performing task j

$$\text{Minimize:} \quad z = \sum_{i=1}^{n} \sum_{j=1}^{n} c_{ij} x_{ij}$$

$$\text{Subject to:} \quad \sum_{j=1}^{n} x_{ij} = 1, \qquad i = 1, 2, \ldots, n$$

$$\sum_{i=1}^{n} x_{ij} = 1, \qquad j = 1, 2, \ldots, n$$

$$x_{ij} \geq 0, \qquad i, j = 1, 2, \ldots, n$$

It can be seen that the above formulation is a transportation problem with $m = n$ and all supplies and demands equal to 1. Note that we have used the nonnegativity constraint, $x_{ij} \geq 0$ instead of the integer constraint, $x_{ij} = 0$ or 1. However, the solution of the model will still be integer-valued (Cooper and Steinberg 1974). Hence, the assignment problem is a special case of the transportation problem with $m = n$, $S_i = 1$, and $D_i = 1$ for all i. Conversely (Dantzig 1963), the transportation problem can also be viewed as a special case of the assignment problem. A transportation problem can be modeled as an assignment problem and vice versa. The basic requirements of an assignment problem are

1. There must be two or more tasks to be completed.
2. There must be two or more resources that can be assigned to the tasks.
3. The cost of using any of the resources to perform any of the tasks must be known.
4. Each resource is to be assigned to one and only one task.

If the number of tasks to be performed is greater than the number of workers available, we will need to add *dummy workers* to balance the problem. Similarly, if the number of workers is greater than the number of tasks, we will need to add *dummy tasks* to balance the problem. If there is no problem of overlapping, a worker's time may be split into segments so that the worker can be assigned more than one task. In this case, each segment of the worker's time will be modeled as a separate resource in the assignment problem. Thus, the assignment problem can be extended to consider partial allocation of resource units to multiple tasks.

Although the assignment problem can be formulated for and solved by the simplex method or the transportation method, a more efficient algorithm has been developed specifically for the assignment problem. The method, known as the *Hungarian method*

(Kuhn 1956), is a simple iterative technique. The method is based on the assignment theory developed by E. Egervary, a Hungarian mathematician. It is based on properties of matrices and relationships between primal and dual problems. Kuhn (1956) discusses several modifications of the Hungarian method. Further details on the method can also be found in Bazaraa and Jarvis (1977). The steps of the assignment method are summarized as follows:

Step 1: Develop the $n \times n$ cost matrix, in which rows represent workers and columns represent tasks.

Step 2: For each row of the cost matrix, subtract the smallest number in the row from each and every number in the row. For each column of the resulting matrix, subtract the smallest number from each and every number in the column. The resulting matrix is called the *matrix of reduced costs*.

Step 3: Find the minimum number of lines through the rows and columns of the reduced-cost matrix such that all zeros have a line through them. If the number of lines is n, the optimal solution has been reached; stop. Otherwise, go to step 4.

Step 4: Define a new reduced-cost matrix as follows: Determine the smallest number in the matrix which does not have a line through it. Subtract this number from all the unlined numbers and add it to all the numbers which have two lines through them (i.e., numbers located at the intersection of two lines in the matrix). Go to step 3.

Example of assignment problem. Suppose five workers are to be assigned to five tasks on the basis of the cost matrix presented in Figure 7–37. Let us use the algorithm steps presented above to solve this assignment problem.

For convenience, we can divide each cost element by 100. When the smallest number in each row of the resulting simplified matrix is subtracted from all the elements in the row, we obtain Figure 7–38. When the smallest number in each column of the figure is subtracted from all the elements in the column, we obtain Figure 7–39.

The minimum number of lines required to cover all the zeros is 3. This is shown in Figure 7–40. The smallest uncovered number in the figure is 1. When this 1 is subtracted

Tasks

		1	2	3	4	5
Worker	1	$200	$400	$500	$100	$400
	2	$400	$700	$800	$1,100	$700
	3	$300	$900	$800	$1,000	$500
	4	$100	$300	$500	$100	$400
	5	$700	$100	$200	$100	$200

Figure 7–37. Cost Matrix for Assignment Problem

Tasks

Worker	1	2	3	4	5
1	1	3	4	0	3
2	0	3	4	7	3
3	0	6	5	7	2
4	0	2	4	0	3
5	6	0	1	0	1

Figure 7–38. Tableau 1 for Assignment Problem

Tasks

Worker	1	2	3	4	5
1	1	3	3	0	2
2	0	3	3	7	2
3	0	6	4	7	1
4	0	2	3	0	2
5	6	0	0	0	0

Figure 7–39. Tableau 2 for Assignment Problem

Tasks

Worker	1	2	3	4	5
1	1	3	3	0	2
2	0	3	3	7	2
3	0	6	4	7	1
4	0	2	3	0	2
5	6	0	0	0	0

Figure 7–40. Tableau 3 for Assignment Problem

from the other uncovered numbers and added to the numbers that are covered by 2 lines, we obtain the new reduced-cost matrix shown in Figure 7–41.

The minimum number of lines needed to cover all the zeros in the figure is 4. The smallest uncovered number is 1. After repeating the appropriate steps of the algorithm, we obtain Figure 7–42. Since $n = 5$ lines are needed to cover all the zeros, the optimal solution has been found.

To determine the optimal assignment, first make assignments to the rows and columns with only one zero. In Figure 7–42, if we start with the rows, we will make assignments in cell (1, 4) and cell (2, 1). Then, considering the columns, we will make

Worker \ Tasks	1	2	3	4	5
1	1	2	2	0	1
2	0	2	2	7	1
3	0	5	3	7	0
4	0	1	2	0	1
5	7	0	0	1	0

Figure 7–41. Tableau 4 for Assignment Problem

Worker \ Tasks	1	2	3	4	5
1	1	1	1	0	1
2	0	1	1	7	1
3	0	4	2	7	0
4	0	0	1	0	1
5	8	0	0	2	1

Figure 7–42. Final Tableau for Assignment Problem

assignments in cell (5, 3) and cell (3, 5). The only remaining assignment is for cell (4, 2). These optimal assignments are shown shaded in the figure. The minimum cost is \$1,500.

7.15 TRAVELING RESOURCE FORMULATION

The *traveling salesman problem* (TSP) is a special case of the assignment problem. The problem is stated as follows: Given a set of n cities, numbered 1 through n, a salesman must determine a *minimum-distance* route, called a *tour*, which begins in city 1, goes through each of the other $(n-1)$ cities exactly once, and then returns to city 1. There are $(n-1)!/2$ possible tours. Thus, complete enumeration of the tours to find the minimum distance is impractical for most application scenarios.

The traveling salesman problem may be viewed as a *traveling resource* problem in project scheduling whereby a single resource unit is expected to perform tasks at several sites. Consider the following model where

$$x_{ij} = 1, \text{if route includes going from city } i \text{ to city } j, i,j = 1,2,\ldots,n; i \neq j;$$
$$\quad 0, \text{otherwise}$$

$$c_{ij} = \text{distance between city } i \text{ and city } j$$

$$\text{Minimize:} \quad z = \sum_{\substack{i=1 \\ i \neq j}}^{n} \sum_{j=1}^{n} c_{ij} x_{ij}$$

$$\text{Subject to:} \quad \sum_{j=1}^{n} x_{ij} = 1, \qquad i = 1, 2, \ldots, n$$

$$\sum_{i=1}^{n} x_{ij} = 1, \qquad j = 1, 2, \ldots, n$$

$$x_{ij} = 0, 1, \qquad i, j = 1, 2, \ldots, n; i \neq j$$

The first constraint represents the salesman's departures, while the second constraint represents arrivals. Although the above formulation can be solved as an assignment problem, a feasible assignment solution does not necessarily represent a *valid tour*. Specifically, a solution of the assignment problem may include subtours, which are disconnected tour cycles. One approach to formulating the salesman problem is to add additional constraints to the assignment problem formulation to eliminate the possibility of subtours.

Hence, we need to add a constraint to the above model indicating that a feasible assignment must define a complete tour with precisely one cycle. The additional constraint destroys the simplicity of the above model. In fact, there is no method available which will determine the optimal tour for a TSP problem in polynomial time. However, numerous heuristics exist that will find good, but not necessarily optimal, solutions in a reasonable amount of time. The techniques can be divided into two classes: tour construction techniques and tour improvement techniques. Tour construction techniques spend a considerable amount of time in constructing a tour to make sure that the final tour is a good tour. Tour improvement heuristics start with an arbitrary tour and switch nodes of the tour if the switch results in a reduction in the total tour length.

The simplest and most efficient tour construction heuristic is the nearest neighbor insertion technique (Rosenkratz, Stearns, and Lewis 1972). The simplest and most efficient tour improvement heuristic is the 2-opt technique developed by Lin (1965). An efficient but more complicated technique is by Lin and Kernighan (1973) which is a k-opt heuristic. Golden et al. (1977) and Bozer, Schorn, and Sharp (1990) discuss the benefit of combining tour construction techniques with tour improvement techniques. Another approach to solving a TSP problem involves the use of a simulated annealing technique (Johnson 1990). We will demonstrate the nearest neighbor technique and the 2-opt technique next. The simulated annealing technique is explained in Section 7.18.

7.15.1 Nearest Neighbor Algorithm

1. Choose any node as the first node in the tour.
2. Locate an unchosen node that is closest to the recently chosen node.
3. Enter this node as the next node to visit in the tour. The node becomes the recently chosen node.
4. Repeat steps 2 and 3 until all the nodes are chosen.
5. Close the tour by connecting the first node and the last node.

Example

Suppose a project involves a crane picking up material from five different locations. Table 7–6 shows the distance between locations i and j. Determine the sequence in which the crane should visit the five locations and return to the original position in a way that minimizes the total distance traveled.

     Starting with location 1 and choosing the nearest location each time, one will get the tour 1-3-2-4-5-1 with the total distance of 49 units. Since the nearest neighbor algorithm is a heuristic, it does not guarantee optimality, but provides good tours.

     The 2-opt improvement technique searches for tour reversals that will lead to a reduction in the tour length. In 2-opt switches, two tour connections are broken and nodes are connected in a way that a complete tour is defined as shown in Figure 7–43, where connections 3-2 and 4-5 are replaced by new connections 3-4 and 2-5.

     The tour length of 1-3-4-2-5-1 is 55. Since this 2-opt switch did not improve the tour length, the new tour will be discarded.

7.15.2 The 2-Opt Technique

The 2-opt technique can be summarized as follows:

1. Pick any 2 connections. Perform the 2-opt switch.
2. Retain the new tour if the tour length is improved. Return to step 1. Otherwise, return to step 1 with the tour before the switch.
3. Stop when no 2-opt switch improves the tour length.

     The traveling salesman problem has many applications in real life and is often used to solve storage, retrieval and distribution problems.

Table 7-6 Distance Matrix for Traveling Resource Problem

Location	1	2	3	4	5
1	—	20	10	30	15
2	20	—	5	10	8
3	10	5	—	12	15
4	30	10	12	—	9
5	15	8	15	9	—

Before 2-opt switch

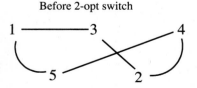

After 2-opt switch

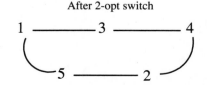

Figure 7–43. 2-opt Example for Traveling Resource Problem

.16 SHORTEST-PATH PROBLEM

The *shortest-path* problem involves finding the shortest path from a specified origin to a specified destination in an acyclic network. A *network* consists of a set of *nodes* (or vertices) and a set of *arcs* (or edges or links) which connect the nodes. Arc (i,j) represents the link from node i to node j. It is often assumed that the arcs are *directed*, that is, each arc has a specific orientation. A *path* is a collection of arcs from one node to another without regard to their directions. A *directed chain* from one node to another is a set of arcs which are all oriented in a specific direction. A directed chain beginning and ending at the same node is a *directed cycle*. A network which has no directed cycles is said to be *acyclic*. In acyclic networks, $i < j$ for all arcs (i,j).

Practical network problems involve assigning costs, profits, distances, capacity restrictions, or other performance measures to arcs. Supply and demand levels can also be associated with nodes in the network. Let us consider a directed network with a distance d_{ij} associated with each arc (i,j). It is assumed that the distances between arcs are nonnegative. When no arc exists from node i to node j, the length of the *dummy arc* (i,j) is set to M, where $M >>> 0$. The shortest-path problem consists of finding the shortest directed chain from a specified origin to a specified destination.

In project planning, transportation and equipment routing among project sites are suitable problems for the application of the shortest-path algorithm. Possible extensions of the basic shortest-path problem may include the following:

> *Smartest path between two project sites*: In some applications, the shortest path may not be the best path in terms of ease of transportation, resource requirement, and other performance considerations. The objective here is to quantify *smartness* and incorporate the measure into the shortest-path procedure.
>
> *Safest path between two project sites*: In this case, the objective is to find the path that offers the highest measure of safety. We will need to quantify *safety* so that the measure can be incorporated into the shortest-path procedure.
>
> *Cheapest path between project sites*: For some problems, the shortest path may turn out to be the most expensive. Commuters using toll roads are familiar with this dilemma. The objective in the cheapest-path problem is to find the least-cost path between two points. Distance-cost relationships can be used as the basis for obtaining quantitative measures to incorporate into the conventional shortest-path procedure so as to minimize cost rather than distance.

Dreyfus (1969) discusses various shortest-path problem scenarios and solution algorithms. Some examples of path-to-path problems of interest are

> Shortest path between the originating node and the terminal node
> Shortest path between the origin and all nodes in the network
> Shortest paths between all nodes in the network
> The n shortest paths from the origin to the terminal node

Dijkstra's algorithm (Lawler 1976) can be used to solve the shortest path problem for networks with nonnegative d_{ij} values. The algorithm can be summarized as

Step 1: Assign to the origin (node 1), the *permanent* label $y_1 = 0$, and assign to every other node the *temporary* label $y_j = M$. Set $i = 1$.

Step 2: From node i, recompute the temporary labels $y_i = \text{Min}\{y_j, y_i + d_{ij}\}$, where node j is temporarily labeled and $d_{ij} < M$.

Step 3: Find the smallest of the temporary labels, say y_i. Label node i permanently with value y_i.

Step 4: If all nodes are permanently labeled, stop. Otherwise, go to step 2.

Example

Figure 7–44 gives the information flow network for the ABC company. The numbers on each arc indicate the probability of successful communication between the two end nodes calculated based on past performances. Select the most reliable path from information source 1 to destination source 7. Let r_i be the reliability of arc i. Then, reliability of path P is defined as $R_p = \prod_{i \in P} r_i$.

The most reliable path is the path that has the maximum product of the reliabilities of the arcs on the path. Taking the logarithm of r_i, one gets

$$R'_p = \log R_p = \sum_{i \in P} \log r_i = \sum_{i \in P} r'_i$$

The problem then reduces to finding the longest path in the network with negative arc distances. Dijkstra's algorithm will determine the path when the minimum operator is changed to the maximum operator in step 2. It is important to note that the arc distances are required to be nonpositive. Otherwise, positive cycles may exist and the algorithm may never terminate. Figure 7–45 indicates r'_i for each arc.

The algorithm starts by assigning a permanent label 0 to node 1, and temporary labels M to all the other nodes. Nodes 2 and 3 are directly connected to node 1. Hence, $y_2 = y_1 + d_{12} = -0.0513$ and $y_3 = y_1 + d_{13} = -0.0202$. Since $-0.0202 > -0.0513$, node 3 receives the permanent label of -0.0202. Nodes 2, 4 and 6 are adjacent to node 3. The new labels are $y_2 = \max(y_2, y_3 + d_{32}) - \max(-0.0513, -0.0202 \pm 0.1054) = -0.0513, y_4 =$

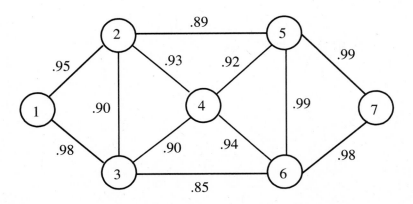

Figure 7–44. Information Flow Network

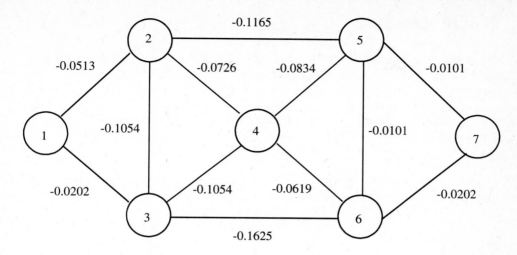

Figure 7–45.   The Network with r_i' Values

$\max(M, -0.0202 \pm 0.1054) = -0.1256, y_6 = \max(M, -0.0202 \pm 0.1625) = -0.1827$. Since $y_2 > y_4 > y_6, y_2$ receives the permanent label -0.0513. The rest of the steps are summarized in Table 7–7. The permanent labels are indicated by shading. The longest path from node 1 to node 7 is found as path 1–2–5–7 with a value of -0.1779. Taking the antilogarithm, one gets 0.84 as the reliability of the path.

Table 7-7 Summary of the Dijkstra's Algorithm
for the Example Problem

| Step | \multicolumn{7}{c}{Node} |
	1	2	3	4	5	6	7
1	0	M	M	M	M	M	M
2	0	−.0513	−.0202	M	M	M	M
3	0	−.0513	−.0202	−.1256	M	−.1827	M
4	0	−.0513	−.0202	−.1238	−.1678	M	M
5	0	−.0513	−.0202	−.1238	−.1678	−.1827	M
6	0	−.0513	−.0202	−.1238	−.1678	−.1779	−.1779
7	0	−.0513	−.0202	−.1238	−.1678	−.1779	−.1779

7.17 GOAL PROGRAMMING

One major shortcoming of linear programming is that only one objective can be considered at a time. In many real-world situations, the decision maker is faced with problems involving more than one objective. *Goal programming* is an extension of linear programming that permits decision makers to set and prioritize multiple goals. It is applicable to situations where not all goals can be satisfied equally and the aim is to minimize the overall dissatisfaction. An assignment of relative priorities to the goals being considered is required. The multiple goals do not have to be on the same measurement scale. For example, cost minimization, revenue maximization, and minimization of project duration may all coexist as goals in a goal programming problem. Goal programming solution techniques choose the values of the decision variables in such a way that deviations from the goals are minimized. If all the goals cannot be satisfied, goal programming will attempt to satisfy them in order of priority. Goal programming has been applied to various problems ranging from activity planning to resource allocation (Tingley and Liebman 1984; Goodman 1974; Hannan 1978; Ignizio 1976; Kornbluth 1973; Lee 1972, 1979). Moore et al. (1978) presented a goal programming project crashing model for the installation of a paper processing system at a bank. Many similar practical applications of goal programming can be found in the literature.

The formulation of goal programming problems is similar to that of linear programming problems except for the following requirements:

1. Explicit consideration of multiple goals
2. A measure of deviation from desired goals
3. Relative prioritization of the multiple goals

The objective of goal programming is to minimize the deviations from the desired goals. The priorities assigned to the goals are considered to be *preemptive*. Lower-priority goals are satisfied only after higher-priority goals have been satisfied. The general formulation of the goal programming model with preemptive weights is presented below. Let

n = number of goals to be considered.
x_i = value of the ith decision variable in the problem.
d_i^+ = amount by which goal i is exceeded.
d_i^- = amount by which goal i is underachieved.
P_i = priority factor for the goal having the ith priority. (The goal with the highest priority has a priority factor of P_1.)
$P_i >>> P_{i+1}$ so that there is no number $k > 0$ such that $nP_{i+1} \geq P_i$. That is, P_i is infinitely larger than P_{i+1}.

The priority factors are included in the objective function with the appropriate deviational variables.

z is the objective function.

$x_1, x_2, \ldots, x_n$ are the n *decision variables.*

$c_1, c_2, \ldots, c_n$ are coefficients of decision variables in the *objective function.*

$a_{i1}, a_{i2}, \ldots, a_{in}$ are *coefficients* of decision variables in the ith constraint.

b_i are the right-hand-side constants of the ith constraint ($i = 1, 2, \ldots, m$).

$$
\begin{aligned}
\text{Minimize:} \quad & z = \sum_k \sum_i P_k w_{i,k}^+ d_i^+ + \sum_s \sum_i P_s w_{i,s}^- d_i^- \\
\text{Subject to:} \quad & \sum_{j=1}^n m_{ij} x_j - d_i^+ + d_i^- = g_i, \qquad i = 1, 2, \ldots, p \\
& \sum_{j=1}^n a_{ij} x_j \le b_i, i = p + 1, \ldots, p + m \\
& x_j, d_i^+, d_i^- \ge 0, \qquad j = 1, \ldots, n; i = 1, \ldots, p
\end{aligned}
$$

where

p = number of goals

m = number of nongoal constraints

n = number of decision variables

d_i^+, d_i^- = deviations from i th goal

P_k, P_s = priority factors

$w_{i,k}^+$ = relative weight of d_i^+ in the kth ranking

$w_{i,k}^-$ = relative weight of d_i^- in the sth ranking

The formulation of a goal programming model requires a very careful analysis. The formulation can be better seen by illustrative examples.

Consider the time-cost trade-off problem discussed in Section 7.8. Suppose that rather than finding a schedule with minimum crashing cost, the management is interested in finding a schedule that satisfies a set of goals as much as possible. The goals have been identified as

1. The crashing cost should not exceed a certain budget.
2. The project should be crashed to T_p days.
3. Certain set of nodes should be realized latest by certain dates.

For example, for the problem given in Figure 7–5, suppose the goals in the order of decreasing priority are

1. The crashing cost should not exceed \$225.
2. The project should be completed by day 9.
3. Node 4 should be realized exactly on day 8.

The goal programming model for the time/cost trade-off problem can then be formulated as

$$\text{Minimize} : P_1 d_1^+ + P_2 d_2^+ + P_3(d_3^+ + d_3^-)$$

$$\sum_{(i,j) \in A} (u_{ij} - y_{ij}) \cdot a_{ij} - d_1^+ + d_1^- = 3000$$

$$t_5 - d_2^+ + d_2^- = 9$$

$$t_4 - d_3^+ + d_3^- = 8$$

$$t_i + y_{ij} - t_j \leq 0 \qquad \forall (i,j) \in A$$

$$d_{ij} \leq y_{ij} \leq u_{ij} \qquad \forall (i,j) \in A$$

It is suggested that the reader should compare this model with the model defined by equations (7.5) through (7.8). More specifically the model can be written as

Minimize:
$$P_1 d_1^+ + P_2 d_2^+ + P_3(d_3^+ + d_3^+)$$

Subject to:
$$60(6 - y_6) + 250(4 - y_c) + 60(3 - y_2) +$$
$$100(5 - y_e) + 35(4 - y_f) + 25(2 - y_g) -$$
$$d_1^+ + d_1 = 225$$
$$t_5 - d_2^+ + d_2^- = 9$$
$$t_4 - d_3^+ + d_3^- = 8$$
$$t_1 + y_a - t_2 \leq 0$$
$$t_1 + y_c - t_4 \leq 0$$
$$t_1 + y_b - t_3 \leq 0$$
$$t_2 + y_d - t_4 \leq 0$$
$$t_3 + y_e - t_4 \leq 0$$
$$t_2 + y_f - t_5 \leq 0$$
$$t_4 + y_g - t_5 \leq 0$$
$$2 \leq y_a \leq 2$$
$$4 \leq y_b \leq 6$$
$$3 \leq y_c \leq 4$$
$$2 \leq y_d \leq 3$$
$$3 \leq y_e \leq 5$$
$$3 \leq y_f \leq 4$$
$$1 \leq y_g \leq 2$$

where P_1, P_2, P_3 are the priority weights for the three goals, respectively. Assuming that $P_1 = 100, P_2 = 10$ and $P_3 = 1$, the LINDO solution for the above problem is given in Table 7–8. Interested readers may refer to Wu and Coppins (1981), Goodman (1974), Hannan (1978), Ignizio (1976), Kornbluth (1973), and Lee (1972) for additional examples.

Note that a value of zero for the objective function indicates that all the goals are satisfied. This is an infrequent occurrence in practice.

Table 7-8 LINDO solution for Goal
Programming Example

Variable	Value	Reduced cost
d_1^+	0.0000	100.0000
d_2^+	0.0000	10.0000
d_3^+	0.0000	1.0000
d_3^-	0.0000	0.0000
y_b	5.0000	0.0000
y_c	3.0000	0.0000
y_d	3.0000	0.0000
y_e	3.0000	0.0000
y_f	4.0000	0.0000
y_g	1.0000	0.0000
d_1^-	310.0000	0.0000
t_5	9.0000	0.0000
d_2^-	0.0000	0.0000
t_4	8.0000	0.0000
t_1	0.0000	0.0000
y_a	2.0000	0.0000
t_2	5.0000	0.0000
t_3	5.0000	0.0000

Objective function value = 0.0000.

18 RESOURCE ALLOCATION USING SIMULATED ANNEALING

In Chapter 5, we discussed several priority orders and performed resource allocation based on the order. Simulated annealing (Eglese 1990) is an iterative process which starts with an arbitrary order and performs a switch in the order and reallocates the resource with respect to the new order. If the project completion time is reduced, then the new order is kept and another switch is performed. Otherwise, the new order is kept with some probability which decreases as the iteration number increases. The process stops when the order has not changed for a given number of iterations. The annealing step prevents the process from converging to a local minimum quickly and enables the process to get out of a local minimum point in the search for a global minimum. The simulated annealing procedure has been successfully applied to several combinatorial optimization problems such as the traveling salesman problem, sequencing, scheduling, and facility layout problems. We will show its application to the resource allocation problem. The discussions will assume one resource problem in order to simplify the illustration. However, it can easily be extended to cover the multiple-resource allocation problem.

Simulated annealing method has its roots in thermodynamics and is specifically related to the way liquids freeze and crystallize and metals cool and anneal. The slow cooling process allows ample time for atoms to move around and achieve perfect crystallization. Similarly, in discrete minimization problems, the objective is to prevent rapid descent and allow the procedure to escape the local minimum and continue the descent process. This is only possible if the method allows ascent, that is, accepts solutions which increase the objective function value. The procedure controls the ascent process using a probability function defined as

$$P(z_i) = e^{-(z_i - z_{i-1})/KT}$$

where z_i is the ith solution value, k is a constant and T is the temperature. When $z_i < z_{i-1}$, the probability is greater than one which is then set equal to one. When $z_i > z_{i-1}$ indicating an ascent, the solution z_i is only accepted with probability $P(z_i)$. The temperature T is decreased (cooling) as the number of iterations increase which in turn decreases the probability of ascent at the later stages. This enables the process to converge to a good local minimum. The initial value of T and the annealing schedule differ with the problem and are generally defined after experimentation.

The initial parameters needed for simulated annealing are

1. $T(0)$, initial temperature value and a temperature function. We will assume that $T(0) = 65$ and decrease it by $k = 0.9$ after every iteration.
2. Stopping criterion which can be a time limit or a test on some variable. We will stop when the sequence is not updated for 5 iterations.

The algorithm as applied to the resource allocation problem can be outlined as follows:

Step 1: Select an initial order S_0. S_0 must satisfy precedence relationship. Set $i = 0$. Select an initial temperature T. Set $n = 0$.

Step 2: Allocate the resource with respect to the priority order S_0. Basically, the first feasible job in the order is scheduled if the available resource is sufficient for day k. When no feasible allocation exists for day k, day $k + 1$ is considered. Let z_0 be the project completion time under S_0.

Step 3: Generate S_j. S_j must differ from S_i in only two positions, and also must satisfy the precedence relationship. S_j is referred to as the neighbor of S_i. If $z_j \leq z_i$, then set $i = j$. Set $n = 0$, $T = 0.9T$; return to start of step 3. Otherwise, go to step 4.

Step 4: Let $r = \text{random } (0, 1)$. If, $r < e^{-(z_j - z_i)/T}$, then set $i = j, n = 0, T = 0.9T$. Return to step 3. Otherwise, set $n = n + 1, T = 0.9T$. If $n \geq 5$, stop. Otherwise, return to step 3.

Consider the resource allocation example problem discussed earlier. The maximum daily resource available is 10 units. The set of jobs to be performed, the predecessors, durations, and resource requirements are given in Table 7–9.

Suppose the procedure starts with the arbitrary feasible initial order ABCDEFG. When the resource is allocated according to this order, $z_0 = 13.17$ days. Possible switches with A are B and C. Job B can be switched with jobs A, C, D, F. Job C can be switched with jobs F and D. Similarly job D with jobs C, F, E, job E with jobs D and F, job F with jobs, C, D, E and G, and job G with job F. Suppose S_1 is defined by switching jobs A and C. Therefore, the new order is CBADEFG. Resource allocation results in $z_1 = 13.17$. The new order is accepted with probability 1 since $z_1 = z_0$.

Table 7-9 Problem Data for Simulated Annealing Example

Jobs	Immediate predecessors	Time	Resource requirement
A	—	2.17	3
B	—	6.00	5
C	—	3.83	4
D	A	2.83	2
E	C	5.17	4
F	A	4.00	2
G	B,D,E	2.00	6

The feasible switches on S_2 to define a neighbor order S_3 are

Job C with job A or B
Job B with jobs C, A or E
Job A with job C
Job D with jobs E or F
Job E with job D or F
Job F with jobs D, E or G
Job G with job F

Note that job B can be switched with job E since the successor of B (which is job G) comes after job E in the sequence. Suppose job D is switched with job F. The new order is CBAFEDG which results in $z_3 = 13.17$. Hence, the order is accepted. Suppose job F is switched with job E next, which leads to CBAEFDG with $z_4 = 13.17$. Switching job A with job E, one gets the order CBEAFDG and $z_5 = 13.00$. The procedure continues in this fashion until no new order is accepted in 5 consecutive iterations. For this problem $z = 13$ is the optimal solution. Figure 7–46 is the schedule chart for the order CBEAFDG.

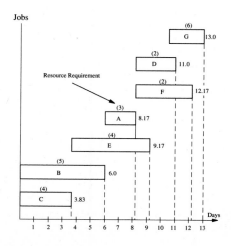

Figure 7–46. Schedule Chart for CBEAFDG

7.19 EXERCISES

7.1. Repeat the activity scheduling example presented in this chapter with the assumption that a 15-minute break is required between the completion of one activity and the start of the next activity.

7.2. A deliveryperson makes deliveries to two project sites, site A and site B. He receives $1,000 per delivery to site A and $200 per delivery to site B. He is required to make at least 5 deliveries per week to site B. Each delivery to either site A or site B takes him 3 hours to complete. There is an upper limit of 35 hours of delivery time per week. The objective of the deliveryperson is to increase his delivery income as much as possible. Formulate his problem as an LP model.

7.3. Solve the LP model of the example presented in this chapter on the number of workers to assign to a project workstation. Discuss how you would go about implementing the LP solution in a practical setting.

7.4. Develop and solve the knapsack formulation to schedule the project presented below. Draw the Gantt chart for the schedule. Use the following notation in your formulation:

C_t = the set of activities eligible for scheduling at time t
CAF_i = relative priority weight for activity i
R_{jt} = units of resource type j available at time t
$\overline{R}_{jt}$ = units of resource type j in use at time t
r_{ji} = units of resource type j required by activity i
x_{it} = indicator variable for scheduling activity i at time t
    $x_{it} = 1$ if activity i is scheduled at time t
    $x_{it} = 0$ if activity i is not scheduled at time t

Activity	Predecessor	Duration (days)	Resource requirement (type 1, type 2)	Priority weight
A	—	6	2, 5	50.1
B	—	5	1, 3	57.1
C	—	2	3, 1	56.8
D	A, B	4	2, 0	100
E	B	5	1, 4	53.2
F	C	6	3, 1	51.8
G	B, F	1	4, 3	44.9
H	D, E	7	2, 3	54.9

There are two resource types. Resource availability varies from day to day based on the schedule below:

Time period	Units available (type 1, type 2)
0 to 5	3, 5
5 + to 10	2, 3
10 + to 14	3, 4
14 + to 20	4, 4
20 + to 22	5, 3
22 + to 30	4, 5
30 + and up	6, 6

7.5. Solve the resource assignment problem with the following cost matrix:

$$\begin{bmatrix} 3 & 7 & 2 & 4 \\ 3 & 6 & 5 & 7 \\ 6 & 4 & 5 & 6 \\ 4 & 3 & 4 & 5 \end{bmatrix}$$

7.6. Starting with Vogel's approximation as the starting solution, solve the transportation problem example presented in this chapter. The example was previously solved with the northwest-corner starting solution.

7.7. Boomer-Sooner, Inc., produces three styles of portable stadium seats: high-rider style, low-rider style, and line-hugger style. The seats are made of high-impact composite material and are produced in two steps, assembly and painting. The schedule for labor and material inputs for each seat style are as follows:

Assembly labor

1 hour per high-rider seat
3 hours per low-rider seat
1.5 hours per line-hugger seat

Painting labor

1.5 hour per high-rider seat
1.5 hours per low-rider seat
4 hours per line-hugger seat

Material requirement

4 pounds per high-rider seat
3 pounds per low-rider seat
6 pounds per line-hugger seat

2,000 hours of assembly labor hours are available per month.
1,000 hours of painting labor hours are available per month.

The contributions to profit for the styles are $35, $40, and $65, respectively. The company has two equally desirable goals: minimization of idle time in painting, and making a monthly profit of $20,000. Set up this problem as a goal programming problem and show the initial tableau.

7.8. A project manager has four crews which can be hired on a temporary basis during peak construction periods. The manager currently has three projects more than can be handled with his regular crews. It is desired to determine the crew assignments that will minimize the total project cost. The costs of using certain crews for certain projects are presented below:

Crew 1

Project A: $4,000
Project B: $3,000
Project C: $9,000

Crew 2

Project A: $7,000
Project B: $1,000
Project C: $8,000

Crew 3

Project A: $2,000
Project B: $6,000
Project C: $4,000

Crew 4

Project A: $9,000
Project B: $5,000
Project C: $5,000

7.9. Find the shortest path between the originating node and the terminal node for the following acyclic network:

Arc	Duration
1–2	7
1–3	6
1–4	10
2–3	2
2–4	4
3–4	2
3–5	7
4–5	3

7.10. Develop a computer program for finding the shortest path between all pairs of nodes in a directed acyclic network. Use the program to verify your solution to Exercise 7.9.

7.11. Develop a general LP formulation for scheduling the following project. Resource requirements and precedence relationships must be satisfied in the model ($R_1 = 5$, $R_2 = 3$).

Activity	Predecessor	Duration (days)	Resource requirement (type 1, type 2)
A	—	3	0, 3
B	A	4	2, 1
C	A	2	1, 1
D	B, C	7	5, 2

7.12. Use any software tool available to you to solve the LP model in Exercise 7.11.

7.13. White Water Fun Park is planning to add another ride which consists of three main parts A, B and C. The manager anticipates that it will take 14 days to receive the material. It should then take 7 days to assemble the components of part A which is then to be tested separately which will take 3 days. It should take 6 and 4 days to complete the assembly of the components of parts B and C, respectively. Then, parts B and C have to be assembled (2 days) and attached to part A (5 days). The completed ride has to be tested for 21 days before it can be opened to the public. Draw the project network (AOA representation) and determine the critical path. Suppose on day 18 the project progress is evaluated, and the following report has been generated. Assembly of part A is on time, part B will be ready two days early, part C will be ready a day late. Will the critical path remain the same? If not, find the new critical path.

7.14. Develop an integer programming model for the resource allocation problem in Figure 7–47 assuming that daily resource availability is 5 units. Use LINDO or any other integer programming software to determine the optimal solution to your model. Discuss the results. Draw the project schedule chart (Gantt chart).

7.15. Consider the project network given in Figure 7–48.
   (a) Determine the critical path for the network when each activity is realized at its normal time.

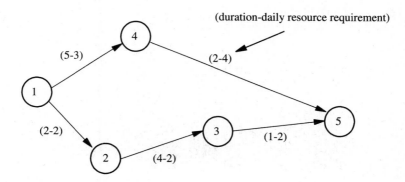

Figure 7–47.   Resource Allocation Problem for Exercise 7.14

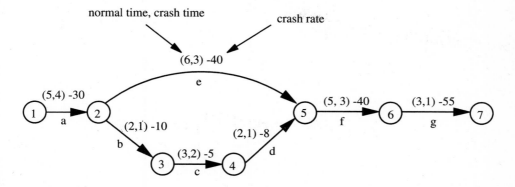

Figure 7–48.   Project Network for Exercise 7–15

(b) Suppose the project must be completed in 15 days. Formulate a linear programming model that will minimize the cost of meeting the deadline.

(c) Solve the model in (b) using a linear programming software.

(d) Generate the time-cost trade-off curve using the network flow approach.

7.16. A fashion clothing warehouse is located in Oklahoma City and serves five stores located in Norman, Tulsa, Oklahoma City, Muskogee and Stillwater. Each week a truck delivers clothes form the warehouse to the stores and returns to the warehouse. The distances between each pair of locations (in miles) is tabulated below.

	Warehouse	Norman	Oklahoma City	Tulsa	Muskogee	Stillwater
Warehouse	0	45	10	105	157	54
Norman	45	0	35	130	140	80
Oklahoma City	10	35	0	115	90	64
Tulsa	105	130	115	0	52	64
Muskogee	157	140	90	52	0	85
Stillwater	54	80	64	64	85	0

Determine the order in which the truck should visit the five stores so that the total distance traveled is minimized.

7.17. Solve Exercise 7.14 (integer programming) using the simulated annealing procedure. Use $T = 60, k = 0.8$ and stop when the order does not change in 3 iterations.

8

Financial and Economic Analyses

This chapter discusses financial and economic analyses for project management. The next chapter discusses general quantitative models for project management. The contents of the two chapters provide essential inputs for the techniques presented in the preceding chapters. Good financial and economic analyses and practical quantitative representation of a project facilitate better project management in terms of setting goals, planning, organizing, scheduling, allocating resources, controlling, optimizing, and terminating the project.

SET GOALS → PLAN → ORGANIZE → SCHEDULE →
ALLOCATE RESOURCES → CONTROL → OPTIMIZE →
TERMINATE

8.1 FUNDAMENTAL COST CONCEPTS

Cost management in a project environment refers to the functions required to maintain effective financial control of the project throughout its life cycle. There are several cost concepts that influence the economic aspects of managing projects. Within a given scope of analysis, there may be a combination of different types of cost aspects to consider. These cost aspects are discussed next.

Life-cycle cost. This is the sum of all costs, recurring and nonrecurring, associated with a project during its entire life cycle.

First cost. This is the total initial investment required to initiate a project or the total initial cost of the equipment needed to start the project.

Budgeted cost for work performed. This is the sum of the budgets for completed work plus the appropriate portion of the budgets for level of effort and apportioned effort. Apportioned effort is effort that by itself is not readily divisible into short-span work packages but is related in direct proportion to measured effort.

Budgeted cost for work scheduled. This is the sum of budgets for all work packages and planning packages scheduled to be accomplished (including work in progress) plus the amount of level of effort and apportioned effort scheduled to be accomplished within a given period of time.

Operating cost. This is a recurring cost needed to keep a project in operation during its life cycle. Operating costs may consist of such items as labor cost, material cost, and energy cost.

Maintenance cost. This is a cost that occurs intermittently or periodically for the purpose of keeping project equipment in good operating condition.

Overhead cost. This is a cost incurred for activities performed in support of the operations of a project. The activities that generate overhead costs support the project efforts rather than contribute directly to the project goal. The handling of overhead costs varies widely from company to company. Typical overhead items are electric power cost, insurance premiums, cost of security, and inventory carrying cost.

Sunk cost. This is a cost that occurred in the past and cannot be recovered under the present analysis. Sunk costs should have no bearing on the prevailing economic analysis and project decisions. Ignoring sunk costs is always a difficult task for analysts. For example, if $950,000 was spent four years ago to buy a piece of equipment for a technology-based project, a decision on whether or not to replace the equipment now should not consider that initial cost. But uncompromising analysts might find it difficult to ignore that much money. Similarly, an individual making a decision on selling a personal automobile would typically try to relate the asking price to what was paid for the automobile when it was acquired. This is wrong under the strict concept of sunk costs.

Opportunity cost. This is the cost of forgoing the opportunity to invest in a venture that would have produced an economic advantage. Opportunity costs are usually incurred due to limited resources that make it impossible to take advantage of all investment opportunities. It is often defined as the cost of the best rejected opportunity. Opportunity costs can also be incurred due to a missed opportunity rather than due to an intentional rejection. In many cases, opportunity costs are hidden or implied because they typically relate to future events that cannot be accurately predicted.

Direct cost. This is a cost that is directly associated with actual operations of a project. Typical sources of direct costs are direct material costs and direct labor costs. Direct costs are those that can be reasonably measured and allocated to a specific component of a project.

Applied direct cost. This represents the amounts recognized in the time period associated with the consumption of labor, material, and other direct resources without regard to the date of commitment or the date of payment. These amounts are to be charged to work in process (WIP) when resources are actually consumed, material resources are withdrawn from inventory for use, or material resources are received and scheduled for use within 60 days.

Indirect cost.   This is a cost that is indirectly associated with project operations. Indirect costs are those that are difficult to assign to specific components of a project. An example of an indirect cost is the cost of computer hardware and software needed to manage project operations. Indirect costs are usually calculated as a percentage of a component of direct costs. For example, the direct costs in an organization may be computed as 10 percent of direct labor costs.

Standard cost.   This is a cost that represents the normal or expected cost of a unit of the output of an operation. Standard costs are established in advance. They are developed as a composite of several component costs such as direct labor cost per unit, material cost per unit, and allowable overhead charge per unit.

Fixed cost.   This is a cost incurred irrespective of the level of operation of a project. Fixed costs do not vary in proportion to the quantity of output. Examples of costs that make up the fixed cost of a project are administrative expenses, certain types of taxes, insurance cost, depreciation cost, and debt servicing cost. These costs usually do not vary in proportion to quantity of output.

Variable cost.   This is a cost that varies in direct proportion to the level of operation or quantity of output. For example, the costs of material and labor required to make an item will be classified as variable costs since they vary with changes in level of output.

Total cost.   This is the sum of all the variable and fixed costs associated with a project.

Incremental cost.   This refers to the additional cost of changing the production output from one level to another. Incremental costs are normally variable costs.

Marginal cost.   This is the additional cost of increasing production output by one additional unit. The marginal cost is equal to the slope of the total cost curve or line at the current operating level.

Actual cost of work performed.   This represents the cost actually incurred and recorded in accomplishing the work performed within a given time period.

Estimated cost at completion.   This is the actual direct cost plus indirect cost that can be allocated to the contract, plus the estimate of costs (direct and indirect) for authorized work remaining.

Economies of scale.   This refers to a reduction of the relative weight of the fixed cost in total cost by increasing output quantity. This helps reduce the final unit cost of a product. Economies of scale is often simply referred to as the savings due to *mass production*.

In a typical project, several cost elements intermingle. A project analyst must be careful to analyze all the relevant cost elements and include them as appropriate in the decision-making process.

8.2 FINANCING STRATEGIES

Financing a project means raising capital for the project. Capital is a resource consisting of funds available to execute a project. Capital includes not only privately owned production facilities but also public investment. Public investments provide the infrastructure of the economy such as roads, bridges, water supply, and so on. Other public capital that indirectly supports production and private enterprise includes schools, police stations, a central financial institution, and postal facilities.

If the physical infrastructures of the economy are lacking, the incentive for private entrepreneurs to invest in production facilities is likely to be lacking also. The government or community leaders can create the atmosphere for free enterprise by constructing better roads, providing better public safety, facilities, and encouraging ventures that assure adequate support services.

As far as project investment is concerned, what can be achieved with project capital is very important. The avenues for raising capital funds include banks, government loans or grants, business partners, cash reserves, and other financial institutions. The key to the success of the free enterprise system is the availability of capital funds and the availability of sources to invest the funds in ventures that yield products needed by the society. Some specific ways that funds can be made available for business investments are discussed next.

8.2.1 Commercial Loans

Commercial loans are the most common sources of project capital. Banks should be encouraged to loan money to entrepreneurs, particularly those just starting a business. Government guarantees may be provided to make it easier for the enterprise to obtain the needed funds.

8.2.2 Bonds and Stocks

Bonds and stocks are also common sources of capital. National policies regarding the issuance of bonds and stocks can be developed to target specific project types to encourage entrepreneurs.

8.2.3 Interpersonal Loans

Interpersonal loans are unofficial means of raising capital. In some cases, there may be individuals with enough personal funds to provide personal loans to aspiring entrepreneurs. But presently, there is no official mechanism that handles the supervision of interpersonal business loans. If a supervisory body existed at a national level, wealthy citizens would be less apprehensive about loaning money to friends and relatives for business purposes. The wealthy citizens could thus become a strong source of business capital.

8.2.4 Foreign Investment

Foreign investments can be attracted for local enterprises through government incentives. The incentives may be attractive zoning permits, foreign exchange permits, or tax breaks.

8.2.5 Investment Banks

The operations of investment banks are often established to raise capital for specific projects. Investment banks buy securities from enterprises and resell them to other investors. Proceeds from these investments may serve as a source of business capital.

8.2.6 Mutual Funds

Mutual funds represent collective funds from a group of individuals. The collective funds are often large enough to provide capital for business investments. Mutual funds may be established by individuals or under the sponsorship of a government agency. Encouragement and support should be provided for the group to spend the money for business investment purposes.

8.2.7 Supporting Resources

A clearing house of potential goods and services that a new project can provide may be established by the government. New entrepreneurs interested in providing the goods and services should be encouraged to start relevant enterprises. They should be given access to technical, financial, and information resources to facilitate starting production operations. As an example, the state of Oklahoma, under the auspices of the Oklahoma Center for the Advancement of Science and Technology (OCAST), has established a resource data base system. The system, named TRAC (Technical Resource Access Center), provides information about resources and services available to entrepreneurs in Oklahoma. The system is linked to the statewide economic development information system. This is a clearing house arrangement that will facilitate access to resources for project management.

The time value of money is an important factor in project planning and control. This is particularly crucial for long-term projects that are subject to changes in several cost parameters. Both the timing and quantity of cash flows are important for project management. The evaluation of a project alternative requires consideration of the initial investment, depreciation, taxes, inflation, economic life of the project, salvage value, and cash flows.

.3 CASH FLOW ANALYSIS

The basic reason for performing economic analysis is to make a choice between mutually exclusive projects that are competing for limited resources. The cost performance of each project will depend on the timing and levels of its expenditures. The techniques of computing cash flow equivalence permit us to bring competing project cash flows to a common basis for comparison. The common basis depends on the prevailing interest rate. Two cash flows that are equivalent at a given interest rate will not be equivalent at a different interest rate. The basic techniques for converting cash flows from one point in time to another are presented in the next section.

8.3.1 Cash Flow Conversion Factors

Cash flow conversion involves the transfer of project funds from one point in time to another. The following notation is used for the variables involved in the conversion process:

i = interest rate per period
n = number of interest periods
P = a present sum of money
F = a future sum of money
A = a uniform end-of-period cash receipt or disbursement
G = a uniform arithmetic gradient increase in period-by-period payments or disbursements

In many cases, the interest rate used in performing economic analysis is set equal to the *minimum attractive rate of return (MARR)* of the decision maker. The MARR is also sometimes referred to as the hurdle rate, required internal rate of return (IRR), return on investment (ROI), or discount rate. The value of the MARR is chosen with the objective of maximizing the economic performance of a project.

Compound amount factor.   The procedure for the single payment compound amount factor finds a future sum of money, F, that is equivalent to a present sum of money, P, at a specified interest rate, i, after n periods. This is calculated as

$$F = P(1 + i)^n$$

A graphic representation of the relationship between P and F is shown in Figure 8–1.

Example

A sum of $5,000 is deposited in a project account and left there to earn interest for 15 years. If the interest rate per year is 12%, the compound amount after 15 years can be calculated as

$$F = \$5,000(1 + 0.12)^{15}$$
$$= \$27,367.85$$

Present worth factor.   The present worth factor computes P when F is given. The present worth factor is obtained by solving for P in the equation for the compound amount factor. That is,

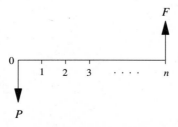

Figure 8-1.   Single-Payment Compound Amount Cash Flow

$$P = F(1 + i)^{-n}$$

Suppose it is estimated that $15,000 would be needed to complete the implementation of a project five years from now. How much should be deposited in a special project fund now so that the fund would accrue to the required $15,000 exactly five years from now? If the special project fund pays interest at 9.2% per year, the required deposit would be

$$P = \$15,000(1 + 0.092)^{-5}$$
$$= \$9,660.03$$

Uniform series present worth factor.   The uniform series present worth factor is used to calculate the present worth equivalent, P, of a series of equal end-of-period amounts, A. Figure 8–2 shows the uniform series cash flow. The derivation of the formula uses the finite sum of the present worth of the individual amounts in the uniform series cash flow as shown below. Some formulas for series and summation operations are presented in appendix D.

$$P = \sum_{t=1}^{n} A(1 + i)^{-t}$$
$$= A\left[\frac{(1 + i)^n - 1}{i(1 + i)^n}\right]$$

Example

Suppose the sum of $12,000 must be withdrawn from an account to meet the annual operating expenses of a multiyear project. The project account pays interest at 7.5% per year compounded on an annual basis. If the project is expected to last 10 years, how much must be deposited in the project account now so that the operating expenses of $12,000 can be withdrawn at the end of every year for 10 years? The project fund is expected to be depleted to zero by the end of the last year of the project. The first withdrawal will be made one year after the project account is opened, and no additional deposits will be made in the account during the project life cycle. The required deposit is calculated to be

$$P = \$12,000\left[\frac{(1 + 0.075)^{10} - 1}{0.075(1 + 0.075)^{10}}\right]$$
$$= \$82,368.92$$

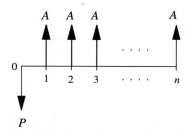

Figure 8-2.   Uniform Series Cash Flow

Uniform series capital recovery factor. The capital recovery formula is used to calculate the uniform series of equal end-of-period payments, A, that are equivalent to a given present amount, P. This is the converse of the uniform series present amount factor. The equation for the uniform series capital recovery factor is obtained by solving for A in the uniform series present amount factor. That is,

$$A = P\left[\frac{i(1+i)^n}{(1+i)^n - 1}\right]$$

Example

Suppose a piece of equipment needed to launch a project must be purchased at a cost of \$50,000. The entire cost is to be financed at 13.5% per year and repaid on a monthly installment schedule over 4 years. It is desired to calculate what the monthly loan payments will be. It is assumed that the first loan payment will be made exactly one month after the equipment is financed. If the interest rate of 13.5% per year is compounded monthly, then the interest rate per month will be 13.5%/12 = 1.125% per month. The number of interest periods over which the loan will be repaid is 4(12) = 48 months. Consequently, the monthly loan payments are calculated to be

$$A = \$50,000\left[\frac{0.01125(1+0.01125)^{48}}{(1+0.01125)^{48} - 1}\right]$$

$$= \$1,353.82$$

Uniform series compound amount factor. The series compound amount factor is used to calculate a single future amount that is equivalent to a uniform series of equal end-of-period payments. The cash flow is shown in Figure 8–3. Note that the future amount occurs at the same point in time as the last amount in the uniform series of payments. The factor is derived as

$$F = \sum_{t=1}^{n} A(1+i)^{n-t}$$

$$= A\left[\frac{(1+i)^n - 1}{i}\right]$$

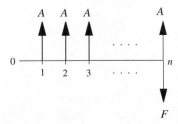

Figure 8-3. Uniform Series Compound Amount Cash Flow

Example

If equal end-of-year deposits of $5,000 are made to a project fund paying 8% per year for 10 years, how much can be expected to be available for withdrawal from the account for capital expenditure immediately after the last deposit is made?

$$F = \$5,000 \left[\frac{(1 + 0.08)^{10} - 1}{0.08} \right]$$

$$= \$72,432.50$$

Uniform series sinking fund factor.   The sinking fund factor is used to calculate the uniform series of equal end-of-period amounts, A, that are equivalent to a single future amount, F. This is the reverse of the uniform series compound amount factor. The formula for the sinking fund is obtained by solving for A in the formula for the uniform series compound amount factor. That is,

$$A = F \left[\frac{i}{(1 + i)^n - 1} \right]$$

Example

How large are the end-of-year equal amounts that must be deposited into a project account so that a balance of $75,000 will be available for withdrawal immediately after the twelfth annual deposit is made? The initial balance in the account is zero at the beginning of the first year. The account pays 10% interest per year. Using the formula for the sinking fund factor, the required annual deposits are

$$A = \$75,000 \left[\frac{0.10}{(1 + 0.10)^{12} - 1} \right]$$

$$= \$3,507.25$$

Capitalized cost formula.   *Capitalized* cost refers to the present value of a single amount that is equivalent to a perpetual series of equal end-of-period payments. This is an extension of the series present worth factor with an infinitely large number of periods. This is shown graphically in Figure 8–4.

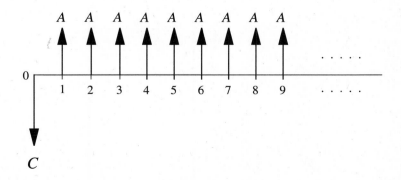

Figure 8-4.   Capitalized Cost Cash Flow

Using the limit theorem from calculus as n approaches infinity, the series present worth factor reduces to the following formula for the capitalized cost:

$$P = \lim_{n\to\infty} A\left[\frac{(1+i)^n - 1}{i(1+i)^n}\right]$$
$$= A\left\{\lim_{n\to\infty}\left[\frac{(1+i)^n - 1}{i(1+i)^n}\right]\right\}$$
$$= A\left(\frac{1}{i}\right)$$

Example

How much should be deposited in a general fund to service a recurring public service project at a cost of \$6,500 per year forever if the fund yields an annual interest rate of 11%? Using the capitalized cost formula, the required one-time deposit to the general fund is

$$P = \frac{\$6,500}{0.11}$$
$$= \$59,090.91$$

The formulas presented above represent the basic cash flow conversion factors. The factors are widely tabulated, for convenience, in engineering economy books (White et al. 1989; Park and Sharp-Bette 1990; Newnan 1991; Thuesen et al. 1977; Steiner 1992; DeGarmo et al. 1988; Grant et al. 1982). Several variations and extensions of the factors are available. Such extensions include the arithmetic gradient series factor and the geometric series factor. Variations in the cash flow profiles include situations where payments are made at the beginning of each period rather than at the end and situations where a series of payments contain unequal amounts. Conversion formulas can be derived mathematically for those special cases by using the basic factors presented above.

Arithmetic Gradient Series.   The gradient series cash flow involves an increase of a fixed amount in the cash flow at the end of each period. Thus, the amount at a given point in time is greater than the amount at the preceding period by a constant amount. This constant amount is denoted by G. Figure 8–5 shows the basic gradient series in which the base amount at the end of the first period is zero. The size of the cash flow in the gradient series at the end of period t is calculated as

$$A_t = (t-1)G, \qquad t = 1, 2, \ldots, n$$

The total present value of the gradient series is calculated by using the present amount factor to convert each individual amount from time t to time 0 at an interest rate of $i\%$ per period and summing up the resulting present values. The finite summation reduces to a closed form as follows:

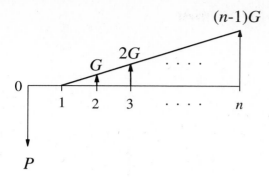

P

Figure 8-5. Arithmetic Gradient Cash Flow with Zero Base Amount

$$P = \sum_{t=1}^{n} A_t(1 + i)^{-t}$$

$$= \sum_{t=1}^{n} (t - 1)G(1 + i)^{-t}$$

$$= G \sum_{t=1}^{n} (t - 1)(1 + i)^{-t}$$

$$= G \left[\frac{(1 + i)^n - (1 + ni)}{i^2(1 + i)^n} \right]$$

Example

The cost of supplies for a 10-year project increases by $1,500 every year starting at the end of year 2. There is no supplies cost at the end of the first year. If the interest rate is 8% per year, determine the present amount that must be set aside at time zero to take care of all the future supplies expenditures. We have $G = \$1,500, i = 0.08$, and $n = 10$. Using the arithmetic gradient formula, we obtain

$$P = \$1,500 \left[\frac{1 - (1 + 10(0.08))(1 + 0.08)^{-10}}{(0.08)^2} \right]$$

$$= \$1,500(25.9768)$$

$$= \$38,965.20$$

In many cases, an arithmetic gradient starts with some base amount at the end of the first period and then increases by a constant amount thereafter. The nonzero base amount is denoted as A_1. Figure 8–6 shows this type of cash flow.

The calculation of the present amount for such cash flows requires breaking the cash flow into a uniform series cash flow of amount A_1 and an arithmetic gradient cash flow with zero base amount. The uniform series present worth formula is used to calculate the present worth of the uniform series portion while the basic gradient series formula is used to calculate the gradient portion. The overall present worth is then calculated as

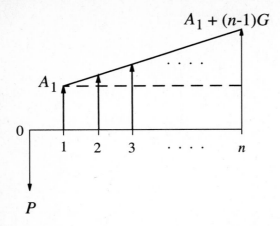

Figure 8-6. Arithmetic Gradient Cash Flow with Nonzero Base Amount

$$P = P_{\text{uniform series}} + P_{\text{gradient series}}$$
$$= A_1 \left[\frac{(1+i)^n - 1}{i(1+i)^n} \right] + G \left[\frac{(1+i)^n - (1+ni)}{i^2(1+i)^n} \right]$$

Increasing Geometric Series Cash Flow. In an increasing geometric series cash flow, the amounts in the cash flow increase by a constant percentage from period to period. There is a positive base amount, A_1, at the end of period 1. Figure 8–7 shows an increasing geometric series. The amount at time t is denoted as

$$A_t = A_{t-1}(1 = j), \qquad t = 2, 3, \ldots, n$$

where j is the percentage increase in the cash flow from period to period. By doing a series of back substitutions, we can represent A_t in terms of A_1 instead of in terms of A_{t-1} as shown below:

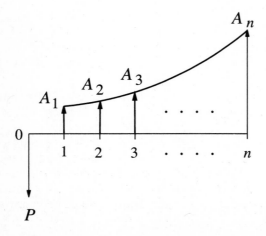

Figure 8-7. Increasing Geometric Series Cash Flow

$$A_2 = A_1(1 + j)$$
$$A_3 = A_2(1 + j) = A_1(1 + j)(1 + j)$$
$$\cdots$$
$$A_t = A_1(1 + j)^{t-1}, \qquad t = 1, 2, 3, \ldots, n$$

The formula for calculating the present worth of the increasing geometric series cash flow is derived by summing the present values of the individual cash flow amounts. That is,

$$
\begin{aligned}
P &= \sum_{t=1}^{n} A_t(1 + i)^{-t} \\
&= \sum_{t=1}^{n} [A_1(1 + j)^{t-1}](1 + i)^{-t} \\
&= \frac{A_1}{(1 + j)} \sum_{t=1}^{n} \left(\frac{1 + j}{1 + i}\right)^t \\
&= A_1 \left[\frac{1 - (1 + j)^n(1 + i)^{-n}}{(i - j)}\right], \qquad i \neq j
\end{aligned}
$$

If $i = j$, the formula above reduces to the limit as ij, shown below:

$$P = \frac{nA_1}{(1 + i)}, \qquad i = j$$

Example

Suppose funding for a 5-year project is to increase by 6% every year with an initial funding of $20,000 at the end of the first year. Determine how much must be deposited into a budget account at time zero in order to cover the anticipated funding levels if the budget account pays 10% interest per year. We have $j = 6\%, i = 10\%, n = 5, A_1 = \$20,000$. Therefore,

$$P = \$20,000 \left[\frac{1 - (1 + 0.06)^5(1 + .10)^{-5}}{(0.10 - 0.06)}\right]$$
$$= \$20,000(4.2267)$$
$$= \$84,533.60$$

Decreasing Geometric Series Cash Flow. In a decreasing geometric series cash flow, the amounts in the cash flow decrease by a constant percentage from period to period. The cash flow starts at some positive base amount, A_1, at the end of period 1. Figure 8–8 shows a decreasing geometric series. The amount at time t is denoted as

$$A_t = A_{t-1}(1 - j), t = 2, 3, \ldots, n$$

where j is the percentage decrease in the cash flow from period to period. As in the case of the increasing geometric series, we can represent A_t in terms of A_1.

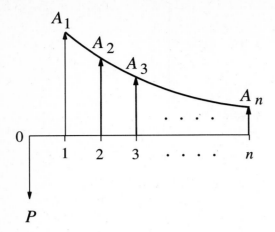

Figure 8-8. Decreasing Geometric Series Cash Flow

$$A_2 = A_1(1 - j)$$
$$A_3 = A_2(1 - j) = A_1(1 - j)(1 - j)$$
$$\ldots$$
$$A_t = A_1(1 - j)^{t-1}, \qquad t = 1, 2, 3, \ldots, n$$

The formula for calculating the present worth of the decreasing geometric series cash flow is derived by finite summation as in the case of the increasing geometric series. The final formula is

$$P = A_1 \left[\frac{1 - (1 - j)^n (1 + i)^{-n}}{(i + j)} \right]$$

Example

The contract amount for a three-year project is expected to decrease by 10% every year with an initial contract of $100,000 at the end of the first year. Determine how much must be available in a contract reservoir fund at time zero in order to cover the contract amounts. The fund pays 10% interest per year. Since $j = 10\%, i = 10\%, n = 3, A_1 = \$100,000$, we should have

$$P = \$100,000 \left[\frac{1 - (1 - 0.10)^3 (1 + 0.10)^{-3}}{(0.10 + 0.10)} \right]$$

$$= \$100,000(2.2615)$$

$$= \$226,150$$

Internal rate of return. The internal rate of return (IRR) for a cash flow is defined as the interest rate that equates the future worth at time n or present worth at time 0 of the cash flow to zero. If we let i^* denote the internal rate of return, then we have

$$FW_{t=n} = \sum_{t=0}^{n} (\pm A_t)(1 + i^*)^{n-t} = 0$$

$$PW_{t=0} = \sum_{t=0}^{n} (\pm A_t)(1 + i^*)^{-t} = 0$$

where $+$ is used in the summation for positive cash flow amounts or receipts and $-$ is used for negative cash flow amounts or disbursements. A_t denotes the cash flow amount at time t, which may be a receipt ($+$) or a disbursement ($-$). The value of i^* is referred to as the *discounted cash flow rate of return, internal rate of return,* or *true rate of return*. The procedure above essentially calculates the net future worth or the net present worth of the cash flow. That is,

net future worth = (future worth of receipts) $-$ (future worth of disbursements)

$$NFW = FW_{(receipts)} - FW_{(disbursements)}$$

net present worth = (present worth of receipts) $-$ (present worth of disbursements)

$$NPW = PW_{(receipts)} - PW_{(disbursements)}$$

Setting the NPW or NFW equal to zero and solving for the unknown variable i, determines the internal rate of return of the cash flow.

Benefit-cost ratio.   The benefit cost ratio of a cash flow is the ratio of the present worth of benefits to the present worth of costs. This is defined as

$$B/C = \frac{\sum\limits_{t=0}^{n} B_t(1 + i)^{-t}}{\sum\limits_{t=0}^{n} C_t(1 + i)^{-t}}$$

$$= \frac{PW_{benefits}}{PW_{costs}}$$

where B_t is the benefit (receipt) at time t and C_t is the cost (disbursement) at time t. If the benefit-cost ratio is greater than one, then the investment is acceptable. If the ratio is less than one, the investment is not acceptable. A ratio of one indicates breakeven situation for the project.

Simple payback period.   Payback period refers to the length of time it will take to recover an initial investment. The approach does not consider the impact of the time value of money. Consequently, it is not an accurate method of evaluating the worth of an investment. However, it is a simple technique that is used widely to perform a "quick-and-dirty" assessment of investment performance. Also, the technique considers only the initial cost. Other costs that may occur after time zero are not included in the calculation. The payback period is defined as the smallest value of $n(n_{min})$ that satisfies the following expression:

$$\sum_{t=1}^{n_{min}} R_t \geq C_0$$

where R_t is the revenue at time t and C_0 is the initial investment. The procedure calls for a simple addition of the revenues period by period until enough total has been accumulated to offset the initial investment.

Example

An organization is considering installing a new computer system that will generate significant savings in material and labor requirements for order processing. The system has an initial cost of $50,000. It is expected to save the organization $20,000 a year. The system has an anticipated useful life of 5 years with a salvage value of $5,000. Determine how long it would take for the system to pay for itself from the savings it is expected to generate. Since the annual savings are uniform, we can calculate the payback period by simply dividing the initial cost by the annual savings. That is,

$$n_{min} = \frac{\$50,000}{\$20,000}$$
$$= 2.5 \, years$$

Note that the salvage value of $5,000 is not included in the above calculation since the amount is not realized until the end of the useful life of the asset (i.e., after 5 years). In some cases, it may be desired to consider the salvage value. In that case, the amount to be offset by the annual savings will be the net cost of the asset. In that case, we would have

$$n_{min} = \frac{\$50,000 - \$5,000}{\$20,000}$$
$$= 2.25 \, years$$

If there are tax liabilities associated with the annual savings, those liabilities must be deducted from the savings before calculating the payback period.

Discounted payback period.   In this book, we introduce the *discounted payback period* approach in which the revenues are reinvested at a certain interest rate. The payback period is determined when enough money has been accumulated at the given interest rate to offset the initial cost as well as other interim costs. In this case, the calculation is done by the following expression:

$$\sum_{t=1}^{n_{min}} R_t(1 + i)^{n_{min}-t} \geq \sum_{t=0}^{n_{min}} C_t$$

Example

A new solar cell unit is to be installed in an office complex at an initial cost of $150,000. It is expected that the system will generate annual cost savings of $22,500 on the electricity bill. The solar cell unit will need to be overhauled every 5 years at a cost of $5,000 per overhaul. If the annual interest rate is 10%, find the *discounted payback period* for the solar cell unit considering the time value of money. The costs of the overhaul are to be considered in calculating the discounted payback period.

Solution.   Using the single payment compound amount factor for one period iteratively, the following solution is obtained:

Time	Cumulative savings
1	$22,500
2	$22,500 + $22,500(1.10)^1 = $47,250
3	$22,500 + $47,250(1.10)^1 = $74,475
4	$22,500 + $74,475(1.10)^1 = $104,422.50
5	$22,500 + $104,422.50(1.10)^1 - $5,000 = $132,364.75
6	$22,500 + $132,364.75(1.10)^1 = $168,101.23

The initial investment is $150,000. By the end of period 6, we have accumulated $168,101.23, more than the initial cost. Interpolating between period 5 and period 6, we obtain

$$n_{\min} = 5 + \frac{(150,000 - 132,364.75)}{(168,101.25 - 132,364.75)}(6 - 5)$$

$$= 5.49$$

That is, it will take 5.49 years or 5 years and 6 months to recover the initial investment.

Investment life for multiple returns.   The time it takes an amount to reach a certain multiple of its initial level is often of interest in many investment scenarios. The *Rule of 72* is one simple approach to calculating how long it will take an investment to double in value at a given interest rate per period. The Rule of 72 gives the following formula for estimating the doubling period:

$$n = \frac{72}{i}$$

where i is the interest rate expressed in a percentage. Referring to the single-payment compound amount factor, we can set the future amount equal to twice the present amount and then solve for n, the number of periods. That is, $F = 2P$. Thus,

$$2P = P(1 + i)^n$$

Solving for n in the above equation yields an expression for calculating the exact number of periods required to double P :

$$n = \frac{\ln(2)}{\ln(1 + i)}$$

where i is the interest rate expressed in decimals. In the general case, for exact computation, the length of time it would take to accumulate m multiples of P is expressed as

$$n = \frac{\ln(m)}{\ln(1 + i)}$$

where m is the desired multiple. For example, at an interest rate of 5% per year, the time it would take an amount, P, to double in value ($m = 2$) is 14.21 years. This, of course, assumes that the interest rate will remain constant throughout the planning horizon. Table 8–1 presents a tabulation of the values calculated from both approaches. Figure 8–9 shows a graphical comparison of the Rule of 72 to the exact calculation.

8.3.2 Effects of Inflation

Inflation is a major player in the financial and economic analyses of projects. Multiyear projects are particularly subject to the effects of inflation. Inflation can be defined as the decline in the purchasing power of money. Some of the most common causes of inflation are

- Increase in amount of currency in circulation
- Shortage of consumer goods
- Escalation of the cost of production
- Arbitrary price increases by resellers

The general effects of inflation are felt in terms of an increase in the prices of goods and a decrease in the worth of currency. In cash flow analysis, return on investment (ROI) for a project will be affected by the time value of money as well as inflation. The *real interest rate* (d) is defined as the desired rate of return in the absence of inflation. When we talk of today's dollars or constant dollars, we are referring to the use of the real interest rate. *Combined interest rate* (i) is the rate of return combining the real interest rate and inflation rate. If we denote the *inflation rate* as j, then the relationship between the different rates can be expressed as

$$(1 + i) = (1 + d)(1 + j)$$

Table 8–1 Evaluation of the Rule of 72

i %	n (Rule of 72)	n (exact value)
0.25	$288.00	$277.61
0.50	144.00	138.98
1.00	72.00	69.66
2.00	36.00	35.00
5.00	14.20	17.67
8.00	9.00	9.01
10.00	7.20	7.27
12.00	6.00	6.12
15.00	4.80	4.96
18.00	4.00	4.19
20.00	3.60	3.80
25.00	2.88	3.12
30.00	2.40	2.64

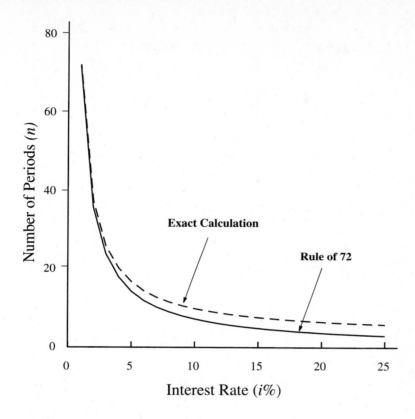

Figure 8-9. Evaluation of Investment Life for Double Return

Thus, the combined interest rate can be expressed as

$$i = d + j + dj$$

Note that if $j = 0$ (i.e., no inflation), then $i = d$. We can also define *commodity escalation rate* (g) as the rate at which individual commodity prices escalate. This may be greater than or less than the overall inflation rate. In practice, several measures are used to convey inflationary effects. Some of these are the *Consumer Price Index*, *Producer Price Index*, and *Wholesale Price Index*. A *market basket rate* is defined as the estimate of inflation based on a weighted average of the annual rates of change in the costs of a wide range of representative commodities. A *then-current* cash flow is a cash flow that explicitly incorporates the impact of inflation. A *constant worth* cash flow is a cash flow that does not incorporate the effect of inflation. The real interest rate, d, is used for analyzing constant worth cash flows. Figure 8–10 shows constant worth and then-current cash flows.

The then-current cash flow in the figure is the equivalent cash flow considering the effect of inflation. C_k is what it would take to buy a certain basket of goods after k time periods if there was no inflation. T_k is what it would take to buy the same

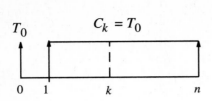
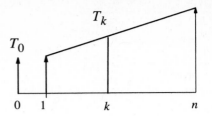

Figure 8-10. Cash Flows for Effects of Inflation

basket in k time period if inflation were taken into account. For the constant worth cash flow, we have

$$C_k = T_0, \qquad k = 1, 2, \ldots, n$$

and for the then-current cash flow, we have

$$T_k = T_0(1 + j)^k, \qquad k = 1, 2, \ldots, n$$

where j is the inflation rate. If $C_k = T_0 = \$100$ under the constant worth cash flow, then we mean $100 worth of buying power. If we are using the commodity escalation rate, g, then we will have

$$T_k = T_0(1 + g)^k, \qquad k = 1, 2, \ldots, n$$

Thus, a then-current cash flow may increase based on both a regular inflation rate (j) and a commodity escalation rate (g). We can convert a then-current cash flow to a constant worth cash flow by using the following relationship:

$$C_k = T_k(1 + j)^{-k}, \qquad k = 1, 2, \ldots, n$$

If we substitute T_k from the commodity escalation cash flow into the expression for C_k above, we get

$$\begin{aligned} C_k &= T_k(1 + j)^{-k} \\ &= T_0(1 + g)k(1 + j)^{-k} \\ &= T_0[(1 + g)/(1 + j)]^k, \qquad k = 1, 2, \ldots, n \end{aligned}$$

Note that if $g = 0$ and $j = 0$, then $C_k = T_0$. That is, no inflationary effect. We now define effective commodity escalation rate (v) as

$$v = [(1 + g)/(1 + j)] - 1$$

and we can express the commodity escalation rate (g) as

$$g = v + j + vj$$

Inflation can have a significant impact on the financial and economic aspects of a project. Inflation may be defined, in economic terms, as the increase in the amount of currency in circulation, resulting in a relatively high and sudden fall in its value. To a producer, inflation means a sudden increase in the cost of items that serve as inputs for the production process (equipment, labor, materials, etc). To the retailer, inflation implies an imposed higher cost of finished products. To an ordinary citizen, inflation portends an unbearable escalation of prices of consumer goods. All these views are interrelated in a project management environment.

The amount of money supply, as a measure of a country's wealth, is controlled by the government. Faced with no other choice, governments often feel impelled to create more money or credit to take care of old debts and pay for social programs. When money is generated at a faster rate than the growth of goods and services, it becomes a surplus commodity, and its value (purchasing power) will fall. This means that there will be too much money available to buy only a few goods and services. When the purchasing power of a currency falls, each individual in a product's life cycle has to dispense more of the currency in order to obtain the product. Some of the classic concepts of inflation are discussed next.

1. Increases in producer's costs are passed on to consumers. At each stage of the product's journey from producer to consumer, prices are escalated disproportionately in order to make a good profit. The overall increase in the product's price is directly proportional to the number of intermediaries it encounters on its way to the consumer. This type of inflation is called cost-driven (or cost-push) inflation.

2. Excessive spending power of consumers forces an upward trend in prices. This high spending power is usually achieved at the expense of savings. The law of supply and demand dictates that the more the demand, the higher the price. This type of inflation is known as demand-driven (or demand-pull) inflation.

3. The impact of international economic forces can induce inflation in a local economy. Trade imbalances and fluctuations in currency values are notable examples of international inflationary factors.

4. Increasing base wages of workers generate more disposable income and, hence, higher demand for goods and services. The high demand, consequently, creates a pull on prices. Coupled with this, employers pass on the additional wage cost to consumers through higher prices. This type of inflation is, perhaps, the most difficult to solve because wages set by union contracts and prices set by producers almost never fall, at least not permanently. This type of inflation may be referred to as wage-driven (or wage-push) inflation.

5. Easy availability of credit leads consumers to buy now and pay later and, thereby, create another loophole for inflation. This is a dangerous type of inflation because the credit not only pushes prices up, but it also leaves consumers with less money later on to pay for the credit. Eventually, many credits become uncollectible debts, which may then drive the economy into recession.

6. Deficit spending results in an increase in the money supply and, thereby, creates less room for each dollar to get around. The popular saying which indicates that "a dollar does not go far anymore" simply refers to inflation in laymen's terms.

The different levels of inflation may be categorized as discussed next.

Mild inflation. When inflation is mild (2 to 4 percent) the economy actually prospers. Producers strive to produce at full capacity in order to take advantage of the high prices to the consumer. Private investments tend to be brisk and more jobs become available. However, the good fortune may be only temporary. Prompted by the prevailing success, employers are tempted to seek larger profits and workers begin to ask for higher wages. They cite their employer's prosperous business as a reason to bargain for bigger shares of the business profit. So, we end up with a vicious cycle where the producer asks for higher prices, the unions ask for higher wages, and inflation starts an upward trend.

Moderate inflation. Moderate inflation occurs when prices increase at 5 to 9 percent. Consumers start purchasing more as an edge against inflation. They would rather spend their money now than watch it decline further in purchasing power. The increased market activity serves to fuel further inflation.

Severe inflation. Severe inflation is indicated by price escalations of 10 percent or more. Double-digit inflation implies that prices rise much faster than wages do. Debtors tend to be the ones who benefit from this level of inflation because they repay debts with money that is less valuable than the one borrowed.

Hyperinflation. When each price increase signals the increase in wages and costs, which again sends prices up further, the economy has reached a stage of malignant galloping inflation or hyperinflation. Rapid and uncontrollable inflation destroys the economy. The currency becomes economically useless as the government prints it excessively to pay for obligations.

Inflation can affect any project in terms of raw materials procurement, salaries and wages, and/or a cost tracking dilemma. Some effects are immediate and easily observable. Other effects are subtle and pervasive. Whatever form it takes, inflation must be taken into account in long-term project planning and control. Large projects may be adversely affected by the effects of inflation in terms of cost overruns and poor resource utilization. The level of inflation determines the severity of the impact on projects.

8.4 BREAK-EVEN ANALYSIS

Break-even analysis refers to the determination of the balanced performance level where project income is equal to project expenditure. The total cost of an operation is expressed as the sum of the fixed and variable costs with respect to output quantity. That is,

$$TC(x) = FC + VC(x)$$

where x is the number of units produced, $TC(x)$ is the total cost of producing x units, FC is the total fixed cost, and $VC(x)$ is the total variable cost associated with producing x units. The total revenue resulting from the sale of x units is defined as

$$TR(x) = px$$

where p is the price per unit. The profit due to the production and sale of x units of the product is calculated as

$$P(x) = TR(x) - TC(x)$$

The break-even point of an operation is defined as the value of a given parameter that will result in neither profit nor loss. The parameter of interest may be the number of units produced, the number of hours of operation, the number of units of a resource type allocated, or any other measure of interest. At the break-even point, we have the following relationship:

$$TR(x) = TC(x)$$

or

$$P(x) = 0$$

In some cases, there may be a known mathematical relationship between cost and the parameter of interest. For example, there may be a linear cost relationship between the total cost of a project and the number of units produced. The cost expressions facilitate straightforward break-even analysis. Figure 8–11 shows an example of a break-even point for a single project. Figure 8–12 shows examples of multiple break-even points that exist when multiple projects are compared. When two project alternatives are compared, the break-even point refers to the point of indifference between the two alternatives. In Figure 8–12, x_1 represents the point where projects A and B are equally desirable, x_2 represents where A and C are equally desirable, and x_3 represents where B and C are equally desirable. The figure shows that if we are operating below a production level of x_2 units, then project C is the preferred project among the three. If we are operating at a level of more than x_2 units, then project A is the best choice.

Example

Three project alternatives are being considered for producing a new product. The required analysis involves determining which alternative should be selected on the basis of how many units of the product are produced per year. Based on past records, there is a known relationship between the number of units produced per year, x, and the net annual profit, $P(x)$, from each alternative. The level of production is expected to be between 0 and 250 units per year. The net annual profits (in thousands of dollars) are given below for each alternative:

Project A : $P(x) = 3x - 200$
Project B : $P(x) = x$
Project C : $P(x) = (1/50)x_2 - 300.$

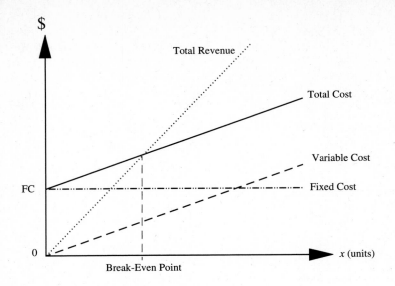

Figure 8-11. Break-Even Point for a Single Project

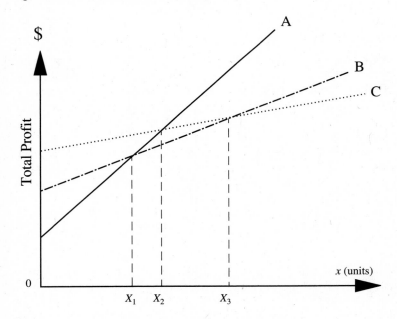

Figure 8-12. Break-Even Points for Multiple Projects

   This problem can be solved mathematically by finding the intersection points of the profit functions and evaluating the respective profits over the given range of product units. It can also be solved by a graphical approach. Figure 8–13 shows a plot of the profit functions. Such a plot is called a break-even chart. The plot shows that project B should be selected if between 0 and 100 units are to be produced. Project A should be selected if between 100 and 178.1 units (178 physical units) are to be produced. Project C should be selected if more than 178 units are to be produced. It should be noted that if less than 66.7 units (66 physical units) are produced, project A will generate net loss rather than net profit. Similarly, project C will generate losses if less than 122.5 units (122 physical units) are produced.

8.4.1 Profit Ratio for Project Investment

Break-even charts offer opportunities for several different types of analysis. In addition to the break-even points, other measures of worth or criteria may be derived from the charts. A measure, called *profit ratio* (Badiru 1991a), is presented here for the purpose of obtaining a further comparative basis for competing projects. Profit ratio is defined as the ratio of the profit area to the sum of the profit and loss areas in a break-even chart. That is,

$$\text{profit ratio} = \frac{\text{area of profit region}}{\text{area of profit region} + \text{area of loss region}}$$

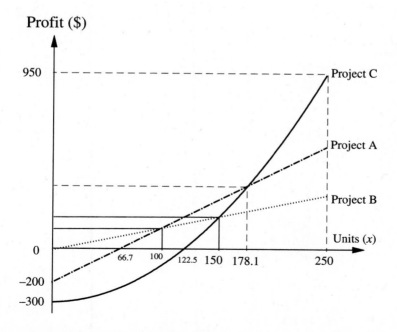

Figure 8-13.   Plot of Profit Functions

For example, suppose the expected revenue and the expected total cost associated with a project are given, respectively, by the following expressions:

$$R(x) = 100 + 10x$$
$$TC(x) = 2.5x + 250$$

where x is the number of units produced and sold from the project. Figure 8–14 shows the break-even chart for the project. The break-even point is shown to be 20 units. Net profits are realized from the project if more than 20 units are produced, and net losses are realized if less than 20 units are produced. It should be noted that the revenue function in Figure 8–14 represents an unusual case where a revenue of $100 is realized when 0 units are produced.

Suppose it is desired to calculate the profit ratio for this project if the number of units that can be produced is limited to between 0 and 100 units. From Figure 8–14, the surface area of the profit region and the area of the loss region can be calculated by using the standard formula for finding the area of a triangle: area = (1/2)(base)(height).

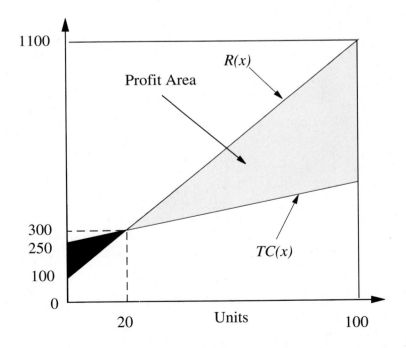

Figure 8-14. Area of Profit versus Area of Loss

Using this formula, we have the following:

$$\text{area of profit region} = \frac{1}{2}(\text{base})(\text{height})$$
$$= \frac{1}{2}(1{,}100 - 500)(100 - 20)$$
$$= 24{,}000 \text{ square units}$$
$$\text{area of loss region} = \frac{1}{2}(\text{base})(\text{height})$$
$$= \frac{1}{2}(250 - 100)(20)$$
$$= 1{,}500 \text{ square units}$$

Thus, the profit ratio is computed as

$$\text{profit ratio} = \frac{24{,}000}{24{,}000 + 1{,}500}$$
$$= 0.9411$$
$$= 94.11\%$$

The profit ratio may be used as a criterion for selecting among project alternatives. If this is done, the profit ratios for all the alternatives must be calculated over the same values of the independent variable. The project with the highest profit ratio will be selected as the desired project. For example, Figure 8–15 presents the break-even chart for an alternate project, say project II. It is seen that both the revenue and cost functions for the project are nonlinear. The revenue and cost are defined as follows:

$$R(x) = 160x - x^2$$
$$TC(x) = 500 + x^2$$

If the cost and/or revenue functions for a project are not linear, the areas bounded by the functions may not easily be determined. For those cases, it may be necessary to use techniques such as definite integrals to find the areas. Figure 8–15 indicates that the project generates a loss if less than 3.3 units (3 actual units) or more than 76.8 units (76 actual units) are produced. The respective profit and loss areas on the chart are calculated as

$$\text{area 1 (loss)} = \int_{0}^{3.3} [(500 + x^2) - (160x - x^2)] \, dx$$
$$= 802.8 \text{ unit-dollars}$$
$$\text{area 2 (profit)} = \int_{3.3}^{76.8} [(160 - x^2) - (500x + x^2)] \, dx$$
$$= 132{,}272.08 \text{ unit-dollars}$$
$$\text{area 3 (loss)} = \int_{76.8}^{100} [(500 + x^2) - (160x - x^2)] \, dx$$
$$= 48{,}135.98 \text{ unit-dollars}$$

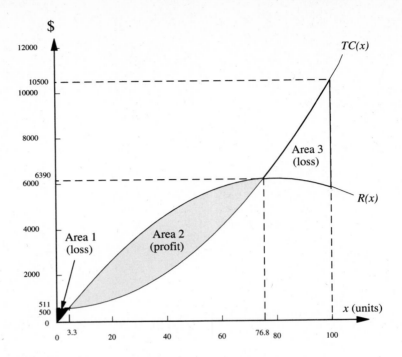

Figure 8-15. Break-Even Chart for Revenue and Cost Functions

Consequently, the profit ratio for project II is computed as

$$\text{profit ratio} = \frac{\text{total area of profit region}}{\text{total area of profit region} + \text{total area of loss region}}$$

$$= \frac{132{,}272.08}{802.76 + 132{,}272.08 + 48{,}135.98}$$

$$= 0.7299$$

$$= 72.99\%$$

The profit ratio approach evaluates the performance of each alternative over a specified range of operating levels. Most of the existing evaluation methods use single-point analysis with the assumption that the operating condition is fixed at a given production level. The profit ratio measure allows an analyst to evaluate the net yield of an alternative given that the production level may shift from one level to another. An alternative, for example, may operate at a loss for most of its early life, while it may generate large incomes to offset the losses in its later stages. Conventional methods cannot easily capture this type of transition from one performance level to another. In addition to being used to compare alternate projects, the profit ratio may also be used for evaluating the economic feasibility of a single project. In such a case, a decision rule may be developed. An example of such a decision rule is

If the profit ratio is greater than 75%, accept the project.
If the profit ratio is less than or equal to 75%, reject the project.

.5 AMORTIZATION SCHEDULE FOR PROJECT FINANCE

Many capital investment projects are financed with external funds. A careful analysis must be conducted to ensure that the amortization schedule can be handled by the organization involved. A computer program such as GAMPS (Graphic Evaluation of Amortization Payments) might be used for this purpose (Badiru 1988c). The program analyzes the installment payments, the unpaid balance, principal amounts paid per period, total installment payment, and current cumulative equity. It also calculates the *equity break-even point* for the debt being analyzed. The equity break-even point indicates the time when the unpaid balance on a loan is equal to the cumulative equity on the loan. With the output of this program, the basic cost of servicing the project debt can be evaluated quickly. A part of the output of the program presents the percentage of the installment payment going into equity and interest charge respectively. The computational procedure for analyzing project debt is as follows:

1. Given a principal amount, P, a periodic interest rate, i (in decimals), and a discrete time span of n periods, the uniform series of equal end-of-period payments needed to amortize P is computed as

$$A = \frac{P[i(1 + i)^n]}{(1 + i)^n - 1}$$

It is assumed that the loan is to be repaid in equal monthly payments. Thus, $A(t) = A$, for each period t throughout the life of the loan.

2. The unpaid balance after making t installment payments is given by

$$U(t) = \frac{A[1 - (1 + i)^{t-n}]}{i}$$

3. The amount of equity or principal amount paid with installment payment number t is given by

$$E(t) = A(1 + i)^{t-n-1}$$

4. The amount of interest charge contained in installment payment number t is derived to be

$$I(t) = A[1 - (1 + i)^{t-n-1}]$$

where $A = E(t) + I(t)$.

5. The cumulative total payment made after t periods is denoted by

$$C(t) = \sum_{k=1}^{t} A(k)$$

$$= \sum_{k=1}^{t} A$$

$$= A \cdot t$$

6. The cumulative interest payment after t periods is given by

$$Q(t) = \sum_{x=1}^{t} I(x)$$

7. The cumulative principal payment after t periods is computed as

$$S(t) = \sum_{k=1}^{t} E(k)$$

$$= A \sum_{k=1}^{t} (1 + i)^{-(n-k+1)}$$

$$= A \left[\frac{(1 + i)^t - 1}{i(1 + i)^n} \right]$$

where

$$\sum_{n=1}^{t} x^n = \frac{x^{t+1} - x}{x - 1}$$

8. The percentage of interest charge contained in installment payment number t is:

$$f(t) = \frac{I(t)}{A}(100\%)$$

9. The percentage of cumulative interest charge contained in the cumulative total payment up to and including payment number t is

$$F(t) = \frac{Q(t)}{C(t)}(100\%)$$

10. The percentage of cumulative principal payment contained in the cumulative total payment up to and including payment number t is

$$H(t) = \frac{S(t)}{C(t)}$$
$$= \frac{C(t) - Q(t)}{C(t)}$$
$$= 1 - \frac{Q(t)}{C(t)}$$
$$= 1 - F(t)$$

Example

Suppose a manufacturing productivity improvement project is to be financed by borrowing $500,000 from an industrial development bank. The annual nominal interest rate for the loan is 10%. The loan is to be repaid in equal monthly installments over a period of 15 years. The first payment on the loan is to be made exactly one month after financing is approved. It is desired to perform a detailed analysis of the loan schedule. The GAMPS computer program (Badiru 1988c) was used to analyze this problem. Table 8–2 presents a partial output of GAMPS for the loan repayment schedule.

The tabulated result shows a monthly payment of $5,373.04 on the loan. Considering time $t = 10$ months, one can see the following results:

Table 8-2   Amortization Schedule for Financed Project

t	$U(t)$	$A(t)$	$E(t)$	$I(t)$	$C(t)$	$S(t)$	$f(t)$	$F(t)$
1	498794.98	5373.04	1206.36	4166.68	5373.04	1206.36	77.6	77.6
2	497578.56	5373.04	1216.42	4156.62	10746.08	2422.78	77.4	77.5
3	496352.01	5373.04	1226.55	4146.49	16119.12	3649.33	77.2	77.4
4	495115.24	5373.04	1236.77	4136.27	21492.16	4886.10	76.9	77.3
5	493868.16	5373.04	1247.08	4125.96	26865.20	6133.18	76.8	77.2
6	492610.69	5373.04	1257.47	4115.57	32238.24	7390.65	76.6	77.1
7	491342.74	5373.04	1267.95	4105.09	37611.28	8658.61	76.4	76.9
8	490064.22	5373.04	1278.52	4094.52	42984.32	9937.12	76.2	76.9
9	488775.05	5373.04	1289.17	4083.87	48357.36	11226.29	76.0	76.8
10	487475.13	5373.04	1299.91	4073.13	53730.40	12526.21	75.8	76.7
⋮	⋮	⋮	⋮	⋮	⋮	⋮	⋮	⋮
170	51347.67	5373.04	4904.27	468.77	913416.80	448656.40	8.7	50.9
171	46402.53	5373.04	4945.14	427.90	918789.84	453601.54	7.9	50.6
172	41416.18	5373.04	4986.35	386.69	924162.88	458587.89	7.2	50.4
173	36388.27	5373.04	5027.91	345.13	929535.92	463615.80	6.4	50.1
174	31318.47	5373.04	5069.80	303.24	934908.96	468685.60	5.6	49.9
175	26206.42	5373.04	5112.05	260.99	940282.00	473797.66	4.9	49.6
176	21051.76	5373.04	5154.65	218.39	945655.04	478952.31	4.1	49.4
177	15854.15	5373.04	5197.61	175.43	951028.08	484149.92	3.3	49.1
178	10613.23	5373.04	5240.92	132.12	956401.12	489390.84	2.5	48.8
179	5328.63	5373.04	5284.60	88.44	961774.16	494675.44	1.7	48.6
180	0.00	5373.04	5328.63	44.41	967147.20	500004.07	0.8	48.3

$U(10) = \$487,475.13$ (unpaid balance)
$A(10) = \$5,373.04$ (monthly payment)
$E(10) = \$1,299.91$ (equity portion of the tenth payment)
$I(10) = \$4,073.13$ (interest charge contained in the tenth payment)
$C(10) = \$53,730.40$ (total payment to date)
$S(10) = \$12,526.21$ (total equity to date)
$f(10) = 75.81\%$ (percentage of the tenth payment going into interest charge)
$F(10) = 76.69\%$ (percentage of the total payment going into interest charge)

Thus, over 76% of the sum of the first ten installment payments goes into interest charges. The analysis shows that by time $t = 180$, the unpaid balance has been reduced to zero. That is, $U(180) = 0.0$. The total payment made on the loan is \$967,148.40 and the total interest charge is $\$967,148.20 - \$500,000 = \$467,148.20$. So, 48.30% of the total payment goes into interest charges. The information about interest charges might be very useful for tax purposes. The tabulated output shows that equity builds up slowly while the unpaid balance decreases slowly. Note that very little equity is accumulated during the first three years of the loan schedule. This is shown graphically in Figure 8–16. The effects of inflation, depreciation, property appreciation, and other economic factors are not included in the analysis presented above. A project analyst should include such factors whenever they are relevant to the loan situation.

The point at which the curves intersect is referred to as the equity break-even point. It indicates when the unpaid balance is exactly equal to the accumulated equity or the cumulative principal payment. For the example, the equity break-even point is 120.9 months (over 10 years). The importance of the equity break-even point is that any equity accumulated after that point represents the amount of ownership or equity that the debtor is entitled to after the unpaid balance on the loan is settled with project collateral. The implication of this is very important, particularly in the case of mortgage loans. *Mortage* is a word of French origin meaning *death pledge*—perhaps, a sarcastic reference to the

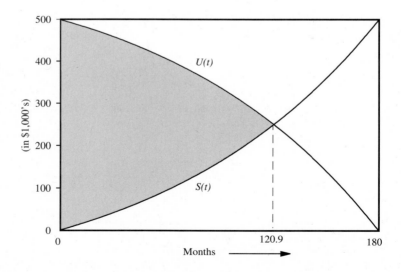

Figure 8-16. Plot of Unpaid Balance and Cumulative Equity

burden of mortgage loans. The equity break-even point can be calculated directly from the following formula.

Let the equity break-even point, x, be defined as the point where $U(x) = S(x)$. That is,

$$A\left[\frac{1 - (1 + i)^{-(n-x)}}{i}\right] = A\left[\frac{(1 + i)^x - 1}{i(1 + i)^n}\right]$$

Multiplying both the numerator and denominator of the left-hand side of the above expression by $(1 + i)^n$ and simplifying yields

$$\frac{(1 = i)^n - (1 + i)^x}{i(1 + i)^n}$$

on the left-hand side. Consequently, we have

$$(1 + i)^n - (1 + i)^x = (1 + i)^x - 1$$
$$(1 + i)^x = \frac{(1 + i)^n + 1}{2}$$

which yields the equity break-even expression

$$x = \frac{\ln[0.5(1 + i)^n + 0.5]}{\ln(1 + i)}$$

where

ln is the natural log function.
n is the number of periods in the life of the loan.
i is the interest rate per period.

Figure 8–17 presents a plot of the total loan payment and the cumulative equity with respect to time. The total payment starts from $0.0 at time 0 and goes up to $967,147.20 by the end of the last month of the installment payments. Since only $500,000 was borrowed, the total interest payment on the loan is $967,147.20 − $500,000 = $467,147.20. The cumulative principal payment starts at $0.0 at time 0 and slowly builds up to $500,001.34, which is the original loan amount. The extra $1.34 is due to a round-off error in the calculations.

Figure 8–18 presents a plot of the percentage of interest charge in the monthly payments and the percentage of interest charge in the total payment. The percentage of interest charge in the monthly payments starts at 77.55% for the first month and decreases to 0.83% for the last month. By comparison, the percentage of interest in the total payment also starts at 77.55% for the first month and slowly decreases to 48.30% by the time the last payment is made at time 180. Table 8–2 and Figure 8–18 show that an increasing proportion of the monthly payment goes into the principal payment as time goes on. If the interest charges are tax deductible, the decreasing values of $f(t)$ mean that there would be decreasing tax benefits from the interest charges in the later months of the loan.

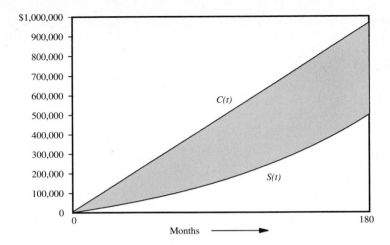

Figure 8-17.   Plot of Total Loan Payment and Total Equity

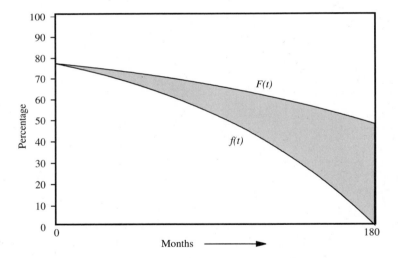

Figure 8-18.   Plot of Percentage of Interest Charge

8.6 REVENUE REQUIREMENT ANALYSIS

Companies evaluating capital expenditures for proposed projects must weigh the expected benefits against the initial and expected costs over the life cycle of the project. One method that is often used is the minimum annual revenue requirements (MARR) analysis. Using the information about costs, interest payments, recurring expenditures, and other project-related financial obligations, the minimum annual revenue required by a project can be calculated. We can compute the break-even point of the project. The break-even point is then used to determine the level of revenue that must be produced by the project in order for it to be profitable. The analysis can be done with either the flow-through method or the normalizing method (Stevens 1979).

The factors to be included in the analysis are initial investment cost, book salvage value, tax salvage value, annual project costs, useful life for bookkeeping purposes, book depreciation method, tax depreciation method, useful life for tax purposes, rate of return on equity, rate of return on debt, capital interest rate, debt ratio, and investment tax credit. The minimum annual revenue requirement for any year n may be determined by means of the net cash flows expected for that year: net cash flow = income − taxes − principal paid. That is,

$$X_n = (G - C - I) - t - P$$

where

G = gross income for year n
C = expenses for year n
I = interest payment for year n
t = taxes for year n
P = principal payment for year n

Rewriting the equation yields

$$G = X_n + C + I + t + P$$

This equation assumes that there are no capital requirements, salvage value considerations, or working capital changes in year n. For the minimum annual gross income, the cash flow, X_n, must satisfy the following relationship:

$$X_n = D_e + f_n$$

where

D_e = recovered portion of the equity capital
f_n = return on the unrecovered equity capital

It is assumed that the total equity and debt capital recovered in a year are equal to the book depreciation, D_b, and that the principal payments are a constant percentage of the book depreciation. That is,

$$P = c(D_b)$$

where c is the debt ratio. The recovery of equity capital is, therefore, given by

$$D_e = (1 - c)D_b$$

and the annual return on equity, f_n, and interest, I, are based on the unrecovered balance as

$$f_n = (1 - c)k_e(BV_{n-1})$$
$$I = ck_d(BV_{n-1})$$

where

c = debt ratio
k_e = required rate of return on equity
k_d = required rate of return on debt capital
BV_{n-1} = book value at the beginning of year n

Based on the preceding equations, the minimum annual gross income, or revenue requirement, for year n can be described as

$$G = D_b + f_n + C + I + t$$

An expression for taxes, t, is given by

$$t = (G - C - D_t - I)T$$

where

D_t = depreciation for tax purposes
T = tax rate

If the expression for R is substituted for G in the above equation, the following alternate expression for t can be obtained:

$$t = \frac{T}{1 - T}(D_b + f_n - D_t)$$

The calculated minimum annual revenue requirement can be used to evaluate the economic feasibility of a project. An example of a decision criterion that may be used for that purpose is presented next.

Decision criterion.   If expected gross incomes are greater than the minimum annual revenue requirements, then the project is considered to be economically acceptable and the project investment is considered to be potentially profitable. Economic acceptance should be differentiated from technical acceptance. If, of course, other alternatives being considered have similar results, a comparison based on the margin of difference (i.e., incremental analysis) between the expected gross incomes and minimum annual revenue requirements must be made. There are two extensions to the basic analysis procedure presented above. They are the *flow-through* and *normalizing methods* (Stevens 1979).

8.6.1 Flow-Through Method of MARR

The flow-through method of the basic revenue requirement analysis allocates credits and costs in the year that they occur. That is, there are no deferred taxes and the investment tax credit is not amortized. Capitalized interest is taken as an expense in the first year. The resulting equation for calculating the minimum annual revenue requirements is

$$R = D_b + f_e + I + gP + C + t$$

where the required return on equity is given by

$$f_e = k_e(1 - c)K_{n-1}$$

where

k_e = implied cost of common stock
c = debt ratio
K_{n-1} = chargeable investment for the preceding year
$K_n = K_{n-1} - D_b$(with K_0 = initial investment)
g = capitalized interest rate

The capitalized interest rate is usually set by federal regulations. The debt interest is given by

$$I = (c)k_d K_{n-1}$$

where

k_d = after-tax cost of capital

The investment tax credit is calculated as

$$C_t = i_t P$$

where i_t is the investment tax credit. Costs, C, are estimated totals that include such items as ad valorem taxes, insurance costs, operation costs, and maintenance costs. The taxes for the flow-through method are calculated as

$$t = \frac{T}{1 - T}(f_e + D_b - D_t) - \frac{C}{1 - T}$$

8.6.2 Normalizing Method of MARR

The normalizing method differs from the flow-through method in that deferred taxes are utilized. These deferred taxes are sometimes included as expenses in the early years of the project and then as credits in later years. This *normalized* treatment of the deferred taxes is often used by public utilities to minimize the potential risk of changes in tax rules that may occur before the end of the project but are unforeseen at the start of the project. Also, the interest paid on the initial investment cost is capitalized. That is, it is taken as a tax deduction in the first year of the project and then amortized over the life of the project to spread out the interest costs. The resulting minimum annual revenue requirement is expressed as

$$R = D_b + d_t + C_t - A_t + I + f_e + t + C$$

where the depreciation schedules are based on the following capitalized investment cost:

$$K = P + gP$$

with P and g as previously defined. The deferred taxes, d_t, are the differences in taxes that result from using an accelerated depreciation model instead of a straight line rate over the life of the project. That is,

$$d_t = (D_t - D_s)T$$

where

D_t = accelerated depreciation for tax purposes
D_s = straight line depreciation for tax purposes

The amortized investment tax credit, A_t, is spread over the life of the project, n, and is calculated as

$$A_t = \frac{C_t}{n}$$

The debt interest is similar to the earlier equation for capitalized interest. However, the chargeable investment differs by taking into account the investment tax credit, deferred taxes, and the amortized investment tax credit. The resulting expressions are

$$I = k_d(c)K_{n-1}$$
$$K_n = K_{n-1} - D_b - C_t - d_t - A_t$$

In this case, the expression for taxes, t, is given by

$$t = \frac{T}{1-T}(f_e + D_b + d_t + C_t - A_t - D_t - gP) - \frac{C_t}{1-T}$$

The differences between the procedures for calculating the minimum annual revenue requirements for the flow-through and the normalizing methods yield some interesting and important details (Badiru and Russell 1987). If the MARRs are converted to uniform annual amounts (leveled), a better comparison between the effects of the calculations for each method can be made. For example, the MARR data calculated by using each method are presented in Table 8–3.

Table 8-3 Normalizing versus Flow-Through Revenue Analysis

Year	Normalizing		Flow-through	
	R_n	R_u	R_n	R_u
1	$7,135	$5,661	$5,384	$5,622
2	6,433	5,661	6,089	5,622
3	5,840	5,661	5,913	5,622
4	5,297	5,661	5,739	5,622
5	4,812	5,661	5,565	5,622
6	4,380	5,661	5,390	5,622
7	4,005	5,661	5,214	5,622
8	3,685	5,661	5,040	5,622

The annual MARR values are denoted by R_n, and the uniform annual amounts are denoted by R_u. The uniform amounts are found by calculating the present value for each yearly amount and then converting that total amount to equal yearly amounts over the same span of time. For a given investment, the flow-through method will produce a lower levelized minimum annual revenue requirement. This is because the normalized data includes an amortized investment tax credit and also deferred taxes. The yearly data for the flow-through method should give values closer to the actual cash flows, because credits and costs are assigned in the year in which they occur and not upfront as in the normalizing method.

The normalizing method, however, provides for a faster recovery of the project investment. For this reason, this method is often used by public utility companies when establishing utility rates. The normalizing method also agrees more in practice with required accounting procedures used by utility companies than does the flow-through method. Return on equity also differs between the two methods. For a given internal rate of return, the normalizing method will give a higher rate of return on equity than the flow-through method. This difference occurs because of the inclusion of deferred taxes in the normalizing method.

Example

Suppose we have the following data for a project. It is desired to perform a revenue requirement analysis using both the flow-through and normalizing methods.

Initial project cost	= $100,000
Book salvage value	= $10,000
Tax salvage value	= $10,000
Book depreciation model	= Straight line
Tax depreciation model	= Sum-of-years' digits
Life for book purposes	= 10 years
Life for tax purposes	= 10 years
Total costs per year	= $4,000
Debt ratio	= 40%
Required return on equity	= 20%
Required return on debt	= 10%
Tax rate	= 52%

Capitalized interest   = 0%
Investment tax credit   = 0%.

Tables 8-4, 8-5, 8-6, and 8-7 show the differences between the normalizing and flow-through methods for the same set of data. The different treatments of capital investment produced by the investment tax credit can be seen.

There is a big difference in the distribution of taxes, since most of the taxes are paid early in the investment period with the normalizing method and taxes are deferred with the flow-through method. The resulting minimum annual revenue requirements are larger for the normalizing method early in the period. However, there is a more gradual decrease with the flow-through method. Therefore, the use of the flow-through method does not put as great a demand on the project to produce high revenues early in the project life cycle as does the normalizing method. Also, the normalizing method produces a lower rate of return on equity. This fact may be of particular interest to shareholders.

Table 8-4 Part One of MARR Analysis

	Tax depreciation		Deferred taxes	
Year	Normalizing	Flow-through	Normalizing	Flow-through
1	$16,363.64	$16,363.64	$3,829.09	None
2	14,727.27	14,727.27	2,978.18	
3	13,090.91	13,090.91	2,127.27	
4	11,454.55	11,454.55	1,276.36	
5	9,818.18	9,818.18	425.45	
6	8,181.82	8,181.82	−425.45	
7	6,545.45	6,545.45	−1,276.36	
8	4,909.09	4,909.09	−2,127.27	
9	3,272.73	3,272.73	−2,978.18	
10	1,636.36	1,636.36	−3,829.09	

Table 8-5 Part Two of MARR Analysis

	Capitalized investment		Taxes	
Year	Normalizing	Flow-Through	Normalizing	Flow-Through
	$100,000.00	$100,000	—	—
1	87,170.91	91,000	$9,170.91	$5,022.73
2	75,192.73	92,000	8,354.04	5,625.46
3	64,065.46	73,000	7,647.78	6,228.18
4	53,789.90	64,000	7,052.15	6,830.91
5	44,363.64	55,000	6,567.13	7,433.64
6	35,789.09	46,000	6,192.73	8,036.36
7	28,065.45	37,000	5,928.94	8,639.09
8	21,192.72	28,000	5,775.78	9,241.82
9	15,170.90	19,000	5,733.24	9,844.55
10	10,000.00	10,000	5,801.31	1,0447.27

Table 8-6   Part Three of MARR Analysis

Year	Return on debt		Return of equity	
	Normalizing	Flow-through	Normalizing	Flow-through
1	$4,000.00	$4,000	$12,000.00	$12,000
2	3,486.84	3,640	10,460.51	10,920
3	3,007.71	3,280	9,023.13	9,840
4	2,562.62	2,920	7,687.86	8,760
5	2,151.56	2,560	6,454.69	7,680
6	1,774.55	2,200	5,323.64	6,600
7	1,431.56	1,840	4,294.69	5,520
8	1,122.62	1,480	3,367.85	4,440
9	847.71	1,120	2,543.13	3,360
10	606.84	760	1,820.51	2,280

Table 8-7   Part Four of MARR Analysis

Year	Minimum annual revenues	
	Normalizing	Flow-through
1	$42,000.00	$34,022.73
2	38,279.56	33,185.45
3	34,805.89	32,348.18
4	31,578.98	31,510.91
5	28,598.84	30,673.64
6	25,865.45	29,836.36
7	23,378.84	28,999.09
8	21,138.98	28,161.82
9	19,145.89	27,324.55
10	17,399.56	26,487.27

.7 COST ESTIMATION

Cost estimation and budgeting help establish a strategy for allocating resources in project planning and control. There are three major categories of cost estimation for budgeting. These are based on the desired level of accuracy. The categories are *order-of-magnitude estimates*, *preliminary cost estimates*, and *detailed cost estimates*. Order-of-magnitude cost estimates are usually gross estimates based on the experience and judgment of the estimator. They are sometimes called "ballpark" figures. These estimates are typically made without a formal evaluation of the details involved in the project. The level of accuracy associated with order-of-magnitude estimates can range from −50% to +50% of the actual cost. These estimates provide a quick way of getting cost information during the initial stages of a project.

50%(actual cost) ≤ order-of-magnitude estimate ≤ 150%(actual cost)

Preliminary cost estimates are also gross estimates, but with a higher level of accuracy. In developing preliminary cost estimates, more attention is paid to some selected details of the project. An example of a preliminary cost estimate is the estimation of expected labor cost. Preliminary estimates are useful for evaluating project alternatives before final commitments are made. The level of accuracy associated with preliminary estimates can ranges from −20% to +20% of the actual cost.

$$80\%(\text{actual cost}) \leq \text{preliminary estimate} \leq 120\%(\text{actual cost})$$

Detailed cost estimates are developed after careful consideration is given to all the major details of a project. Typically considerable time is needed to obtain detailed cost estimates. Because of the amount of time and effort needed to develop detailed cost estimates, the estimates are usually developed after there is firm commitment that the project will take off. Detailed cost estimates are important for evaluating actual cost performance during the project. The level of accuracy associated with detailed estimates normally range from −5% to +5% of the actual cost.

$$95\%(\text{actual cost}) \leq \text{detailed cost} \leq 105\%(\text{actual cost})$$

There are two basic approaches to generating cost estimates. The first one is a variant approach, in which cost estimates are based on variations of previous cost records. The other approach is the generative cost estimation, in which cost estimates are developed from scratch without taking previous cost records into consideration.

8.7.1 Optimistic and Pessimistic Cost Estimates

Using an adaptation of the **PERT** formula, we can combine optimistic and pessimistic cost estimates. Let

O = optimistic cost estimate
M = most likely cost estimate
P = pessimistic cost estimate

Then, the estimated cost can be estimated as

$$E[C] = \frac{O + 4M + P}{6}$$

and the cost variance can be estimated as

$$V[C] = \left[\frac{P - O}{6}\right]^2$$

Several quantitative techniques are available for cost estimation. Some of these techniques are discussed in Chapter 9. They include forecasting, learning curve analysis, regression, and cost simulation.

8.8 BUDGETING AND CAPITAL RATIONING

Budgeting involves sharing limited resources among several project groups or functions in a project environment. Budget analysis can serve any of the following purposes:

- Plan for resources expenditure
- Project selection criterion
- Projection of project policy
- Basis for project control
- Performance measure
- Standardization of resource allocation
- Incentive for improvement

8.8.1 Top-Down Budgeting

Top-down budgeting involves collecting data from upper-level sources such as top and middle managers. The figures supplied by the managers may come from their personal judgment, past experience, or past data on similar project activities. The cost estimates are passed to lower-level managers, who then break the estimates down into specific work components within the project. These estimates may, in turn, be given to line managers, supervisors, and lead workers to continue the process until individual activity costs are obtained. Top management provides the global budget, while the functional level worker provides specific budget requirements for project items.

8.8.2 Bottom-Up Budgeting

In this method, elemental activities and their schedules, descriptions, and labor skill requirements are used to construct detailed budget requests. Line workers familiar with specific activities are requested to provide cost estimates. Estimates are made for each activity in terms of labor time, materials, and machine time. The estimates are then converted to an appropriate cost basis. The dollar estimates are combined into composite budgets at each successive level up the budgeting hierarchy. If estimate discrepancies develop, they can be resolved through the intervention of senior management, middle management, functional managers, project manager, accountants, or standard cost consultants. Figure 8–19 shows the breakup of a project into phases and parts to facilitate bottom-up budgeting and improve both schedule and cost control.

    Elemental budgets may be developed on the basis of the timed progress of each part of the project. When all the individual estimates are gathered, we obtain a composite budget estimate. Figure 8–20 shows an example of the various components that may be involved in an overall budget. The bar chart appended to a segment of the pie chart indicates the individual cost components making up that particular segment. Analytical tools such as learning curve analysis, work sampling, and statistical estimation may be employed in the cost estimation and budgeting processes.

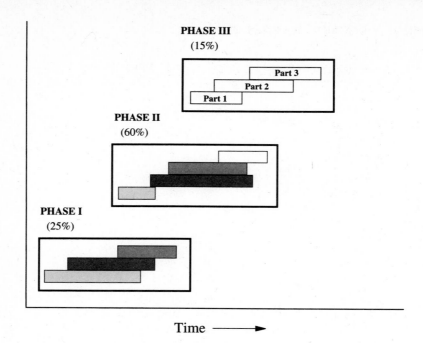

Figure 8-19.   Budgeting by Project Phases

8.8.3 LP Formulation for Capital Rationing

Capital rationing involves selecting a combination of projects that will optimize the return on investment. A mathematical formulation of the capital budgeting problem is presented as follows:

$$\text{Maximize:} \quad z = \sum_{i=1}^{n} v_i x_i$$

$$\text{Subject to:} \quad \sum_{i=1}^{n} c_i x_i \leq B$$

$$x_i = 0, 1, \qquad i = 1, \ldots, n$$

where

n = number of projects
v_i = measure of performance for project i (e.g., present value)
c_i = cost of project i
x_i = indicator variable for project i
B = budget availability level

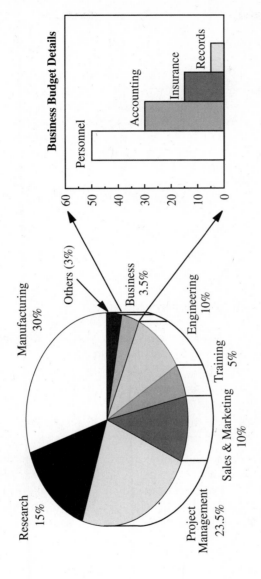

Figure 8-20. Budget Breakdown and Distribution

A solution of the above model will indicate which projects should be selected in combination with which projects. The example that follows illustrates a capital rationing problem.

8.8.4 Example of a Capital Rationing Problem

Planning a portfolio of projects is essential in resource-limited projects. Scheinberg (1992) presents a planning model for implementing a number of projects simultaneously. The capital rationing example presented here (Pulat and Badiru 1990) involves the determination of the optimal combination of project investments so as to maximize total return on investment. Suppose a project analyst is given N projects, $X_1, X_2, X_3, \ldots, X_N$, with the requirement to determine the level of investment in each project so that total investment return is maximized subject to a specified limit on available budget. The projects are not mutually exclusive.

The investment in each project starts at a base level $b_i (i = 1, 2, \ldots, N)$ and increases by a variable increments $k_{ij}(j = 1, 2, 3, \ldots, K_i)$, where K_i is the number of increments used for project i. Consequently, the level of investment in project X_i is defined as

$$x_i = b_i + \sum_{j=1}^{K_i} k_{ij}$$

where

$$x_i \geq 0 \qquad \forall i$$

For most cases, the base investment will be 0. In those cases, we will have $b_i = 0$. In the modeling procedure used for this problem,

$$X_i = \begin{cases} 1 & \text{if the investment in project } i \text{ is greater than zero} \\ 0 & \text{otherwise} \end{cases}$$

and

$$Y_{ij} = \begin{cases} 1 & \text{if } j\text{th increment of alternative } i \text{ is used.} \\ 0 & \text{otherwise.} \end{cases}$$

The variable x_i is the actual level of investment in project i, while X_i is an indicator variable indicating whether or not project i is one of the projects selected for investment. Similarly, k_{ij} is the actual magnitude of the jth increment while Y_{ij} is an indicator variable that indicates whether or not the jth increment is used for project i. The maximum possible investment in each project is defined as M_i such that

$$b_i \leq x_i \leq M_i$$

There is a specified limit, B, on the total budget available to invest such that

$$\sum_i x_i \leq B$$

There is a known relationship between the level of investment, x_i, in each project and the expected return, $R(x_i)$. This relationship will be referred to as the *utility function*, $f(.)$, for the project. The utility function may be developed through historical data, regression analysis, and forecasting models. For a given project, the utility function is used to determine the expected return, $R(x_i)$, for a specified level of investment in that project. That is,

$$R(x_i) = f(x_i)$$

$$= \sum_{j=1}^{K_i} r_{ij} Y_{ij}$$

where r_{ij} is the incremental return obtained when the investment in project i is increased by k_{ij}. If the incremental return decreases as the level of investment increases, the utility function will be *concave*. In that case, we will have the following relationship:

or

$$r_{ij} - r_{i,j+1} \geq 0$$

Thus,

$$Y_{ij} \geq Y_{i,j+1}$$

or

$$Y_{ij} - Y_{i,j+1} \geq 0$$

so that only the first n increments ($j = 1, 2, \ldots, n$) that produce the highest returns are used for project i. Figure 8–21 shows an example of a concave investment utility function.

If the incremental returns do not define a concave function, $f(x_i)$, then one has to introduce the inequality constraints presented above into the optimization model. Otherwise, the inequality constraints may be left out of the model, since the first inequality, $Y_{ij} \geq Y_{i,j+1}$, is always implicitly satisfied for concave functions. Our objective is to maximize the total return. That is,

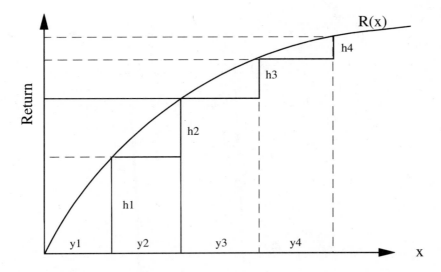

Figure 8-21.   Utility Curve for Investment Yield

$$\text{Maximize}: Z = \sum_i \sum_j r_{ij}\, Y_{ij}$$

subject to the following constraints:

$$x_i = b_i + \sum_j k_{ij}\, Y_{ij} \qquad \forall i$$

$$b_i \le x_i \le M_i \qquad \forall i$$

$$Y_{ij} \ge Y_{i,j+1} \qquad \forall i,j$$

$$\sum_i x_i \le B$$

$$x_i \ge 0 \qquad \forall i$$

$$Y_{ij} = 0 \text{ or } 1 \qquad \forall i,j$$

Now suppose we are given four projects (i.e., $N = 4$) and a budget limit of \$10 million. The respective investments and returns are shown in Tables 8–8, 8–9, 8–10, and 8–11.

Table 8-8   Investment Data for Project 1
for Capital Rationing

Stage (j)	Incremental investment y_{1j}	Level of investment x_1	Incremental return r_{1j}	Total return $R(x_1)$
0	—	0	—	0
1	0.80	0.80	1.40	1.40
2	0.20	1.00	0.20	1.60
3	0.20	1.20	0.30	1.90
4	0.20	1.40	0.10	2.00
5	0.20	1.60	0.10	2.10

Table 8-9   Investment Data for Project 2
for Capital Rationing

Stage (j)	Incremental investment y_{2j}	Level of investment x_2	Incremental return r_{2j}	Total return $R(x_2)$
0	—	0	—	0
1	3.20	3.20	6.00	6.00
2	0.20	3.40	0.30	6.30
3	0.20	3.60	0.30	6.60
4	0.20	3.80	0.20	6.80
5	0.20	4.00	0.10	6.90
6	0.20	4.20	0.05	6.95
7	0.20	4.40	0.05	7.00

Table 8-10 Investment Data for Project 3
for Capital Rationing

Stage (j)	Incremental investment y_{3j}	Level of investment x_3	Incremental return r_{3j}	Total return $R(x_3)$
0	0	-	-	0
1	2.00	2.00	4.90	4.90
2	0.20	2.20	0.30	5.20
3	0.20	2.40	0.40	5.60
4	0.20	2.60	0.30	5.90
5	0.20	2.80	0.20	6.10
6	0.20	3.00	0.10	6.20
7	0.20	3.20	0.10	6.30
8	0.20	3.40	0.10	6.40

Table 8-11 Investment Data for Project 4
for Capital Rationing

Stage (j)	Incremental investment y_{4j}	Level of investment x_4	Incremental return r_{4j}	Total return $R(x_4)$
0	—	0	—	0
1	1.95	1.95	3.00	3.00
2	0.20	2.15	0.50	3.50
3	0.20	2.35	0.20	3.70
4	0.20	2.55	0.10	3.80
5	0.20	2.75	0.05	3.85
6	0.20	2.95	0.15	4.00
7	0.20	3.15	0.00	4.00

All the values are in millions of dollars. For example, in Table 8–8, if an incremental investment of $0.20 million from stage 2 to stage 3 is made in project 1, the expected incremental return from the project will be $0.30 million. Thus, a total investment of $1.20 million in project 1 will yield a total return of $1.90 million. The question addressed by the optimization model is to determine how many investment increments should be used for each project. That is, when should we stop increasing the investments in a given project? Obviously, for a single project we would continue to invest as long as the incremental returns are larger than the incremental investments. However, for multiple projects, investment interactions complicate the decision so that investment in one project cannot be independent of the other projects. The LP model of the capital rationing example was solved with LINDO software. The model is

Maximize: $Z = 1.4Y11 + .2Y12 + .3Y13 + .1Y14 + .1Y15 + 6Y21 + .3Y22 + .3Y23 + .2Y24$
             $+ .1Y25 + .05Y26 + .05Y27 + 4.9Y31 + .3Y32 + .4Y33 + .3Y34 + .2Y35 + .1Y36$
             $+ .1Y37 + .1Y38 + 3Y41 + .5Y42 + .2Y43 + .1Y44 + .05Y45 + .15Y46$

Subject to:

$.8Y11 + .2Y12 + .2Y13 + .2Y14 + .2Y15 - X1 = 0$
$3.2Y21 + .2Y22 + .2Y23 + .2Y24 + .2Y25 + .2Y26 + .2Y27 - X2 = 0$
$2.0Y31 + .2Y32 + .2Y33 + .2Y34 + .2Y35 + .2Y36 + .2Y37 + .2Y38 - X3 = 0$
$1.95Y41 + .2Y42 + .2Y43 + .2Y44 + .2Y45 + .2Y46 + .2Y47 - X4 = 0$
$X1 + X2 + X3 + X4 <= 10$
$Y12 - Y13 >= 0$
$Y13 - Y14 >= 0$
$Y14 - Y15 >= 0$
$Y22 - Y23 >= 0$
. . . . .
$Y26 - Y27 >= 0$
$Y32 - Y33 >= 0$
$Y33 - Y34 >= 0$
$Y34 - Y35 >= 0$
$Y35 - Y36 >= 0$
$Y36 - Y37 >= 0$
$Y37 - Y38 >= 0$
$Y43 - Y44 >= 0$
$Y44 - Y45 >= 0$
$Y45 - Y46 >= 0$
$X_i >= 0$ for $i = 1, 2, \ldots, 4$
$Y_{ij} = 0, 1$ for all i and j

The solution indicates the following values for Y_{ij}.

Project 1

$$Y11 = 1, Y12 = 1, Y13 = 1, Y14 = 0, Y15 = 0.$$

Thus, the investment in project 1 is $X1 = \$1.20$ million. The corresponding return is $1.90 million.

Project 2

$$Y21 = 1, Y22 = 1, Y23 = 1, Y24 = 1, Y25 = 0, Y26 = 0, Y27 = 0.$$

Thus, the investment in project 2 is $X2 = \$3.80$ million. The corresponding return is $6.80 million.

Project 3

$$Y31 = 1, Y32 = 1, Y33 = 1, Y34 = 1, Y35 = 0, Y36 = 0, Y37 = 0.$$

Thus, the investment in project 3 is $X3 = \$2.60$ million. The corresponding return is $5.90 million.

Project 4

$$Y41 = 1, Y42 = 1, Y43 = 1.$$

Thus, the investment in project 4 is $X4 = \$2.35$ million. The corresponding return is $3.70 million.

The total investment in all four projects is $9,950,000. Thus, the optimal solution indicates that not all of the $10,000,000 available should be invested. The expected return from the total investment is $18,300,000. This translates into 83.92% return on investment. Figure 8–22 presents histograms of the investments and the returns for the four projects. The individual returns on investment from the projects are shown graphically in Figure 8–23.

The optimal solution indicates an unusually large return on total investment. In a practical setting, expectations may need to be scaled down to fit the realities of the project environment. Not all optimization results will be directly applicable to real situations. Possible extensions of the above model of capital rationing include the incorporation of risk and time value of money into the solution procedure. Risk analysis would be relevant, particularly for cases where the levels of returns for the various levels of investment are not known with certainty. The incorporation of time value of money would be useful if the investment analysis is to be performed for a given planning horizon. For example, we might need to make investment decisions to cover the next five years rather than just the current time.

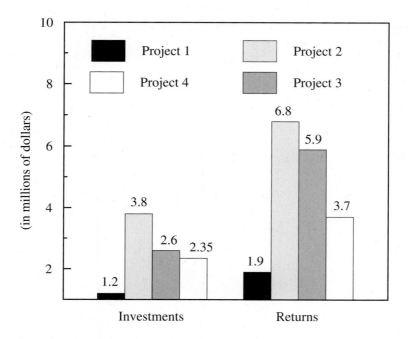

Figure 8-22. Histogram of Capital Rationing Example

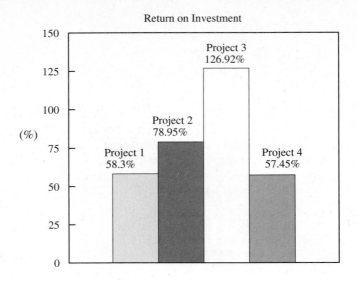

Figure 8-23. Histogram of Returns on Investments

8.9 COST MONITORING

As a project progresses, costs can be monitored and evaluated to identify areas of unacceptable cost performance. Figure 8–24 shows a plot of cost versus time for projected cost and actual cost. The plot permits a quick identification of when cost overruns occur in a project.

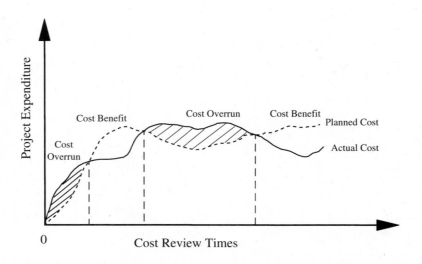

Figure 8-24. Evaluation of Actual and Projected Cost

Plots similar to those presented above may be used to evaluate cost, schedule, and time performance of a project. An approach similar to the profit ratio presented earlier may be used together with the plot to evaluate the overall cost performance of a project over a specified planning horizon. Presented below is a formula for the *cost performance index (CPI)*:

$$CPI = \frac{\text{area of cost benefit}}{\text{area of cost benefit} + \text{area of cost overrun}}$$

As in the case of the profit ratio, the CPI may be used to evaluate the relative performances of several project alternatives or to evaluate the feasibility and acceptability of an individual alternative. In Figure 8–25, we present another cost monitoring tool that we refer to as a cost control pie chart. The chart is used to track the percentage of cost going into a specific component of a project. Control limits can be included in the pie chart to identify out-of-control cost situations. The example in Figure 8–25 shows that 10% of total cost is tied up in supplies. The control limit is located at 12% of total cost. Hence, the supplies expenditure is within control (so far, at least).

8.9.1 Project Balance Technique

One other approach to monitoring cost performance is the project balance technique. The technique helps in assessing the economic state of a project at a desired point in

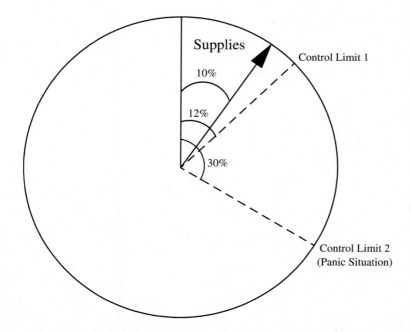

Figure 8-25.   Cost Control Pie Chart

time in the life cycle of the project. It calculates the net cash flow of a project up to a given point in time. The project balance is calculated as

$$B(i)_t = S_t - P(1 + i)^t + \sum_{k=1}^{t} PW_{\text{income}}(i)_k$$

where

$B(i)_t$ = project balance at time t at an interest rate of $i\%$ per period
$PW_{income}(i)_t$ = present worth of net income from the project up to time t
P = initial cost of the project
S_t = salvage value at time t

The project balance at time t gives the net loss or net profit associated with the project up to that time.

8.10 CONTRACT MANAGEMENT AND C/SCSC

Contract management involves the process by which goods and services are acquired, utilized, monitored, and controlled in a project. Contract management addresses the contractual relationships from the initiation of a project to the completion of the project (i.e., completion of services and/or hand over of deliverables). Some of the important aspects of contract management are

- Principles of contract law
- Bidding process and evaluation
- Contract and procurement strategies
- Selection of source and contractors
- Negotiation
- Worker safety considerations
- Product liability
- Uncertainty and risk management
- Conflict resolution

In 1967, the U.S. Department of Defense (U.S. DOD) introduced a set of 35 standards or criteria to which contractors must comply under cost or incentive contracts (U.S. DOD 1967). The system of criteria is referred to as the cost and schedule control systems criteria (C/SCSC). Many government agencies now require compliance with C/SCSC for major contracts. The purpose is to manage the risk of cost overrun to the government. The system presents an integrated approach to cost and schedule management. It is now widely recognized and used in major project environments. It is intended to facilitate greater uniformity and provide advance warning about impending schedule or cost overruns.

The topics covered by C/SCSC include cost estimating and forecasting, budgeting, cost control, cost reporting, earned value analysis, resource allocation and management, and schedule adjustments. The tools and techniques needed to address these topics are presented in the various sections of this book. The important link among all of these is the dynamism of the relationship among performance, time, and cost. Such a relationship is represented in Figure 8–26. This is essentially a multiobjective problem. Since performance, time, and cost objectives cannot be satisfied equally well, concessions or compromises would need to be worked out in implementing C/SCSC.

Another dimension of the performance-time-cost relationship is the U.S. Air Force's R&M 2000 Standard which addresses *reliability* and *maintainability* of systems. R&M 2000 is intended to integrate reliability and maintainability into the performance, cost, and schedule management for government contracts. C/SCSC and R&M 2000 constitute an effective guide for project design. Further details on cost and risk aspects of project management with respect to C/SCSC can be found in Michaels and Wood (1989) and Obradovitch and Stephanou (1990).

To comply with C/SCSC, contractors must use standardized planning and control methods based on *earned value*. Earned value refers to the actual dollar value of work performed at a given point in time compared to planned cost for the work. This is different from the conventional approach of measuring actual versus planned, which is explicitly forbidden by C/SCSC. In the conventional approach, it is possible to misrepresent the actual content (or value) of the work accomplished. The work rate analysis technique presented in Chapter 5 can be useful in overcoming the deficiencies of the conventional approach. C/SCSC is developed on a work content basis using the following factors:

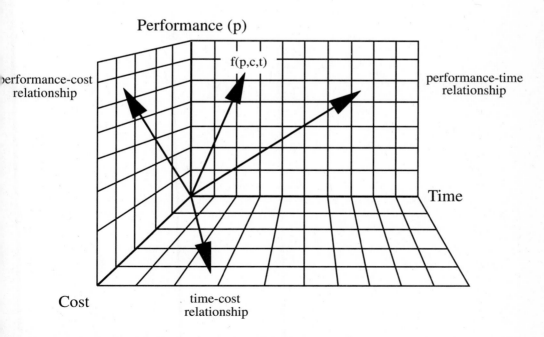

Figure 8-26.   Performance-Cost-Time Relationships for C/SCSC

- The actual cost of work performed (ACWP), which is determined on the basis of the data from cost accounting and information systems.
- The budgeted cost of work scheduled (BCWS) or baseline cost determined by the costs of scheduled accomplishments.
- The budgeted cost of work performed (BCWP) or earned value. This is the actual work of effort completed as of a specific point in time.

The following equations can be used to calculate cost and schedule variances for work package at any point in time:

Cost variance	$= BCWP - ACWP$
Percent cost variance	$= (\text{Cost variance}/BCWP) \cdot 100$
Schedule variance	$= BCWP - BCWS$
Percent schedule variance	$= (\text{Schedule variance}/BCWS) \cdot 100$
ACWP and remaining funds	$= \text{Target cost (TC)}$
ACWP + cost to complete	$= \text{Estimated cost at completion (EAC)}$

8.11 EXERCISES

8.1. A manufacturing company has revenue and cost functions that are defined, respectively, as follows (x is the number of units):

$$R(x) = \frac{3x}{4}$$

$$C(x) = \frac{x^3 - 8x^2 + 25x + 30}{25}$$

(a) Plot the cost and revenue functions on the same graph and show the break-even points.

(b) At what value of x will the company's profit be maximum?

8.2. In performing a make-or-buy analysis for a piece of equipment needed for a project, the project analyst developed the following cost functions:

Alternative 1 (make): $C(x) = 2,500 + 125x - 0.025x^2$
Alternative 2 (buy Type A): $C(x) = 5,000 + 75x$
Alternative 3 (buy Type B): $C(x) = 2,000 + 100x + 0.005x^2$

where x is the number of hours of utilization of the equipment. Determine the ranges of values of x over which each alternative is the preferred choice. Make a plot of the three cost functions on the same graph and indicate the respective preference regions.

8.3. The net present worths, in millions of dollars, for three project alternatives, A, B, and C, are given below:

NPW of $A = 1.25x + 22.5$
NPW of $B = 2.5x + 20$
NPW of $C = 2^x$

where x represents thousands of units of a product generated by the project. Find the range of production units over which each alternative is the preferred choice. Plot the break-even chart for the alternatives on the same graph and indicate the respective preference regions. What is the fixed cost of alternative C?

8.4. Find the capitalized cost of a project fund needed to provide a maintenance expense of $3,000 every two years forever if the first maintenance operation is to be conducted two years after the project fund is started. The interest rate on the fund is 9.25% per year.

8.5. A project involves laying 100 miles of oil distribution pipe in the Alaskan wilderness. It is desired to determine the lowest-cost crew size and composition that can get the job done within 90 days. Three different types of crew are available. Each crew can work independently or work simultaneously with other crews. Crew 1 can lay pipe at the rate of 1 mile per day. If Crew 2 is to do the whole job alone, it can finish it in 120 days. Crew 3 can lay pipes at the rate of 1.7 miles per day. But this crew already has other imminent project commitments that will limit its participation in the piping project to only 35 contiguous days. All three crews are ready, willing, and available to accept their contracts at the beginning of the project. Crew 3 charges $50,000 per mile of pipes laid. Crew 2 charges $35,000 per mile and crew 1 charges $40,000 per mile. Contracts can be awarded only in increments of 10 miles. That is, a crew can get a contract for only 0 mile, 10 miles, 20 miles, 30 miles, and so on. *Required*: Determine the best composition of the project crews and the minimum budget needed to finish the project in the shortest possible time. Assume that the following constraints are applicable:

> There is a limit of $20,000,000 on the budget.
> The project must be completed within 50 days.
> Crew 1 and Crew 3 cannot work together.

8.6. Maintenance expenditures for a 20-year project will occur at periodic outlays of $1,000 at the end of the fifth year, $2,000 at the end of the tenth year, and $3,500 at the end of the fifteenth year. With interest at 10% per year, find the equivalent uniform annual cost over the life of the project.

8.7. Two machines are under consideration for a manufacturing project. There is some uncertainty concerning the hours of usage of the machines. Machine A costs $3,000 and has an efficiency of 90%. Machine B costs $1,400 and has an efficiency of 80%. Each machine has a ten-year life with no salvage value. Electric service for both machines costs $1 per year per kilowatt of demand and $0.01 per kilowatt-hour of energy. The output of the motors used in the machines is expected to be 100 horsepower. The interest rate is 8% per year. At how many hours of operation per year would the machines be equally economical? If annual usage is expected to be less than this amount, which machine is preferable?

8.8. Presented below are the monthly total dollar expenditures on a community service project.

Month	Expenditure
January	$500,000
February	100,000
March	76,000
April	125,125
May	4,072
June	127,000
July	50,000
August	17,000
September	100,000
October	25,000
November	80,000
December	275,000

The total budget allocated for the project is $1,250,000. It is a one-year project, and the work content is evenly distributed over the project duration. That is, the same amount of work (in labor-hours) is expected to be accomplished each month. As a cost control measure, the project manager has stipulated that the cumulative project expenditure must be directly proportional to the amount of work completed.

(a) Develop a cumulative cost control chart for the project. Identify the control limits and the points where cumulative cost is out of control.

(b) If the lower control limit on monthly expenditure is $20,000, identify the months where expenditures are below this control limit. Discuss the potential causes for the low expenditures and what, if any, investigation or control actions should be taken.

8.9. Suppose you borrow $50,000 to start an engineering consulting business. The loan is to be repaid by making end-of-month payments for 48 months at a monthly interest rate of 1%. During the first year, you make monthly payments of $1,000 each. However, due to a drastic change in your consulting business, you negotiate with the bank to change your second-year monthly payments to X per month. At the beginning of the third year, there is another drastic change in your business. This time, you negotiate with the bank to reduce your remaining monthly payments to $X/2$ per month. Find what the value of X should be in order to have the loan totally paid off at the end of the fourth year.

8.10. The chief engineer of a major corporation is evaluating alternatives to supply electricity to the company. Under present conditions, it is anticipated that the company will pay $3 million at the end of this year for electricity purchased from a utility company. It is estimated that this cost will increase by $300,000 per year. The engineer needs to decide if the company should build a 4,000-kilowatt power plant to generate its own electricity. If a power plant is built, the operating costs (excluding cost of fuel) are estimated to be $130,000 per year. Two alternate power plants are under consideration:

(a) Coal power plant: Installed cost of the coal power plant is $1,200 per kilowatt. Coal fuel consumption will be 30,000 tons per year. The cost of coal fuel for the first year is $20 per ton and is estimated to increase at a rate of $2 per ton per year. The plant will have no salvage value.

(b) Oil fuel power plant: Installed cost of the oil-based power plant is $1,000 per kilowatt. Oil consumption will be 46,000 barrels per year. The cost of oil fuel for the first year is $34 per barrel and is estimated to increase at $1 per barrel per year. The plant will have no salvage value.

  If the interest rate is 12% per year and the planning horizon is 10 years, use the equivalent annual cost (EAC) method to determine which alternative the engineer should recommend to the company. Assume that the do-nothing alternative is feasible.

8.11. A local manufacturer has been ordered by the city to stop discharging toxic waste into the city sewer system. As a consulting engineer to the company, you have determined that there are three options for meeting the city's requirements. The options and their associated costs are presented below.

System	Initial cost	Annual operating cost	Salvage value at end of 20 years
Downhill system	$30,000	$6,000	$2,000
Quicksilver system	35,000	5,000	5,000
Rolling camaro system	80,000	1,000	40,000

Each system is expected to last and be fully used for 20 years. If the MARR is 8% per year, use incremental present worth analysis to determine which system should be selected. The do-nothing alternative is not acceptable.

8.12. As an engineering economist, you have developed mathematical functions for the marginal revenue (MR) and total cost (TC) for Whacked Widget Company of Norman. It is assumed that the company can sell all the units it makes. The functions (in thousands of dollars) are

$$MR(x) = 100 - 0.02x$$

$$TC(x) = \frac{2x^2}{10^4} + 10{,}000$$

where x represents the number of units of whacked widgets produced.

(a) Determine the number of units the company should make in order to maximize its revenue.

(b) What is the maximum profit the company can realize from the production of whacked widgets?

8.13. Given a geometric cash flow series defined by

$$A_t C(1 - j)^t, \qquad \text{for } t = 0, 1, 2, 3, 4, \dots, n$$

where

    C is a constant
    j is a constant percent rate of decrease per period
    i is the interest rate per period

(a) Draw the cash flow diagram for A_t.

(b) Derive a general formula to find F_n (i.e., future value at time $t = n$). Derive the formula *from scratch*. Do not use any of the existing formulas in your derivation. Simplify the formula into a closed form in terms of $C, i, j,$ and n that can be easily used for calculations. Do not attempt to oversimplify the formula.

(c) Given $C = \$10{,}000, j = 8\%, i = 10\%$, and $n = 20$, use your formula above to find $\mathbf{F}_{20}$ (i.e., future value at time $t = 20$).

8.14. Suppose you borrow \$100,000 for a project. The loan is to be repaid by making end-of-month payments for 60 months. The payment schedule is arranged as follows:

    First year: Payments of \$1,000 per month
    Second year: Payments of \$X per month
    Third year through fifth year: Payments of $\frac{\$X}{2}$ per month

The first payment is to be made one month after getting the loan. The interest rate charged by the bank is 1% per month. Find the value of X.

8.15. The revenue and total cost functions for an engineering investment are as follows:

$$R(x) = 39.86 + 87.25x + \frac{1{,}615}{x}$$

$$TC(x) = 39.86 + 24.03x + \frac{2{,}463}{x}$$

where $2 <= x <= 10$ and x = number of boxes of widgets sold per year. (Each full box contains 10,000 widgets.)

(a) Determine how many units of widgets must be sold per year in order to break even. Assume that partially filled boxes can be sold.

(b) Make a sketch (plot) of the profit function for this investment. Looking at your plot, determine how sensitive the profit is to changes in x. Very sensitive? Moderately sensitive? Somewhat sensitive? Not sensitive? Explain your choice.

8.16. (a) For the engineering investment in Exercise 8.15, calculate the profit ratio if no more than eight boxes of widgets can be sold.

    (b) Determine where the profit function in Exercise 8.15(b) has a minimum or maximum point. What is the maximum or minimum profit?

8.17. Today is May 6, 1995. Suppose that 10 years ago you had started putting $100 per month in a special business account that pays a nominal interest rate of 6%. The first deposit was made when the account was opened and the subsequent deposits were made on a monthly schedule. You made the deposits for six years and then stopped. You plan to start an engineering consulting company on May 6, 1999. You will withdraw $1,000 per month from the business account to run the company. The first withdrawal will be on May 6, 1999. How many months can you continue the monthly withdrawals before the fund in the account is exhausted? Solve to two decimal places.

8.18. For the cash flow in Figure 8–27, find the future value at time $t = 12$ if the interest rate is 12% per period. Use the arithmetic gradient series factor as appropriate in your solution.

8.19. A project has an initial cost of $50,000 and an economic life of 10 years. The annual incomes from the project over the ten years are $8,000, $10,000, $12,000, $15,000, $20,000, $14,000, $10,000, $8,000, $12,000, and $6,000, respectively. The project has a fixed salvage value of $5,000 regardless of when the project is terminated. Assuming a MARR of 10% per year, compute the project balance for each year. Plot the project balance versus time.

8.20. A new grinding machine is to be purchased at an initial cost of $80,000. The machine is expected to increase productivity and generate annual cost savings of $15,000. The machine will need to be overhauled every two years at a cost of $2,000 per overhaul. If the annual interest rate is 8%, find the *discounted payback period* for the machine considering the time value of money. The costs of overhaul are to be considered in calculating the discounted payback period. What is your recommendation about the proposed machine?

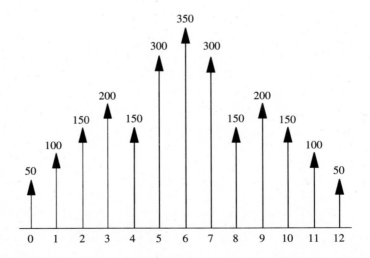

Figure 8-27. Cash Flow for Problem 8.18

9

Decision Analysis for Project Selection

This chapter presents useful techniques for assessing and comparing projects for selection purposes. The techniques presented include the utility models, project value model, polar plots, benchmarking techniques, and the analytic hierarchy process (AHP).

9.1 PROJECT SELECTION PROBLEM

Project selection is an important aspect of project planning. The right project must be undertaken at the right time to satisfy the constraints of time and resources. Projects may be selected on the basis of a combination of several criteria. Some of these criteria are technical merit, management desire, schedule efficiency, cost-benefit ratio, resource availability, critical need, availability of sponsors, and user acceptance.

Many aspects of project selection cannot be expressed in quantitative terms. For this reason, project analysis and selection must be addressed by techniques that permit the incorporation of both quantitative and qualitative factors. Khorramshahgol et al. (1988) and Bidanda (1989) discusses crucial aspects of project selection. Some techniques for project analysis and selection are presented in the sections that follow. The techniques facilitate the marriage of quantitative and qualitative considerations in the project decision process. Other techniques such as net present value, profit ratio, equity break-even point, and learning curves presented in the preceding chapters are also useful for project selection strategies.

9.2 UTILITY MODELS

The concept of utility refers to the rational behavior of a decision maker when faced with making a choice under uncertainty. The overall utility of a project can be measured in terms of both quantitative and qualitative factors. The vast body of literature

397

available on utility theory makes the utility approach to project selection very appealing. This section presents an approach to project assessment based on utility models. The approach fits an empirical utility function to each factor to be included in a multi-attribute selection model. The specific utility values (weights) that are obtained from the utility functions are used as the basis for selecting a project.

Utility theory is a branch of decision analysis that involves the building of mathematical models to describe the behavior of a decision maker when faced with making a choice among alternatives in the presence of risk. Several utility models are available in the literature (Keeney and Raiffa 1976). The utility of a composite set of outcomes of n decision factors is expressed in the general form

$$U(x) = U(x_1, x_2, \ldots, x_n)$$

where x_i = specific outcome of attribute X_i, $i = 1, 2, \ldots, n$ and $U(x)$ is the utility of the set of outcomes to the decision maker. The basic assumption of utility theory is that people make decisions with the objective of maximizing their *expected utility*. Drawing on an example presented by Park and Sharp-Bette (1990), we may consider a decision maker whose utility function with respect to project selection is represented by

$$u(x) = 1 - e^{-0.0001x}$$

where x represents a measure of the benefit derived from a project. Benefit, in this sense, may be a combination of several factors (e.g., quality improvement, cost reduction, productivity improvement) that can be represented in dollar terms. Suppose this decision maker is faced with a choice between two project alternatives with benefits as specified below.

Project I: Probabilistic levels of project benefits

Benefit, x	−$10,000	$0	$10,000	$20,000	$30,000
Probability, $P(x)$	0.2	0.2	0.2	0.2	0.2

Project II: A definite benefit of $5,000

Assuming an initial benefit of zero and identical levels of required investment, the decision maker is required to choose between the two projects. For project I, the expected utility is computed as

$$E[u(x)] = \sum u(x)\{P(x)\}$$

Benefit, x	Utility, $u(x)$	$P(x)$	$u(x)P(x)$
−$10,000	−1.7183	0.2	−0.3437
0	0	0.2	0
10,000	0.6321	0.2	0.1264
20,000	0.8647	0.2	0.1729
30,000	0.9502	0.2	0.1900
		Sum	0.1456

Thus, $E[u(x)_1] = 0.1456$. For project II, we have $u(x)_2 = u(\$5000) = 0.3935$. Consequently, the project providing the certain amount of \$5,000 is preferred to the more risky project I even though project I has a higher expected benefit of $\sum xP(x) = \$10,000$. A plot of the utility function used in this example is presented in Figure 9–1.

If the expected utility of 0.1456 is set equal to the decision maker's utility function, we obtain

$$0.1456 = 1 - e^{-0.0001x^*}$$

which yields $x^* = \$1,574$, which is referred to as the *certainty equivalent (CE)* of project I $(CE_1 = 1,574)$. The certainty equivalent of an alternative with variable outcomes is a *certain amount (CA)*, which a decision maker will consider to be desirable to the same degree as the variable outcomes of the alternative. In general, if CA represents the certain amount of benefit that can be obtained from project II, then the criteria for making a choice between the two projects can be summarized as follows:

If $CA < \$1,754$, select project I.
If $CA = \$1,754$, select either project.
If $CA > \$1,574$, select project II.

The key point in using utility theory for project selection is the proper choice of utility models. The sections that follow present two simple but widely used utility models: the *additive utility model* and the *multiplicative utility model*.

9.2.1 Additive Utility Model

The additive utility of a combination of outcomes of n factors $(X_1, X_2, \ldots, X_n)$ is expressed as

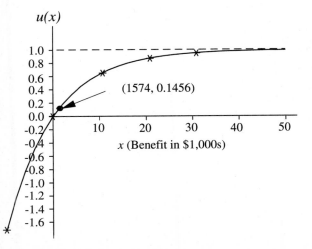

Figure 9-1. Utility Function and Certainty Equivalent

$$U(x) = \sum_{i=1}^{n} U(x_i, \overline{x}_i^0)$$

$$= \sum_{i=1}^{n} k_i U(x_i)$$

where:

x_i = measured or observed outcome of attribute i
n = number of factors to be compared
x = combination of the outcomes of the n factors
$U(x_i)$ = utility of the outcome for attribute i, x_i
$U(x)$ = combined utility of the set of outcomes, x
k_i = weight or scaling factor for attribute $i (0 < k_i < 1)$
X_i = variable notation for attribute i
x_i^0 = worst outcome of attribute i
x_i^* = best outcome of attribute i
$\overline{x}_i^0$ = set of worst outcomes for the complement of x_i.
$U(x_i, \overline{x}_i^0)$ = utility of the outcome of attribute i and the set of worst outcomes for the complement of attribute i
$k_i = U(x_i^*, \overline{x}_i^0)$
$\sum_{i=1}^{n} k_i = 1.0$ (required for the additive model)

For example, let **A** be a collection of four project attributes defined as **A** = {Profit, Flexibility, Quality, Productivity}. Now, define **X** = {Profit, Flexibility} as a subset of **A**. Then, $\overline{\mathbf{X}}$ is the complement of **X** defined as $\overline{\mathbf{X}}$ = {Quality, Productivity}. An example of the comparison of two projects under the additive utility model is summarized in Table 9–1 and yields the following results:

$$U(x)_A = \sum_{i=1}^{n} k_i U_i(x_i) = .4(.95) + .2(.45) + .3(.35) + .1(.75) = 0.650$$

$$U(x)_B = \sum_{i=1}^{n} k_i U_i(x_i) = .4(.90) + .2(.98) + .3(.20) + .1(.10) = 0.626$$

Table 9-1 Example for Additive Utility Model

Attribute (i)	k_i	Project A $U_i(x_i)$	Project B $U_i(x_i)$
Profitability	0.4	0.95	0.90
Flexibility	0.2	0.45	0.98
Quality	0.3	0.35	0.20
Throughput	0.1	0.75	0.10
	1.00		

Since $U(x)_A > U(x)_B$, project A is selected.

9.2.2 Multiplicative Utility Model.

Under the multiplicative utility model, the utility of a combination of outcomes of n factors $(X_1, X_2, \ldots, X_{n1})$ is expressed as

$$U(x) = \frac{1}{C}\left[\prod_{i=1}^{n}(Ck_i\,U_i(x_i) + 1) - 1\right]$$

where C and k_i are scaling constants satisfying the following conditions:

$$\prod_{i=1}^{n}(1 + Ck_i) - C = 1.0$$
$$-1.0 < C < 0.0$$
$$0 < k_i < 1$$

The other variables are as defined previously for the additive model. Using the multiplicative model for the data in Table 9–1 yields $U(x)_A = 0.682$ and $U(x)_B = 0.676$. Thus, project A is selected.

9.2.3 Fitting a Utility Function

An approach presented in this section for multiattribute project selection is to fit an empirical utility function to each factor to be considered in the selection process. The specific utility values (weights) that are obtained from the utility functions may then be used in any of the available justification methodologies discussed by Canada and Sullivan (1989). One way to develop an empirical utility function for a project attribute is to plot the *best* and the *worst* outcomes expected from the attribute and then fit a reasonable approximation of the utility function using concave, convex, linear, S-shaped, or any other logical functional form.

Alternately, if an appropriate probability density function can be assumed for the outcomes of the attribute, then the associated cumulative distribution function may yield a reasonable approximation of the utility values between 0 and 1 for corresponding outcomes of the attribute. In that case, the cumulative distribution function gives an estimate of the cumulative utility associated with increasing levels of attribute outcome. Simulation experiments, histogram plotting, and goodness-of-fit tests may be used to determine the most appropriate density function for the outcomes of a given attribute. For example, the following five attributes are used to illustrate how utility values may be developed for a set of project attributes. The attributes are return on investment (ROI), productivity improvement, quality improvement, idle time reduction, and safety improvement.

Suppose we have historical data on the return on investment (ROI) for investing in a particular project. Assume that the recorded ROI values range from 0% to 40%. Thus, the worst outcome is 0% and the best outcome is 40%. A frequency distribution of the observed ROI values is developed and an appropriate probability density function (*pdf*) is fitted to the data. For our example, suppose the ROI is found to be exponentially distributed with a mean of 12.1%. That is,

$$f(x) = \begin{cases} \frac{1}{\beta}e^{-x/\beta}, & \text{if } x \geq 0 \\ 0, & \text{otherwise} \end{cases}$$

$$F(x) = \begin{cases} 1 - e^{-x/\beta}, & \text{if } x \geq 0 \\ 0, & \text{otherwise} \end{cases}$$

$$\approx U(x)$$

where $\beta = 12.1$. The probability density function and cumulative distribution function are shown graphically in Figure 9–2. The utility of any observed ROI within the applicable range may be read directly from the cumulative distribution function.

For the productivity improvement attribute, suppose it is found (based on historical data analysis) that the level of improvement is normally distributed with a mean of 10% and a standard deviation of 5%. That is,

$$f(x) = \frac{1}{\sqrt{2\pi}\sigma}e^{-\frac{1}{2}\left(\frac{x-\mu}{\sigma}\right)^2}, \qquad -\infty < x < \infty$$

where $\mu = 10$ and $\sigma = 5$. Since the normal distribution does not have a closed form expression for $F(x)$, $U(x)$ is estimated by plotting representative values based on the standard normal table. Figure 9–3 shows $f(x)$ and the estimated utility function for productivity improvement. The utility of productivity improvement may also be evaluated on the basis of cost reduction.

Suppose quality improvement is subjectively assumed to follow a beta distribution with shape parameters $\alpha = 1.0$ and $\beta = 2.9$. That is,

$$f(x) = \frac{\Gamma(\alpha + \beta)}{\Gamma(\alpha)\Gamma(\beta)} \cdot \frac{1}{(b-a)^{\alpha+\beta-1}} \cdot (x-a)^{\alpha-1}(b-x)^{\beta-1},$$

$$\text{for } a \leq x \leq b \quad \text{and} \quad \alpha > 0, \beta > 0$$

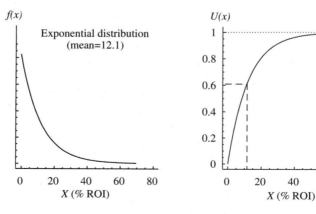

Figure 9-2. Estimated Utility Function for Project ROI

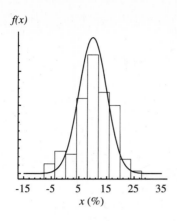
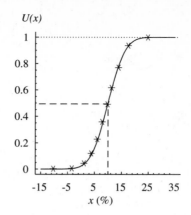

Figure 9-3.   Utility Function for Productivity Improvement

where

a = lower limit for the distribution.
b = upper limit for the distribution.
α and β are the shape parameters for the distribution.

As with the normal distribution, there is no closed form expression for $F(x)$ for the beta distribution. However, if either of the shape parameters is a positive integer, then a binomial expansion can be used to obtain $F(x)$. Figure 9–4 shows a plot of $f(x)$ and the estimated $U(x)$ for quality improvement due to the proposed project.

Based on work study observations, suppose idle time reduction is found to be best described by a lognormal distribution with a mean of 10% and standard deviation of 5%. This is represented as

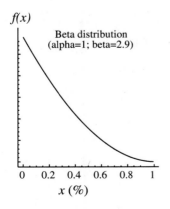
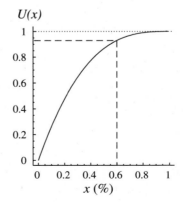

Figure 9-4.   Utility Function for Quality Improvement

$$f(x) = \frac{1}{x\sqrt{2\pi\sigma^2}} e^{\left(\frac{-(\ln x - \mu)^2}{2\sigma^2}\right)}, \qquad x > 0$$

There is no closed form expression for $F(x)$. Figure 9–5 shows $f(x)$ and the estimated $U(x)$ for idle time reduction due to the project.

For the example, suppose safety improvement is assumed to have a previously known utility function defined as

$$U_p(x) = 30 - \sqrt{400 - x^2},$$

where x represents percent improvement in safety. For the expression, the unscaled utility values range from 10 (for 0% improvement) to 30 (for 20% improvement). To express any particular outcome of an attribute i, on a scale of 0.0 to 1.0, it is expressed as a proportion of the range of best to worst outcomes as

$$X = \frac{x_i - x_i^0}{x_i^* - x_i^0}$$

where

X = outcome expressed on a scale of 0.0 to 1.0
x_i = measured or observed raw outcome of attribute i
x_i^0 = worst raw outcome of attribute i
x_i^* = best raw outcome of attribute i

The utility of the outcome may then be represented as $U(X)$ and read off the empirical utility curve. Using the above scaling approach, the utility function for safety improvement is scaled from 0.0 to 1.0. This is shown in Figure 9–6. The numbers within parentheses represent the scaled values.

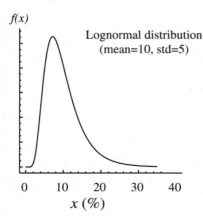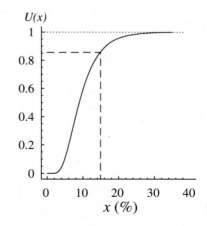

Figure 9-5.   Utility Function for Idle Time Reduction

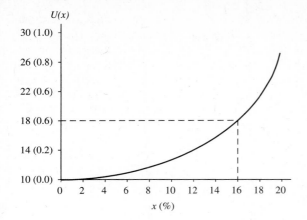

Figure 9-6. Utility Function for Safety Improvement

The respective utility values for the five attributes may be viewed as relative weights for comparing project alternatives. The utility values obtained from the modeled functions can be used in the additive and multiplicative utility models discussed earlier. For example, Table 9–2 shows a composite utility profile for a proposed project.

Using the additive utility model, the *composite utility (CU)* of the project, based on the five attributes, is given by

$$U(X) = \sum_{i=1}^{n} k_i U_i(x_i)$$
$$= .30(.61) + .20(.49) + .25(.93) + .15(.86) + .10(.40) = 0.6825$$

This composite utility value may then be compared with the utilities of other projects. On the other hand, a single project may be evaluated independently on the basis of some *minimum acceptable level of utility (MALU)* desired by the decision maker. The criteria for evaluating a project based on MALU may be expressed by the following rule:

> *Project j is acceptable if its composite utility, $U(X)_j$, is greater than MALU.*
> *Project j is not acceptable if its composite utility, $U(X)_j$, is less than MALU.*

Table 9-2 Composite Utility for a Proposed Project

Attribute (i)	k_i	Value	$U_i(x_i)$
Return on investment	0.30	12.1%	0.61
Productivity improvement	0.20	10.0	0.49
Quality improvement	0.25	60.0	0.93
Idle time reduction	0.15	15.0	0.86
Safety improvement	0.10	15.0	0.40
	1.00		

The utility of a project may be evaluated on the basis of its economic, operational, or strategic importance to an organization. Utility functions can be incorporated into existing justification methodologies. For example, in the analytic hierarchy process, utility functions can be used to generate values that are used to evaluate the relative preference levels of attributes and alternatives. Utility functions can be used to derive component weights when comparing overall effectiveness of projects. Utility functions can generate descriptive levels of project performance as well as indicate the limits of project effectiveness, as shown by the S-Curve in Figure 9–7.

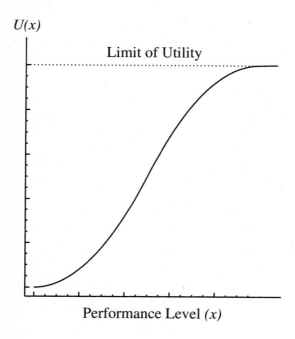

Figure 9-7. S-Curve Model for Project Utility

9.3 PROJECT VALUE MODEL

A technique that is related to the utility modeling is the project value model (PVM), which is an adaptation of the manufacturing system value (MSV) model presented by Troxler and Blank (1989). The model is suitable for the incorporation of utility values. An example of the model is presented below. The model provides a heuristic decision aid for comparing project alternatives. Value is represented as a deterministic vector function that indicates the value of tangible and intangible attributes that characterize an alternative value represented as

$$V = f(A_1, A_2, \ldots, A_p)$$

where V = value, $A = (A_1, \ldots, A_n)$ = vector of quantitative measures or attributes, and p = number of attributes that characterize the project. Examples of project attributes are quality, throughput, capability, productivity, and cost performance. Attributes are considered to be a combined function of factors, x_i, expressed as

$$A_k(x_1, x_2, \ldots, x_{m_k}) = \sum_{i=1}^{m_k} f_i(x_i)$$

where $\{x_i\}$ = set of m factors associated with attribute $A_k (k = 1, 2, \ldots, p)$ and f_i = contribution function of factor x_i to attribute A_k. Examples of factors are market share, reliability, flexibility, user acceptance, capacity utilization, safety, and design functionality. Factors are themselves considered to be composed of indicators, v_i, expressed as

$$x_i(v_1, v_2, \ldots, v_n) = \sum_{j=1}^{n} z_i(v_i)$$

where $\{v_j\}$ = set of n indicators associated with factor $x_i (i = 1, 2, \ldots, m)$ and z_j = scaling function for each indicator variable v_j. Examples of indicators are debt ratio, project responsiveness, lead time, learning curve, and scrap volume. By combining the above definitions, a composite measure of the value of a project is given by

$$\begin{aligned} PV &= f(A_1, A_2, \ldots, A_p) \\ &= f\left\{ \left[\sum_{i=1}^{m_1} f_i \left(\sum_{j=1}^{n} z_j(v_j) \right) \right]_1, \left[\sum_{i=1}^{m_2} f_i \left(\sum_{j=1}^{n} z_j(v_j) \right) \right]_2, \ldots, \left[\sum_{i=1}^{m_k} f_i \left(\sum_{j=1}^{n} z_j(v_j) \right) \right]_p \right\} \end{aligned}$$

where m and n may assume different values for each attribute. A subjective measure to indicate the utility of the decision maker may be included in the model by using an attribute weighting factor, w_i, to obtain the following:

$$PV = f(w_1 A_1, w_2 A_2, \ldots, w_p A_p)$$

where

$$\sum_{k=1}^{P} w_k = 1, \qquad (0 \leq w_k \leq 1)$$

As an example, an analysis using the above model to compare three projects on the basis of four attributes is presented in Table 9–3. The four attributes, *capability, suitability, performance,* and *productivity,* require careful interpretation before developing relative weights for the alternatives.

9.3.1 Capability

Capability refers to the ability of equipment to produce certain features. For example, certain equipment may produce only horizontal or vertical slots, flat finishes, and so on. But a multiaxis machine can produce spiral grooves or remove internal metal from

Table 9-3 Comparison of Technology Values

Alternatives	Suitability ($k = 1$)	Capability ($k = 2$)	Performance ($k = 3$)	Productivity ($k = 4$)
Project A	0.12	0.38	0.18	0.02
Project B	0.30	0.40	0.28	−1.00
Project C	0.53	0.33	0.52	−1.10

prismatic or rotational parts, thus increasing the part variety that can be made. In Table 9–3, the levels of increase in part variety from the three competing projects are 38%, 40%, and 33%, respectively.

9.3.2 Suitability

Suitability refers to the appropriateness of the project to company operations. For example, chemical milling is more suitable for making holes in thin flat metal sheets than drills. Drills need special fixtures to hold the thin metal down and protect it from wrinkling and buckling. The parts that the three projects are suitable for are, respectively, 12%, 30%, and 53% of the current part mix.

9.3.3 Performance

Performance can refer to the proficiency of the project to produce high quality outputs. That is, the ability to meet extra tight performance requirements. In our example, the three projects can, respectively, meet tightened standards on 18%, 28%, and 52% of the normal set of jobs.

9.3.4 Productivity

Productivity can be measured by a simulation of the performance of the current system with the proposed technology at its production rate, quality level, and part application. For the example in Table 9–3, the three projects, respectively, show increases of 0.02, −1.0, and −1.1 on a uniform scale of productivity measurement.

A plot of the histograms of the respective values of the three projects is shown in Figure 9–8. Project C is the best alternative in terms of suitability and performance. Project B shows the best capability measure, but its productivity is too low to justify the needed investment. Project A offers the best productivity, but its suitability measure is low.

The relative weights used in many justification methodologies are based on subjective propositions by the decision maker(s). Some of those subjective weights can be enhanced by the incorporation of utility models. For example, the weights shown in Table 9–3 could be obtained from utility functions.

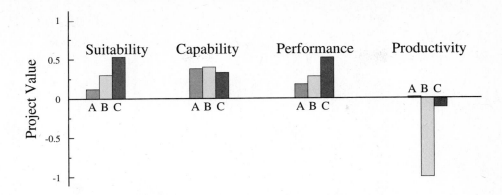

Figure 9-8.   Relative System Value Weights of Three Alternatives

4 POLAR PLOTS

Polar plots provide a means of visually comparing project alternatives (Badiru 1991a). In a conventional polar plot, as shown in Figure 9–9, the vectors drawn from the center of the circle are on individual scales based on the outcome ranges for each attribute. For example, the vector for NPV (net present value) is on a scale of $0 to $500,000 while the scale for quality is from 0 to 10. It should be noted that the overall priority weights for the alternatives are not proportional to the areas of their respective polyhedrons.

A modification of the basic polar plot (Badiru 1991a) is presented in this section. The modification involves a procedure which normalizes the areas of the polyhedrons with respect to the total area of the base circle. With this modification, the normalized areas of the polyhedrons are proportional to the respective priority weights of the alternatives. So, the alternatives can be ranked on the basis of the areas of the polyhedrons. Steps involved in the modified approach are as follows:

1. Let n be the number of attributes involved in the comparison of alternatives such that $n >= 4$. Number the attributes in a preferred order $(1, 2, 3, \ldots, n)$.
2. If the attributes are considered to be equally important (i.e., equally weighted), compute the sector angle associated with each attribute as

$$\theta = \frac{360°}{n}$$

3. Draw a circle with a large enough radius. A radius of 2 inches is usually adequate.
4. Convert the outcome range for each attribute to a standardized scale of 0 to 10 using appropriate transformation relationships.
5. For attribute number 1, draw a vertical vector up from the center of the circle to the edge of the circle.
6. Measure θ clockwise and draw a vector for attribute number 2. Repeat this step for all attributes in the numbered order.
7. For each alternative, mark its standardized relative outcome with respect to each attribute along the attribute's vector. If a 2-inch radius is used for the base circle, we then have the following linear transformation relationship:

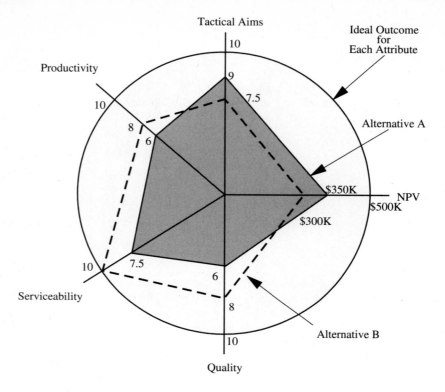

Figure 9-9.   Basic Polar Plot

     0.0 inches = rating score of 0.0
     2.0 inches = rating score of 10.0

8. Connect the points marked for each alternative to form a polyhedron. Repeat this step for all alternatives.

9. Compute the area of the base circle as

$$\Omega = \pi r^2$$

$$= 4\pi \text{ squared inches}$$

$$= 100\pi \text{ squared rating units}$$

10. Compute the area of the polyhedron corresponding to each alternative. This can be done by partitioning each polyhedron into a set of triangles and then calculating the areas of the triangles. To calculate the area of each triangle, note that we will know the lengths of two sides of the triangle and the angle subtended by the two sides. With these three known values, the area of each triangle can be calculated through basic trigonometric formulas.

     For example, the area of each polyhedron may be represented as $\lambda_i (i = 1, 2, \ldots, m)$, where m is the number of alternatives. The area of each triangle in the polyhedron for a given alternative is then calculated as

$$\Delta_t = \frac{1}{2}(L_j)(L_{j+1})(\operatorname{Sin}\theta)$$

where

L_j = standardized rating with respect to attribute j

L_{j+1} = standardized rating with respect to attribute $j + 1$

L_j and L_{j+1} are the two sides that subtend θ

Since $n \geq 4$, θ will be between $0°$ and $90°$ and $\sin \theta$ will be strictly increasing over that interval. The area of the polyhedron for alternative i is then calculated as

$$\lambda_i = \sum_{t(i)=1}^{n} \Delta_{t(i)}$$

Note that θ is constant for a given number of attributes and the area of the polyhedron will be a function of the adjacent ratings (L_j and L_{j+1}) only.

11. Compute the standardized area corresponding to each alternative as

$$w_i = \frac{\lambda_i}{\Omega}(100\%)$$

12. Rank the alternatives in decreasing order of Select the highest ranked alternative as the preferred alternative.

9.4.1 Illustrative Example

The problem solved here is used to illustrate how the modified polar plots can be used to compare project alternatives. Table 9–4 presents the ranges of possible evaluation ratings within which an alternative can be rated with respect to each of five attributes. The evaluation rating of an alternative with respect to attribute j must be between the given range, a_j to b_j. Table 9–5 presents the data for raw evaluation ratings of three alternatives with respect to the five attributes specified in Table 9–4.

The attributes quality level, profit level, and productivity level are quantitative measures that can be objectively determined. The attributes flexibility level and customer satisfaction level are subjective measures that can be intuitively rated by an experienced project analyst. The steps in the solution are as follows:

Table 9–4 Ranges of Raw Evaluation Ratings for Polar Plots

Attribute (j)	Description	Rank (k_j)	Evaluation range	
			Lower limit (a_j)	Upper limit (b_j)
I	Quality	1	0.5	9
II	Profit (× $1,000)	2	0	100
III	Productivity	3	1	10
IV	Flexibility	4	0	12
V	Satisfaction	5	0	10

Table 9–5 Raw Evaluation Ratings for Modified Polar Plots

	Attributes				
Alternatives	I ($j = 1$)	II ($j = 2$)	III ($j = 3$)	IV ($j = 4$)	V ($j = 5$)
A ($i = 1$)	5	50	3	6	10
B ($i = 2$)	1	20	1.5	9	2
C ($i = 3$)	8	75	4	11	1

Step 1: $n = 5$. The attributes are numbered in the following preferred order:

> Quality: Attribute I
> Profit: Attribute II
> Productivity: Attribute III
> Flexibility: Attribute IV
> Satisfaction: Attribute V

Step 2: The sector angle is computed as

$$\theta = 360°/n$$
$$= 72°$$

Step 3: This step is shown in Figure 9–10.

Step 4: Let Y_{ij} be the raw evaluation rating of alternative i with respect to attribute j (see Table 9–5). Let Z_{ij} be the standardized evaluation rating. The standardized evaluation ratings (between 0.0 and 10.0) shown in Table 9–6 were obtained by using the following linear transformation relationship:

$$Z_{ij} = 10 \left[\frac{(Y_{ij} - a_j)}{(b_j - a_j)} \right]$$

Steps 5, 6, 7, 8: These are shown in Figure 9–10.

Step 9: The area of the base circle is $\Omega = 100\pi$ squared rating units. Note that it is computationally more efficient to calculate the areas in terms of rating units rather than inches.

Step 10: Using the expressions presented in step 10, the areas of the triangles making up each of the polyhedrons are computed and summed. The respective areas are

$$\lambda_A = 72.04 \text{ squared units}$$
$$\lambda_B = 10.98 \text{ squared units}$$
$$\lambda_C = 66.14 \text{ squared units}$$

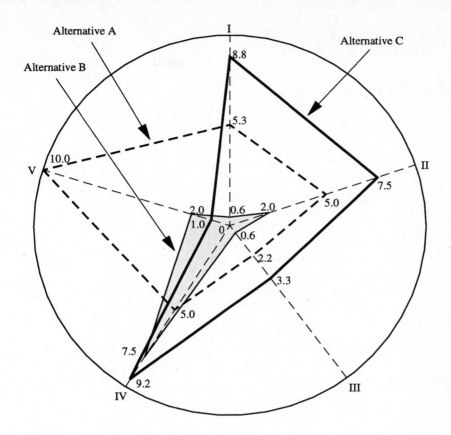

Figure 9-10. Modified Polar Plot

Table 9-6 Standardized Evaluation Ratings for
Modified Polar Plots

	Attributes				
Alternatives	I ($j = 1$)	II ($j = 2$)	III ($j = 3$)	IV ($j = 4$)	V ($j = 5$)
A ($i = 1$)	5.3	5.0	2.2	5.0	10.0
B ($i = 2$)	0.6	2.0	0.6	7.5	2.0
C ($i = 3$)	8.8	7.5	3.3	9.2	1.0

Step 11:   The standardized areas for the three alternatives are

$$w_A = 22.93\%$$
$$w_B = 3.50\%$$
$$w_C = 21.05\%$$

Step 12:   On the basis of the standardized areas in step 11, alternative A is found to be the best choice.

As an extension to the modification presented above, the sector angle may be a variable indicating relative attribute weights while the radius represents the evaluation rating of the alternatives with respect to the weighted attribute. That is, if the attributes are not equally weighted, the sector angles will not all be equal. In that case, the sector angle for each attribute is computed as

$$\theta_j = p_j(360°)$$

where

p_j = relative numeric weight of each of n attributes

$$\sum_{j=1}^{n} p_j = 1.0$$

Suppose the attributes in the example are considered to have unequal weights as shown in Table 9–7.

The resulting polar plots for weighted sector angles are shown in Figure 9–11. The respective weighted areas for the alternatives are

$$\lambda_A = 51.56 \text{ squared units}$$
$$\lambda_B = 9.07 \text{ squared units}$$
$$\lambda_C = 60.56 \text{ squared units}$$

The standardized areas for the alternatives are as follows:

$$w_A = 16.41\%$$
$$w_B = 2.89\%$$
$$w_C = 19.28\%$$

Thus, if the given attributes are weighted as shown in Table 9–7, alternative C will turn out to be the best choice. However, it should be noted that the relative weights of the attributes are too skewed, resulting in some sector angles being greater than 90°. It is preferred to have the attribute weights assigned such that all sector angles are less than

Table 9-7 Relative Weighting of Attributes for Polar Plots

Attribute (i)	Weight (p_j)	Angle (θ_j)
I	0.333	119.88
II	0.267	96.12
III	0.200	72.00
IV	0.133	47.88
V	0.067	24.12
	1.000	360.00

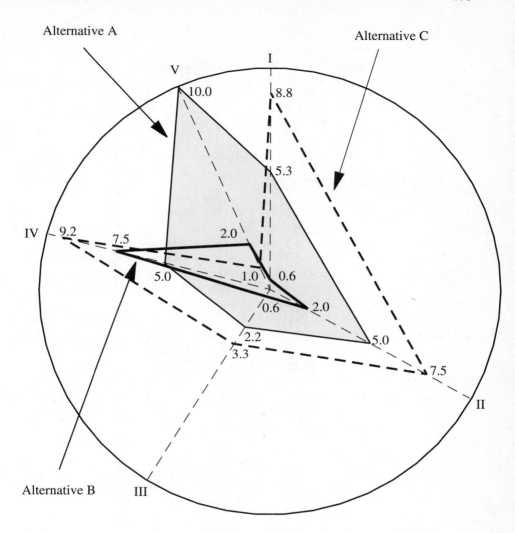

Figure 9-11.   Polar Plot with Weighted Sector Angles

90°. This leads to more consistent evaluation since $\sin(\theta)$ is strictly increasing between 0° and 90°.

It should also be noted that the weighted areas for the alternatives are sensitive to the order in which the attributes are drawn in the polar plot. Thus, a preferred order of the attributes must be defined prior to starting the analysis. The preferred order may be based on the desired sequence in which alternatives must satisfy management goals. For example, it may be desirable to attend to product quality issues before addressing throughput issues. The surface area of the base circle may be interpreted as a measure of global organizational goal with respect to such performance indicators as available capital, market share, capacity utilization, and so on. Thus, the weighted area of the polyhedron associated with an alternative may be viewed as the degree to which that alternative satisfies organizational goals.

Some of the attributes involved in a selection problem might constitute a combination of quantitative and/or qualitative factors or a combination of objective and/or subjective considerations. The prioritizing of the factors and considerations is typically based on the experience, intuition, and subjective preferences of the decision maker. Goal programming is another technique that can be used to evaluate multiple objectives or criteria in decision problems.

9.5 ANALYTIC HIERARCHY PROCESS

The analytic hierarchy process (Saaty 1980) is a practical approach to solving complex decision problems involving the pairwise comparisons of alternatives. The technique, popularly known as AHP, has been used extensively in practice to solve many decision problems. AHP enables decision makers to represent the hierarchical interaction of factors, attributes, characteristics, or alternatives in a multifactor decision-making environment. Figure 9–12 presents an example of a decision hierarchy for project alternatives.

In an AHP hierarchy, the top level reflects the overall objective of the decision problem. The factors or attributes on which the final objective is dependent are listed at intermediate levels in the hierarchy. The lowest level in the hierarchy contains the competing alternatives through which the final objective might be achieved. After the hierarchy has been constructed, the decision maker must undertake a subjective prioritization procedure to determine the weight of each element at each level of the hierarchy. Pairwise comparisons are performed at each level to determine the relative importance of each element at that level with respect to each element at the next higher

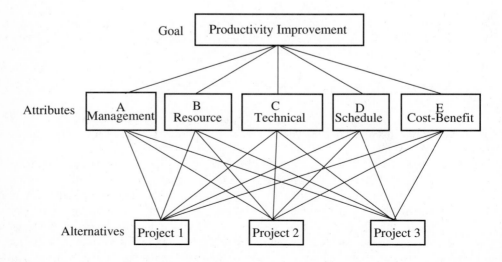

Figure 9-12. AHP for Project Alternatives

level in the hierarchy. In our example, three alternate projects are to be considered. The projects are to be compared on the basis of the following five project attributes:

- Management support
- Resource requirements
- Technical merit
- Schedule effectiveness
- Cost-benefit ratio

The first step in the AHP procedure involves the relative weighting of the attributes with respect to the overall goal. The attributes are compared pairwise with respect to their respective importance to the goal. The pairwise comparison is done through subjective and/or quantitative evaluation by the decision maker(s). The matrix below shows the general layout for pairwise comparisons.

$$\mathbf{F} = \begin{bmatrix} r_{11} & r_{12} & \cdots & r_{1n} \\ r_{21} & r_{22} & \cdots & r_{2n} \\ \cdot & \cdot & \cdots & \cdot \\ \cdot & \cdot & \cdots & \cdot \\ \cdot & \cdot & \cdots & \cdot \\ r_{n1} & r_{n2} & \cdots & r_{nm} \end{bmatrix}$$

where

$\mathbf{F}$ = matrix of pairwise comparisons
r_{ij} = relative preference of the decision maker for i to j
$r_{ij} = 1/r_{ji}$
$r_{ii} = 1$

If $r_{ik}/r_{ij} = r_{jk}$ for all i, j, k, then matrix $\mathbf{F}$ is said to be perfectly consistent. In other words, the transitivity of the preference orders is preserved. Thus, if factor A is preferred to factor B by a scale of 2 and factor A is preferred to factor C by a scale of 6, then factor B will be preferred to factor C by a factor of 3. That is, A = 2B and A = 6. Then, 2B = 6C (i.e., B = 3C). The usual weight scale for AHP is summarized in Table 9–8.

Table 9-8   AHP Weight Scale

Scale	Definition
1	Equal importance
3	Weak importance of one over another
5	Essential importance
7	Demonstrated importance
9	Absolute importance
2,4,6,8	Intermediate weights
Reciprocals	Represented by negative numbers

In practical situations, one cannot expect all pairwise comparison matrices to be perfectly consistent. Thus, *a* tolerance level for consistency was developed by Saaty. The tolerance level, referred to as consistency ratio, is acceptable if it is less then 0.10 (10%). Readers should refer to Saaty (1980) for further details on the procedure for computing the consistency ratio. If a consistency ratio is greater than 10%, the decision maker has the option of going back to reconstruct the comparison matrix or proceeding with the analysis with the recognition that he accepts the potential bias that may exist in the final decision. Once the pairwise comparisons matrix is complete, the relative weights of the factors included in the matrix are obtained from the estimate of the maximum eigenvector of the matrix. This is done by the expression

$$\mathbf{FW} = \lambda_{max}\mathbf{W}$$

where

$\mathbf{F}$ = matrix of pairwise comparisons
λ_{max} = maximum eigenvector of $\mathbf{F}$
$\mathbf{W}$ = vector of relative weights

For the example in Figure 9–12, Table 9–9 shows the tabulation of the pairwise comparison of the project attributes.

Each of the attributes listed along the rows of the table is compared against each of the attributes listed in the columns. Each number in the body of the table indicates the degree of preference or importance of one attribute over the other on a scale of 1 to 9. Typical questions that may be used to arrive at the relative rating are

"With respect to the goal of improving productivity, do you consider project resource requirements to be more important than technical merit?"
"If so, how much more important is it on a scale of 1 to 9?"

Similar questions are asked iteratively until each attribute has been compared with each of the other attributes. For example, in Table 9–9, attribute B (resource requirements) is considered to be more important than attribute C (technical merit) with a degree of 6. In general, the numbers indicating the relative importance of the attributes are based on the following weight scales:

Table 9-9 Pairwise Comparisons of Project Attributes

Attributes	A: Management	B: Resource	C: Technical	D: Schedule	E: Cost-benefit
A: Management	1	1/3	5	6	5
B: Resource	3	1	6	7	6
C: Technical	1/5	1/6	1	3	1
D: Schedule	1/6	1/7	1/3	1	1/4
E: Cost-benefit	1/5	1/6	1	4	1

1:    Equally important
3:    Weakly more important
5:    Strongly more important
7:    Very strongly more important
9:    Absolutely more important

Intermediate ratings are used as appropriate to indicate intermediate levels of importance. If the comparison order is reversed (e.g., B versus A rather than A versus B), then the reciprocal of the importance rating is entered in the pairwise comparison table. The relative evaluation ratings in the table are converted to the matrix of pairwise comparisons shown in Table 9–10.

The entries in Table 9–10 are normalized to obtain Table 9–11. The normalization is done by dividing each entry in a column by the sum of all the entries in the column. For example, the first cell in Table 9–11 (i.e., 0.219) is obtained by dividing 1.000 by 4.567. Note that the sum of the normalized values in each attribute column is 1.

The last column in Table 9–11 shows the normalized average rating associated with each attribute. This column represents the estimated maximum eigenvector of the matrix of pairwise comparisons. The first entry in the column (0.288) is obtained by dividing 1.441 by 5, which is the number of attributes. The averages represent the relative weights (between 0.0 and 1.0) of the attributes that are being evaluated. The relative weights show that attribute B (resource requirements) has the highest importance rating of 0.489. Thus, for this example, resource consideration is seen to be the most important factor in the selection of one of the three alternate projects. It should be emphasized that these attribute weights are valid for only the particular goal specified

Table 9-10   Pairwise Comparisons of Project Attributes

Attributes	A	B	C	D	E
A	1.000	0.333	5.000	6.000	5.000
B	3.000	1.000	6.000	7.000	6.000
C	0.200	0.167	1.000	3.000	1.000
D	0.167	0.143	0.333	1.000	0.250
E	0.200	0.167	1.000	4.000	1.000
Column sum	4.567	1.810	13.333	21.000	13.250

Table 9-11   Normalized Matrix of Pairwise Comparisons

Attributes	A	B	C	D	E	Sum	Average
A: Management support	0.219	0.184	0.375	0.286	0.377	1.441	0.288
B: Resource requirement	0.656	0.551	0.450	0.333	0.454	2.444	0.489
C: Technical merit	0.044	0.094	0.075	0.143	0.075	0.431	0.086
D: Schedule effectiveness	0.037	0.077	0.025	0.048	0.019	0.206	0.041
E: Cost-benefit ratio	0.044	0.094	0.075	0.190	0.075	0.478	0.096
Column sum	1.000	1.000	1.000	1.000	1.000		1.000

in the AHP model for the problem. If another goal is specified, the attributes would need to be reevaluated with respect to that new goal.

After the relative weights of the attributes are obtained, the next step is to evaluate the alternatives on the basis of the attributes. In this step, a relative evaluation rating is obtained for each alternative with respect to each attribute. The procedure for the pairwise comparison of the alternatives is similar to the procedure for comparing the attributes. Table 9–12 presents the tabulation of the pairwise comparisons of the three alternatives with respect to attribute A (management support).

The table shows that project 1 and project 3 have the same level of management support. Examples of questions that may be used in obtaining the pairwise ratings of the alternatives are

"Is project 1 preferred to project 2 with respect to management support?"
"What is the level of preference on a scale of 1 to 9?"

It should be noted that the comparisons shown in Table 9-11 are valid only when management support of the projects is being considered. Separate pairwise comparisons of the project must be done whenever another attribute is being considered. Consequently, for our example, we would have five separate matrices of pairwise comparisons of the alternatives, one matrix for each attribute. Table 9–12 is the first of the five matrices. The other four are not shown. The normalization of the entries in Table 9–12 yields the following relative weights of the projects with respect to management support: project 1 (0.21), project 2 (0.55), and project 3 (0.24). Table 9–13 shows a summary of the normalized relative ratings of the three projects with respect to each of the five attributes.

The attribute weights shown in Table 9–12 are combined with the weights in Table 9–13 to obtain the overall relative weights of the projects as

Table 9-12 Project Ratings Based on Management Support

Alternatives	Project 1	Project 2	Project 3
Project 1	1	1/3	1
Project 2	3	1	2
Project 3	1	1/2	1

Table 9-13 Project Weights Based on Attributes

	Attributes				
	Management support	Resource requirements	Technical merit	Schedule effectiveness	C/B ratio
Project 1	0.21	0.12	0.50	0.63	0.62
Project 2	0.55	0.55	0.25	0.30	0.24
Project 3	0.24	0.33	0.25	0.07	0.14

$$\alpha_j = \sum_i (w_i k_{ij})$$

where

α_j = overall weight for project j

w_i = relative weight for attribute i

k_{ij} = rating (local weight) for project j with respect to attribute i

$w_i k_{ij}$ = global weight of alternative j with respect to attribute i

Table 9–14 shows the summary of the final AHP analysis for the example. The summary shows that project 2 should be selected, since it has the highest overall weight of 0.484. Some commercial computer programs are available for AHP. Expert Choice and NEWTECH from Expert Choice, Inc., are two examples. In the next chapter, we present a computer simulation approach to probability AHP for forecasting project performance. AHP can be used to prioritize multiple projects with respect to several objectives. Table 9–15 shows a generic layout for multiple projects evaluation.

Table 9-14 Summary of AHP Evaluation of
Three Projects

	Attributes					
	A ($i = 1$)	B ($i = 2$)	C ($i = 3$)	D ($i = 4$)	E ($i = 5$)	
$w_i \Rightarrow$	0.288	0.489	0.086	0.041	0.096	
Project j			k_{ij}			α_j
Project 1	0.21	0.12	0.50	0.63	0.62	0.248
Project 2	0.55	0.55	0.25	0.30	0.24	0.484
Project 3	0.24	0.33	0.25	0.07	0.14	0.268
Column sum	1.000	1.000	1.000	1.000	1.000	1.000

Table 9-15 Layout for Multiple Projects Comparison

	Objectives				
	Objective 1	Objective 2	Objective 3	...	Objective k
Project 1					
Project 2					
Project 3					
...					
...					
Project n					

9.6 PROJECT BENCHMARKING

The techniques presented in the preceding sections can be used for benchmarking projects. For example, to develop a baseline schedule, evidence of successful practices from other projects may be needed. Metrics based on an organization's most critical project implementation issues should be developed. *Benchmarking* is a process whereby target performance standards are established based on the best examples available. The objective is to equal or surpass the best example. In simple terms, benchmarking means "learning from the best." The premise of benchmarking is that if an organization replicates the best quality examples, it will become one of the best in the industry. A major approach of benchmarking is to identify performance gaps between projects. Figure 9–13 shows an example of such a gap. Benchmarking requires that an attempt should be made to close the gap by improving performance of the subject project.

Benchmarking requires frequent comparison with the target project. Updates must be obtained from projects already benchmarked and new projects to be benchmarked must be selected on a periodic basis. Measurement, analysis, feedback, and modification should be incorporated into the performance improvement program. The benchmark-feedback model presented in Figure 9–14 is useful for establishing a continuous drive toward performance benchmarks.

The figure shows the block diagram representation of input-output relationships of the components in a benchmarking environment. In the model, $I(t)$ represents the

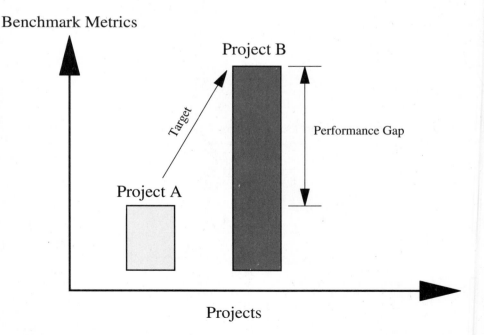

Figure 9-13.   Identification of Benchmark Gaps

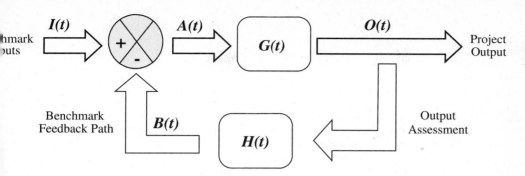

Figure 9-14.   Quality Benchmark-Feedback Model

set of benchmark inputs to the subject project. The inputs may be in terms of data, information, raw material, technical skill, or other basic resources. The index t denotes a time reference. $A(t)$ represents the feedback loop actuator. The actuator facilitates the flow of inputs to the various segments of the project. $G(t)$ represents the forward transfer function which coordinates input information and resources to produce the desired output, $O(t)$. $H(t)$ represents the management control process that monitors the status of improvement and generates the appropriate feedback information, $B(t)$, which is routed to the input transfer junction. The feedback information is necessary to determine what control actions should be taken at the next improvement phase. The primary responsibility of a quality improvement manager should involve ensuring the proper forward and backward flow of information concerning the performance of a project with respect to the benchmarked inputs.

.7 EXERCISES

9.1. Construct a project value model for a super project that involves the construction of a scientific experimentation facility in the vicinity of a densely populated area.

9.2. A multinational company is considering a new manufacturing project which will require some of the scarce resources of the company. Each unit of the product to be produced will require 3 units of resource type 1, 4 units of resource type 2, and 5 units of resource type 3. The company has available 120 units of resource 1, 152 units of resource 2, and 150 units of resource 3.
  (a) What is the maximum quantity of this product that can be made?
  (b) If a $5 profit is expected from each unit, what is the maximum profit that can be made?
  (c) If the cost of tooling up to make the product is $100, should this project be selected for funding by the company? Explain.

9.3. Develop an analytic hierarchy process model for the goal of selecting an industrial development project. List and justify the important factors to be considered.

9.4. The following are evaluation ratings of how well each of three project alternatives satisfies each of four attributes on a scale of 0 to 100.

Attribute	Rank	Alternative 1	Alternative 2	Alternative 3
A	1	25	55	80
B	2	15	73	70
C	3	70	10	70
D	4	85	62	24

   (a) Draw the basic polar plot for the above data. Which alternative do you recommend?
   (b) Draw the modified polar plot for the data. Which alternative do you recommend?
   (c) Draw the polar plot with weighted sector angles for the data. Which alternative do you recommend?

9.5. Using a quantitative example, show how a project analyst might use learning curve analysis as a decision aid for project selection.

9.6. Discuss the differences between project feasibility analysis and project selection analysis.

9.7. Using an appropriate mathematical procedure, find the eigenvector of the matrix below and compare your result to the vector of attribute weights calculated for the AHP example in this chapter.

$$\begin{matrix}
1.000 & 0.333 & 5.000 & 6.000 & 5.000 \\
3.000 & 1.000 & 6.000 & 7.000 & 6.000 \\
0.200 & 0.167 & 1.000 & 3.000 & 1.000 \\
0.167 & 0.143 & 0.333 & 1.000 & 0.250 \\
0.200 & 0.167 & 1.000 & 4.000 & 1.000
\end{matrix}$$

9.8. Provide relative weights to indicate the importance of five to six attributes you would consider in selecting a project manager. Who are the people you should consult in developing the relative weighting of the attributes?

Section 3

Computer Applications

10

Computers and Project Management

This chapter presents computer applications for project management. Topics covered include general applications of project management software, simulation, and other computerized procedures. It is recommended that all the planning functions of a project be completed before software use can begin. Proper planning helps identify the appropriate inputs to provide to a computer program.

10.1 COMPUTER TOOLS AND PROJECT MANAGEMENT

Project management is both a science and an art. It is a science because it uses scientific techniques that have been proven to enhance management processes. It is an art because it relies on the good judgment, expertise, and personal intuition of the project manager. A good project manager will use his or her personal expertise and experience and the tools available to manage projects effectively. Computer hardware and software constitute one group of tools that is available to the project manager. Badiru and Whitehouse (1989) present further details on computer tools for project management.

A large number of procedures and methods have been developed to enhance the application of project management to a vast range of industrial and administrative functions. This rapid spreading of project management is largely due to the availability of computer software packages that make it possible to quickly implement project management techniques. Using a computer should not replace the conventional planning, scheduling, and controlling functions of management. A computer should also not be expected to replace traditional management decision-making processes. The computer should be used only as a tool to facilitate the implementation of validated techniques. The primary value of the computer is the speed at which it will perform the quantitative analysis needed in developing schedules and generating a variety of outputs and reports.

Computer programs for project planning and control have been available for a long time, but the complex designs made general use difficult. The first generation of project management programs was complex, expensive, restrictive, and difficult to use. However, new hardware and software technologies have opened new possibilities. The increasing popularity, acceptance, and use of project management software are evidenced by the number of commercial programs that have been introduced in recent years. Many computer trade publications such as *PC Week*, *PC World*, *Computer Shopper*, and *Software Magazine* frequently carry advertisements, buyers' guides, and reviews on project management software.

Project management software is designed for use not only by the project manager but also by any person involved in managing a group of resources in order to carry out a set of tasks. In larger projects, the overall project leader cannot be aware of all the concurrent activities in the project. A decentralized organization of the project management environment is often more helpful in coordinating the various activities in a large project. The project leader sets the objectives and time frames for accomplishing the objectives and then delegates the actual functions to different groups.

A software package must be able to meet the needs of a decentralized project organization. It must be an aid to both the middle- and lower-level management personnel. Today, it is not necessary to use a mainframe computer to meet these needs. For example, the needs can be met by using flexible microcomputer software tools with facilities that make it possible to send and collect relevant information in a local area network. The personal computer can enhance the productivity of project teams and facilitate better project performance. With computer information systems, bottlenecks in the project environment can be quickly identified.

Instead of spending a lot of time performing conventional tracking, it is much easier to get online facts promptly about project progress rather than making a series of phone calls. Without leaving the project office, it is possible to get detailed project reports quickly via new computer-based technologies such as fax, electronic mail, modems, and local area networks. Instant access to information located at remote project sites is possible with computer networks. Regular software tools can be combined with dedicated project management software to create a versatile information control environment.

Nowadays, many computer innovations are being introduced in the project environment. These innovations facilitate client-server computing, online interactive project consultations, object-oriented computing, multiplatform systems, graphical user interfaces, multimedia output, robust PC LAN, and operator-level computing. Project management should fully exploit all these new tools. Weitz (1993) discusses the new directions in LAN-based project management software. Johnson (1992) presents a survey of the capabilities of commercial project management software for resource-constrained project scheduling.

0.2 PROJECT MANAGEMENT SOFTWARE EVOLUTION

The present stage of project management computer software has been achieved through an evolutionary process that spans several developments. Five distinct generations of software evolution can be identified.

1. First generation (pre-World War II). This was the era of unstructured project analysis when managers executed their project plans (if ever there was one) by mere intuition and the so-called "seat-of-the-pants" approaches.
2. Second generation (from mid-1950s). The emergence of formal operations research techniques led to the introduction of dedicated project scheduling techniques such as PERT and CPM. Project analyses in this generation were mostly performed manually.
3. Third generation (from early 1970s). In this generation, mainframe computer implementations of PERT/CPM became prevalent. However, access to the programs was limited to only those with the necessary hardware. The batch environment of this generation made it impossible to obtain instant information about project status.
4. Fourth generation (from early 1980s). This period marked the development of more accessible project management programs on mini- and microcomputers. The increased access gave managers the tools for timely project monitoring and control.
5. Fifth generation (from mid-1980s). This generation is marked by the introduction of integrated project management packages. These advanced programs combined the traditional PERT/CPM network analysis with project graphics, report generation, spreadsheets, and cost analysis.
6. Sixth generation (the 1990s). The sixth generation promises to incorporate real-time communication and networking capabilities into project analysis. Some of the elements of this sixth generation are already available in some project management packages. Artificial intelligence techniques including expert systems, neural networks, and case-based reasoning for project management will play a major role in project management in this era.

10.3 SELECTING PROJECT MANAGEMENT SOFTWARE

The proliferation of project management software has created an atmosphere whereby every project analyst wants to have and use a software package for every project situation. However, not every project situation deserves the use of project management software. An analyst should first determine whether or not the use of software is justified. If this evaluation is affirmative, then the analyst will need to determine which specific package of the many available should be used. Some of the important factors that may indicate the need for project management software are

1. Multiple projects are to be managed concurrently.
2. A typical project contains more than 20 tasks.
3. The project scheduling environment is very complex.
4. More than five resource types are involved in each project.
5. There is a need to perform complementing numerical analysis to support project management functions.
6. The generation of graphics (e.g., Gantt, PERT charts) is desired to facilitate project communication.
7. Cost analysis is to be performed on a frequent basis.

8. It is necessary to generate forecasts from historical project data.
9. Automated reporting is important to the organization.
10. Computerization is one of the goals of the organization.

A close examination of any modern project environment will reveal that it fits the criteria for using project management software. Only very small and narrowly focused projects will not need the of help software for effective management.

10.3.1 Important Factors

When evaluating project management software, several factors and questions should be evaluated.

1. *Cost.* Can the software be procured for a reasonable amount relative to the organization's budget and intended use? There are some software packages that cost up to $50,000. Such software packages are usually targeted for the large contractors. It would be an overkill and a waste of resources to buy $100,000 worth of software to manage a $10,000 one-time project.
2. *Need.*
   - Is a computer program really needed for the project involved?
   - Is there or will there be a knowledgeable person who can run the software?
   - Are there other projects that may need the same software?
   - Is the organization planning to standardize software choice?
3. *Project plan.* Does the software offer the following capabilities?
   - Ease of entering and storing project plans
   - Flexibility of task precedence structure
   - Capability to handle large project networks
   - Capability for work breakdown structure (WBS)
   - Capability for task splitting
4. *Resource Management.* Does the software offer the following resource management options?
   - Partial allocation of resources
   - Resource leveling
   - Resource-cost analysis
   - Robust resource allocation heuristics
5. *Progress Tracking.* Can the software do the following?
   - Display time-based project schedule
   - Handle project replanning (network editing)
   - Indicate percent completion at specific points in time
   - Handle milestone reporting
6. *Report Generation.* Can the software enhance project communication with the following tools?
   - Network diagram (consider limitation on project size and complexity of the diagram)
   - Gantt chart
   - Milestone schedule

- Resource reports
- Cost reports
- Planned versus actual performance reports

7. *Ease of use*. Does the software have the following?
    - On-line training and help
    - Simplified input format
    - Clear output format
    - Relevant contents of outputs
    - High quality documentation
    - Error handling abilities

8. *Supporting analytical tools*. Does the software offer built-in or interface utility for supporting analytical tools?
    - Economic analysis tools
    - Optimization tools
    - Communication tools
    - Other project specific needs

9. *Hardware requirements*. What computer systems environment is required to run the software?
    - RAM requirement
    - Math coprocessor requirements
    - Input devices (mouse, joystick, keyboard, etc.)
    - Output devices (types of printers supported)
    - Computer display unit requirements

10. *General characteristics*.
    - Software version

Later versions of a program, for example version 3.2, have the tendencies of having fewer bugs and more enhanced features. However, excessively high version numbers, for example version 8.1, could mean that the developer has been slow in implementing improvements promptly. In this case, only minor changes, such as the following, are made from one version to the next:

- Compactness (e.g., storage requirements).
- File import/export capability.
- Availability of demo package.
- Copy protection and backup procedures. Does the program permit limited or unlimited hard disk installations?
- Speed of execution.
- Compatibility with other hardware and software used in the project environment.
- Vendor reputation.
- Access to technical support.

Some of these factors may be more important than others in specific project situations. A careful overall analysis should be done by the project analyst. With more and more new programs and updates to old software coming into the market, a crucial aspect of the project management function is keeping up with what is available and using good judgment in selecting a program. The future directions for software applications in project management may include any of the following:

1. A clearing house for software testing where users can preview specific software of interest
2. Commercial software rental similar to video tapes
3. Proliferation of shareware packages

0.4 COMMERCIAL SOFTWARE PACKAGES

Numerous commercial software packages are available for project management. Because of the dynamic changes in software choices on the market, it will not be effective to include a survey of the available software in an archival book of this nature. Any such review may be outdated before the book is even published. To get the latest on commercial software capabilities and choices, it will be necessary to consult one of the frequently published trade magazines that carry software reviews. Examples of such magazines are *PC Week*, *PC World*, and *Software*. Other professional publications also carry project management software reviews occasionally. Examples of such publications are *Industrial Engineering* and *Project Management Journal*.

The project management software market is a very competitive market. Many packages that were originally developed for specific and narrow applications, such as data analysis and conventional decision support, now offer project management capabilities. For example, SAS/OR, a statistical analysis software package that is popular on mainframe and miniframe computers, has a PC-based version that handles project management. Similarly, AutoCAD, the popular computer-aided design software, now has a project management option within it. The option, called AutoProject, has very good graphics and integrated drafting capabilities. Prospective buyers are often overwhelmed by the range of products available.

As computer-aided project management moves into the twenty-first century, users of project management software can look forward to significant improvements in ease of use and accessibility. The nucleus of the improvements will be the emergence of the graphical user interface (GUI). The GUI technology was first pioneered in the microcomputer industry by Apple Computer, Inc. The release of Microsoft Windows and OS/2 Presentation Manager operating systems have made the GUI a reality for most PC users. Many project management software companies are now incorporating GUI capabilities into their products. This makes all the functionalities of a software package easily accessible to users. With its emphasis on graphical views of projects by using tools such as Gantt charts, PERT networks, and resource diagrams, project management is a major market avenue for graphical user interfaces designs. Important questions to ask when evaluating project management software graphics capabilities are

- Does the graphical interface save keystrokes?
- Does the interface offer more convenient organization and representation of project data?
- Does the interface contain better analytic tools for data analysis?

- Can graphical outputs be customized for different levels of report?
- Does the graphical interface offer more flexibility than conventional user interfaces?
- Does the graphical interface require previous graphics experience?
- Which project charts can be viewed only on-screen?
- Which project charts can be printed in hardcopy form?
- What type of hardware configuration is required by the graphical interface? Although most project management software programs can run on the basic PC systems, users of programs with graphical user interfaces often require a more powerful system.

Some of the most popular project management software packages include InstaPlan 5000, Artemis Project, Microsoft Project for Windows, Plantrac II, Advanced Project Workbench, Qwiknet Professional, Super Project Plus, Time Line, Project Scheduler, Primavera, Texim, Viewpoint, PROMIS, Topdown Project Planner, Harvard Project Manager, PCS (Project Control System), PAC III, VISION, SAS/OR, Autoproject, Visual Planner, Project/2, Timesheet, Task Monitor, Quick Schedule Plus, Suretrak Project Scheduler, Supertime, On Target, Great Gantt, Pro Tracs, Autoplan, AMS Time Machine, MacProject II, Micro Trak, Checkpoint, Maestro II, Cascade, OpenPlan, Vue, and Cosmos.

So prolific are the software offerings that the developers are running out of innovative productive names. At this moment, new products are being introduced while some are being phased out. Prospective buyers of project management software should consult vendors for the latest products. No blanket software recommendation can be offered in this book since product profiles change quickly and frequently.

10.4.1 Special Purpose Software

Special purpose software tools have found a place in project management. Some of these tools are simulation packages, statistical analysis packages, optimization programs, report writers, and others. Computer simulation is a versatile tool that has potential for enhancing project planning and control analysis. The following section presents an application of simulation to project network analysis.

10.5 SIMULATION OF PROJECT NETWORKS

Computer simulation is a tool that can be effectively utilized to enhance project planning, scheduling, and control. At any given time, only a small segment of a project network will be available for direct observation and analysis. The major portion of the project will either have been in the past or will be expected in the future. Such unobservable portions of the project can be studied best by simulation.

Using the historical information from previous segments of a project and the prevailing events in the project environment, projections can be made about future expectations of the project. Outputs of simulation can alert management to real and potential problems. The information provided by simulation can be very helpful in project selection decisions. Simulation-based project analysis may involve the following components:

- Activity time modeling
- Simulation of project schedule
- What-if analysis and statistical modeling
- Management decisions and sensitivity analysis

10.5.1 STARC Software

Schedule simulation is the most obvious avenue for the application of simulation to project analysis. A computer program named STARC (Badiru 1991d) is used in this section to illustrate schedule simulation and analysis. STARC is a project planning aid. It simulates project networks and performs what-if analysis of projects involving probabilistic activity times and resource constraints. The output of STARC can serve as a decision aid for project selection. The effects of different activity time estimates and resource allocation options can be studied with STARC prior to actual project commitments.

Activity time modeling.   The true distribution of activity times will rarely be known. Even when known, the distribution may change from one type to another depending on who is performing the activity, where the activity is performed, and when the activity is performed. Simulation permits a project analyst to experiment with different activity time distributions. Commonly used activity time distributions are beta, normal, uniform, and triangular distributions. The results of simulation experiments can guide the analyst in developing definite action plans. Using the three PERT estimates a (optimistic time), m (most likely time), and b (pessimistic time), a beta distribution with appropriate shape parameters can be modeled for each activity time. The details of the methodology for activity time modeling are presented by Badiru (1991b).

Duration risk coverage factor.   This is a percentage factor (q) which provides a risk coverage for the potential inaccuracies in activity time estimates. A risk factor of 10% ($q = 0.10$), for example, extends the range $[a, b]$ of an activity's duration by 10% over the specified PERT time interval. An example of the extension of the activity duration interval is shown in Figure 10–1.

The extension of the range ensures that there is some probability (greater than zero) of generating activity times below the optimistic estimate or above the pessimistic estimate. Further details of the procedure for the PERT interval extension can be found in Badiru and Whitehouse (1989). If a risk coverage of 0% is specified, there is no interval extension and no adjustments are made to the PERT estimates. By comparison, a risk coverage of 100% yields an extended interval that is twice as wide as the original PERT interval. While a large extension of the PERT interval may be desirable for more simulation flexibility, it does result in a high variance for activity times.

Resource allocation heuristic.   During a simulation run, STARC uses the composite allocation factor (CAF) presented by Badiru (1988a) to prioritize activities

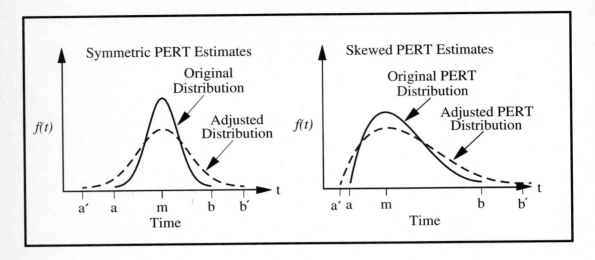

Figure 10-1. Risk Coverage for PERT Estimates

for resource allocation. This heuristic was discussed in detail in Chapter 5. The resource allocation process takes into account both the resource requirements and the variabilities in activity times. Activities with higher values of CAF are given priority during the resource allocation process.

CAF is a weighted and scaled sum of two components: RAF (resource allocation factor) and SAF (stochastic activity duration factor). RAF indicates the degree of resource consumption per unit time while SAF measures the degree of variability in activity durations. For each simulation experiment, a weighting factor, w, is used to specify the relative weights or level of importance to give to RAF and SAF in the computation of CAF. The higher the value of w on a scale of 0 to 1, the higher the importance associated with resource considerations in the scheduling process (see Chapter 5 for further details). Figure 10–2 shows the weighting scale used in STARC.

Simulation examples. The sample project presented in Table 10–1 is used to illustrate project network simulation analysis using STARC. The project network is

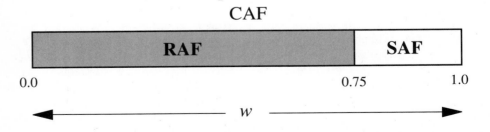

Figure 10-2. CAF Weighting Scale Used in STARC

Table 10-1 Sample Project Data
with Resource Constraint

Activity ID	Activity no.	Preceding activities	a, m, b (in months)	Resource units required
A	1	—	1, 2, 4	3
B	2	—	5, 6, 7	5
C	3	—	2, 4, 5	4
D	4	A	1, 3, 4	2
E	5	C	4, 5, 7	4
F	6	A	3, 4, 5	2
G	7	B, D, E	1, 2, 3	6

shown in Figure 10–3. The sample project consists of seven activities and one resource type. There are 10 units of the resource available.

Simulation runs were made with $w = 0.5$ and $q = 0.15$. Figure 10–4 shows the output of the unconstrained PERT analysis. The expected duration (DUR), earliest start (ES), earliest completion (EC), latest start (LS), latest completion (LC), total slack (TS), free slack (FS), and critical path indicator (CRIT) are presented for each activity. The output shows that the PERT time without resource constraints is 11 months. Activities 3, 5, and 7 are on the critical path.

Figure 10–5 shows the simulated sample averages for the network parameters. The average project duration is 13.23 months (based on a simulation sample size of 100). With the resource constraints, the criticality indices for activities 1 through 7 are 0.97, 0.0, 0.03, 0.84, 0.03, 0.13, and 0.87, respectively. The criticality index of an activity is the probability that the activity will fall on the critical path. The simulation indicates that activity 1 is critical most of the time (probability of 0.97), whereas activity 2 is

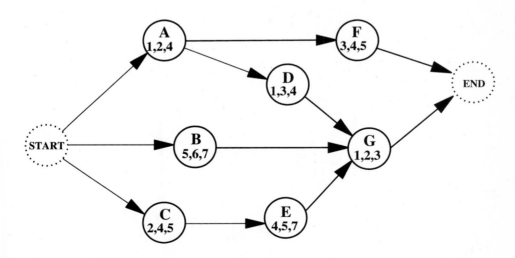

Figure 10-3. Project Network for Simulation Analysis

```
                    UNCONSTRAINED PERT SCHEDULE
     ACT.   DUR.    ES      EC      LS      LC     TS     FS    CRIT
     -------------------------------------------------------------------
      1     2.17    0.00    2.17    4.00    6.17   4.00   0.00   0.000
      2     6.00    0.00    6.00    3.00    9.00   3.00   3.00   0.000
      3     3.83    0.00    3.83    0.00    3.83   0.00   0.00   1.000
      4     2.83    2.17    5.00    6.17    9.00   4.00   4.00   0.000
      5     5.17    3.83    9.00    3.83    9.00   0.00   0.00   1.000
      6     4.00    2.17    6.17    7.00   11.00   4.83   0.00   0.000
      7     2.00    9.00   11.00    9.00   11.00   0.00   0.00   1.000
     UNCONSTRAINED PERT PROJECT COMPLETION TIME =  11.00
```

Figure 10-4. Unconstrained PERT Analysis

```
                       SIMULATED SAMPLE AVERAGES
     ACT.   MEAN   MEAN   MEAN    MEAN   MEAN    MEAN   MEAN   CRIT.
      #     DUR.   ES     EC      LS     LC      TS     FS     INDEX
     -------------------------------------------------------------------
      1     2.28   6.04   8.32    6.04   8.32    0.06   0.00   0.970
      2     6.04   0.00   6.04    5.17  11.21    5.17   5.11   0.000
      3     3.78   0.00   3.78    2.25   6.05    2.25   0.00   0.030
      4     2.81   8.32  11.13    8.40  11.21    0.14   0.08   0.840
      5     5.16   3.78   8.94    6.05  11.21    2.25   2.19   0.030
      6     4.01   8.32  12.33    9.22  13.23    0.91   0.00   0.130
      7     2.01  11.16  13.17   11.22  13.23    0.06   0.00   0.870
     AVERAGE PROJECT DURATION =  13.23
```

Figure 10-5. Simulated Sample Averages

never on the critical path. It should be noted that the average ES, EC, LS, LC, TS, and FS will not necessarily match conventional PERT network calculations. Thus, we cannot draw a PERT network based on the sample averages. This is because each average value (e.g., mean EC) is computed from the results of several independent simulation runs. However, for each separate simulation run, the simulated outputs will fit conventional PERT network calculations.

Figure 10–6 shows a deadline analysis for a set of selected project deadlines. The second column in the figure presents the probabilities calculated analytically on the basis of the *central limit theorem*. The third column presents the sample probabilities based on simulation observations. The larger the number of simulation runs, the closer the analytical and sample probabilities will be. Suppose we are considering a contract deadline of 13 months; we might like to know the probability of being able to finish the project within that time frame. The simulation output indicates a simulated probability of 0.42 for completion within 13 months. So, the chances of finishing the project in 13 months are not so good. There is a very low probability (0.0114) of finishing the project in 11 months, which is the unconstrained PERT duration. A 15-month deadline (with a probability of 0.97) seems quite achievable for this project.

Figure 10–7 shows the simulation output with the shortest project duration of 11.07 months. If plotted on a Gantt chart, this schedule can serve as an operational schedule for the project, provided the simulated activity durations are realistic. It should be recalled, however, that 11 months has a low probability of being achieved.

```
                    PROJECT DEADLINE ANALYSIS
    DEADLINE    CALCULATED PROBABILITY    OBSERVED PROBABILITY
    ------------------------------------------------------------
       10.00          0.0000                    0.0000
       11.00          0.0062                    0.0000
       12.00          0.0836                    0.0700
       13.00          0.3969                    0.4200
       14.00          0.8048                    0.7900
       15.00          0.9761                    0.9700
       16.00          0.9990                    1.0000
       17.00          1.0000                    1.0000
       18.00          1.0000                    1.0000
       19.00          1.0000                    1.0000
```

Figure 10-6. Project Deadline Analysis

```
                      SHORTEST SIMULATED SCHEDULED
    ACT.   DUR.    ES     EC     LS     LC     TS     FS    CRIT
    -------------------------------------------------------------
      1    1.14   5.47   6.61   5.47   6.61   0.00   0.00  1.000
      2    5.47   0.00   5.47   4.31   9.78   4.31   4.16  0.000
      3    4.37   0.00   4.37   0.15   4.52   0.15   0.00  0.000
      4    2.04   6.61   8.65   7.74   9.78   1.13   0.98  0.000
      5    5.26   4.37   9.63   4.52   9.78   0.15   0.00  0.000
      6    4.46   6.61  11.07   6.61  11.07   0.00   0.00  1.000
      7    1.30   9.63  10.92   9.78  11.07   0.15   0.00  0.000
    SHORTEST SIMULATED PROJECT DURATION =  11.07
```

Figure 10-7. Best Simulated Project Schedule

The project network data as organized by STARC are shown in Figure 10–8. The scaled CAF measures for the activities are presented in the last column. Activity 2 has the highest priority (100.0) for resource allocation when activities compete for resources. Activity 4 has the lowest priority (54.0).

A frequency distribution histogram for the project duration based on the simulated sample is presented in Figure 10–9. As expected, the average project duration appears to be approximately normally distributed.

Other portions of the simulation output (not shown) present the sample duration variances, sample duration ranges, and the parameters of the fitted beta distributions

```
                  PROJECT ACTIVITIES DATA
    ACTIVITY      A      M      B     MEAN   VAR.  RANGE   CAF
    ------------------------------------------------------------
       1         1.0    2.0    4.0    2.2    0.3    3.0    55.4
       2         5.0    6.0    7.0    6.0    0.1    2.0   100.0
       3         2.0    4.0    5.0    3.8    0.3    3.0    72.6
       4         1.0    3.0    4.0    2.8    0.3    3.0    54.0
       5         4.0    5.0    7.0    5.2    0.3    3.0    88.0
       6         3.0    4.0    5.0    4.0    0.1    2.0    66.6
       7         1.0    2.0    3.0    2.0    0.1    2.0    75.3
```

Figure 10-8. Project Simulation Data

```
FREQUENCY DISTRIBUTION HISTOGRAM FOR PROJECT DURATION
-----------------------------------------------------
CLASS       INTERVAL          ELEMENTS
-----------------------------------------------------
   1      11.01 to   11.24      3  *********
   2      11.24 to   11.46      0
   3      11.46 to   11.68      5  ***************
   4      11.68 to   11.90      4  ************
   5      11.90 to   12.12      3  *********
   6      12.12 to   12.35      7  *********************
   7      12.35 to   12.57      9  ****************************
   8      12.57 to   12.79     11  **********************************
   9      12.79 to   13.01     14  *******************************************
  10      13.01 to   13.23      7  *********************
  11      13.23 to   13.46     11  **********************************
  12      13.46 to   13.68      8  ************************
  13      13.68 to   13.90      6  ******************
  14      13.90 to   14.12      4  ************
  15      14.12 to   14.35      1  ***
  16      14.35 to   14.57      2  ******
  17      14.57 to   14.79      2  ******
  18      14.79 to   15.01      2  ******
  19      15.01 to   15.23      0
  20      15.23 to   15.46      1  ***_
```

Figure 10-9.   Histogram of Project Duration

for the activity durations. The variances and sample ranges are useful for statistical or analytical purposes such as control charts for activity durations and resource loading diagrams.

What-if analysis.   In the safe environment of simulation, the parameters involved in developing project planning strategies may be varied to determine the resulting effects on the overall project structure. Resource allocation is a major area where simulation-based what-if analysis can be useful. For example, the project analyst can determine the lower and upper bounds for resource allocation in order to achieve the specified project goal. Statistical modeling techniques, such as regression analysis, may be employed to analyze simulation data and study the relationships between factors in the project. The effectiveness of different resource allocation heuristics can be tested in the simulation environment.

A review of the simulation output may indicate what type of what-if analysis may be performed. For example, a change was made in the sample project data. The number of available units of resource was increased from 10 to 15. With the additional resource units available, the average project duration was reduced from 13.23 months to 11.09 months. The deadline analysis revealed that the probability of finishing the project in 13 months increased from 0.42 to 0.9895 after the additional resource allocation.

Another revision of the project data was also analyzed. In this revision, the resource availability was decreased from 10 units to 7 units. It turns out that decreasing the resource availability by 3 units caused the average project duration to increase from 13.23 to 17.56 months. With the decreased resource availability, even a generous deadline of 17 months has a low probability (0.25) of being achieved. Using the revised simulation outputs, management can make better-informed decisions about resource allocation.

Sensitivity analysis.   Management decisions based on a simulation analysis will exhibit more validity than decisions based on absolute subjective reasoning. Simulation simplifies sensitivity analysis so that a project analyst can quickly determine what changes in project parameters will necessitate a change in management decisions. For example, with the information obtained from simulation, we can study the sensitivity of project completion times to changes in resource availability. The potential effects of decisions can be studied prior to making actual resource and time commitments. For the project data presented in Table 10–1, Figure 10–10 shows a plot of the sensitivity of project duration to changes in resource availability. Note that a resource level of 5 units is infeasible since there is one activity (activity G) which requires 6 units of resources. As resource units increase, the project duration decreases until it levels off at around 11 time units. Figure 10–11 shows a plot of the sensitivity of the probability of completion to changes in project due dates. The plot is based on the case of 10 resource units available. Both the analytically calculated and the simulation observed probabilities are shown. Probability of completion increases with an increase in due dates.

Multiple resource constraints.   One additional resource type was added to the sample project data. Table 10–2 presents the revised project data. The simulation outputs for the revised project data are used for the statistical modeling presented later. Several simulation runs of the project network were made. Several combinations of w and q were investigated, and the average project durations were recorded for simulation sample size of 100. Table 10–3 presents a tabulation of the average project durations

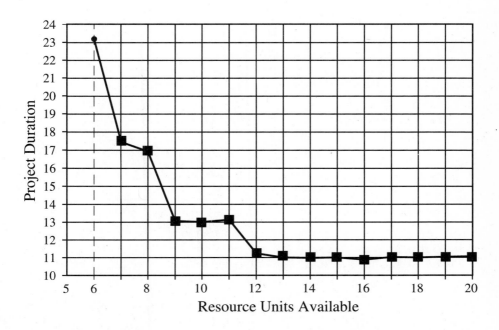

Figure 10-10.   Sensitivity of Project Duration to Resource Availability

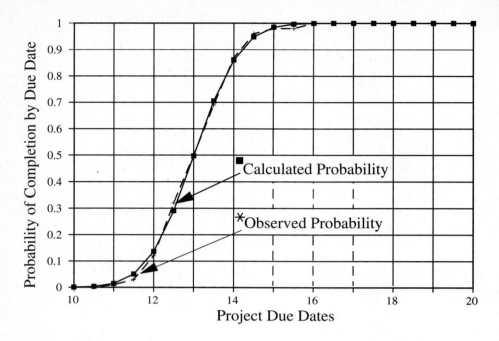

Figure 10-11. Sensitivity of Probability of Completion to Project Due Dates

Table 10-2 Project Data with Multiple Resource Constraints

Activity	Predecessor	a, m, b	Resources r_{i1}, r_{i2}
A	—	1, 2, 4	3, 0
B	—	5, 6, 7	5, 4
C	—	2, 4, 5	4, 1
D	A	1, 3, 4	2, 0
E	C	4, 5, 7	4, 3
F	A	3, 4, 5	2, 7
G	B, D, E	1, 2, 3	6, 2

Units of resource type 1 available = 10
Units of resource type 2 available = 15

based on alternate values of w and q. Values of w range from 0.0 to 1.0, while values of q are 0, 0.1, 0.15, and 0.2.

   Figure 10–12 shows a plot of the simulation output. Note that there is not much difference among the results for risk coverage levels (q) of 0%, 10%, 15%, and 20%. Thus, the project duration appears to be insensitive to risk coverage levels less than or equal to 20%. This preliminary conclusion was confirmed by a formal statistical test. There appear to be differences between the results for different levels of w between 0.0 and 1.0. The increase in the project durations for values of w greater than 0.9 are

Table 10-3 Simulation Output for Project Durations

	Average project duration			
w	$q = 0.0$	$q = 0.1$	$q = 0.15$	$q = 0.2$
0.0	12.98	13.06	13.12	12.60
0.1	13.56	12.88	13.33	13.05
0.2	13.56	12.96	13.03	13.30
0.3	13.48	13.18	12.90	13.03
0.4	13.33	13.08	13.13	13.02
0.5	12.69	13.34	12.51	13.63
0.6	12.76	13.12	13.11	12.91
0.7	13.33	12.10	12.65	12.50
0.8	13.01	13.09	13.45	13.19
0.9	13.25	13.42	13.04	13.23
1.0	16.89	16.77	16.71	17.03

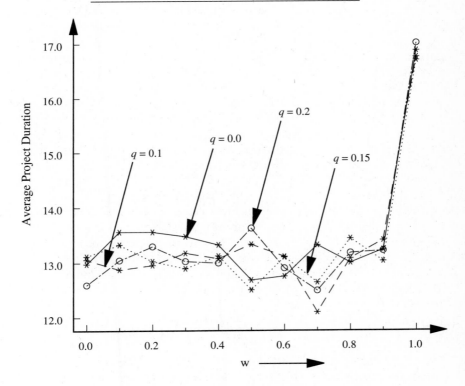

Figure 10-12. Plot of Simulated Average Project Duration

441

particularly significant. Thus, the project duration appears to be sensitive to changes in w. This observation was also confirmed by a formal statistical test.

10.5.2 Regression Metamodels

The use of metamodels (models of a model) is important for achieving an integrated decision model. In conventional use of simulation, we go from the real world to the simulation model to the simulation output and then to the decision. However, the transition from simulation output to actual decision is not necessarily simple. Analysis of the simulation output to generate decisions can be simple or complex depending on the skills of the decision maker. To facilitate a smooth transition and make the best use of the available simulation output, a metamodel may be used. It can be viewed as a transformation model that converts the output in one worldview to another, but simplified, worldview. Graphical representation of tabular data is one common example of metamodeling. Regression metamodels of simulation constitute an effective approach to the use of simulation.

A multifactor analysis of variance (ANOVA) of the simulation results was conducted with STATGRAPHICS software. Two replicates of the simulation experiment were used in the analysis. The data shown in Table 10–3 is for the first replicate. The ANOVA results show that the effect of w on the project duration is significant at the 95% confidence level ($\alpha = 0.05$), while the effect of q is not significant. This confirms graphical observation in Figure 10–12. It was also found that the interaction effect of w and q is significant at the 95% confidence level. Even though q does not seem to have a direct effect on the project duration, it interacts with w to contribute to the observed differences in the project duration.

There is no significant difference between the levels of q for this particular project. There are, however, significant differences between the levels of w. Thus, for this illustrative project, more attention should be directed at the resource constraints. Since $w = 1.0$ yielded the longest average project duration, it is concluded that the duration of this project is likely to increase when emphasis is placed on the resource considerations alone. Thus, variabilities in activity time durations should also be considered in the decision process. A simple linear regression model fitted to the data in Table 10–3 is presented below. The model gives a relationship between the average project duration, D, and w for the case when $q = 0.15$.

$$D_{(q=0.15)} = 12.565 + 1.5936w$$

The R-squared value of 21.40% is very low. It is concluded that the simple linear model does not significantly account for variabilities in the project duration. A multiple linear regression model fitted to the data is presented as

$$f(w, q) = 12.66 + 1.67w - 0.69q$$

with a low R^2 value of 18.46%. The model is shown graphically in Figure 10–13. It is concluded that a multiple linear regression model also does not adequately represent the simulated data.

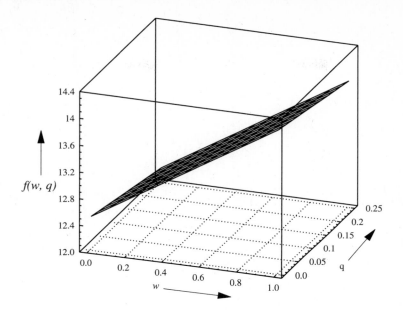

Figure 10-13. Plot of Multiple Linear Regression Model

Multiple nonlinear regression model. A multiple nonlinear regression model is the next logical option in the analysis of the simulation output. We will investigate a multiple nonlinear regression model of the form

$$f(w, q) = \beta_0 + \beta_1 w^{\alpha_1} + \beta_2 q^{\alpha_2} + \epsilon$$

where β_i and α_i are appropriate model parameters and ϵ is the error term. The simulated data in Table 10–3 is once again used for the modeling. The following nonlinear model was obtained:

$$f(w, q) = 13.17 + 3.80 w^{28.37} - 0.20 q^{-0.37}$$

with a high R^2 value of 93.34%. Most of the variabilities in average project duration are explained by the terms in the nonlinear model. A plot of the response surface for the fitted model is shown in Figure 10–14. Note that w accounts for most of the fit in this particular nonlinear model. This confirms the visual assessment obtained earlier from the plot of the simulated data. Alternate nonlinear models were investigated. The alternate models were of the following forms:

$$f(w, q) = \beta_0 + \beta_1 w^{\alpha_1} + \beta_2 q^{\alpha_2} + \beta_3 w^{\alpha^3} q^{\alpha_4} + \epsilon$$
$$f(w, q) = \beta_0 + \beta_1 w + \beta_2 q + \beta_3 w^{\alpha_1} + \beta_4 q^{\alpha_2} + \beta_5 w^{\alpha_3} q^{\alpha_4} + \epsilon$$

These did not yield any feasible model fit. Besides, these alternate models contain too many parameters. Since the objective is to keep the fitted model as parsimonious in parameters as possible, the alternate models were rejected without further investigation. For practical applications, a model must be simple enough to appeal to decision makers.

Using the regression model. Now that the nonlinear model has been fitted and validated, the next step is the application of the model to project decision analysis.

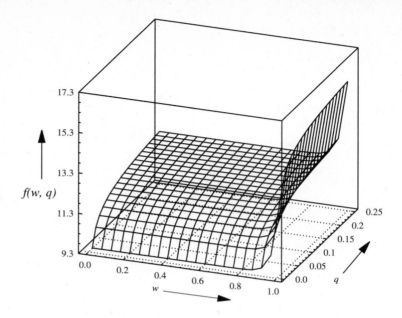

Figure 10-14. Multiple Nonlinear Regression Model

Suppose the ABC classification under a Pareto distribution analysis has been used to determine the relative effects of different factors on the complexity of resource allocation in a large project. Two factors have been identified for analysis. The first factor (factor A) is defined as the "rate of multiple resource consumption per unit time." The second factor (factor B) is defined as "variabilities in task durations." Using the ABC classification, management has assigned a weight of 80% to factor A and 20% to factor B. These relative weights will be used to make resource allocation plans for the project. A project analyst has been asked to analyze the potential effects of the relative weights on the expected number of months it will take to complete the project. Using the fitted model, the analyst would set $w = 0.8$ and $q = 0.20$ and estimate the expected project duration as

$$E[\text{duration}] = 13.17 + 3.80(0.8)^{28.37} - 0.20(0.20)^{-0.37}$$
$$= 12.81 \text{ months}$$

Management would then review this information and probably incorporate it with other qualitative considerations before making final resource allocation decisions. The analyst may try alternate combinations of the values of w and q before making a recommendation to management. If properly used for project planning and control, simulation can yield significant benefits involving better product quality, improved resource utilization, higher project productivity, and increased probability of meeting due dates.

0.6 SIMULATION APPROACH TO AHP

We discussed the analytic hierarchy process (AHP) in Chapter 9. In this section we present a simulation approach using AHP for dynamic decision making in a project environment. It is important to be able to forecast the expected performance of a project. Simulation is useful for this purpose particularly where analytical techniques are not possible or too complex to aid the decision process.

AHP is a technique that has been widely used to solve complex decision problems. However, the conventional AHP takes a static view of the interactions among factors in a decision problem. In real-world problems, decision factors interact in dynamic and probabilistic fashions. The occurrence level of each factor is uncertain and may depend on the occurrence levels of other factors. The implementation of AHP in a simulation environment considering probabilistic factor interactions can significantly enhance the quality of AHP-based decisions. Similarly, conventional forecasting is based on static and certain relationships among the factors in the forecast problem. Forecasting results in single-value predictions. Scenario planning, by contrast, focuses on uncertainties in the forecast environment. In scenario planning, past events, current events, and simulated future events are used to generate probabilistic representations of future events.

10.6.1 DDM Software

Badiru, Pulat, and Kang (1993) present the DDM (Dynamic Decision Making) software. DDM is a computer simulation implementation of probabilistic AHP. The software can be used to forecast possible changes around a project baseline performance. The baseline performance is referred to as a base forecast. The objective is to predict the change in the base forecast as a function of qualitative factors which were not included in the forecast. The model consists of sets of independent events, dependent events, levels at which the events may occur, and alternatives defined as the possible variations in base forecast due to the occurrence of such events.

The occurrence probabilities for the independent events and the conditional probabilities for the dependent events are provided by the user. The user also provides pairwise comparisons for the set of events generated by the procedure. Subevents in a scenario occur according to event probability distributions and the scenario is evaluated using the AHP approach. The AHP approach consists of defining a hierarchy and analyzing the elements of each level in terms of their impact on the elements at the higher level by means of pairwise comparisons. The pairwise comparisons are combined into an overall weight for each element. The weights are then aggregated to derive a set of ratings for the alternatives. The alternative with the highest rating is selected. Figure 10–15 presents a general example of the hierarchy.

The highest hierarchy specifies the objective of the problem, which is to estimate variability in a forecast. The second level contains a set of events which affect the objective. These events are further broken down to subevents contained in the third level. The fourth level contains the decision alternatives, such as the forecasted percent increase (or decrease) from the base forecast. The main events (level 2) may be probabilistic in nature. In Figure 10–15, event 3 depends on event 2 and event 4 depends on event 3.

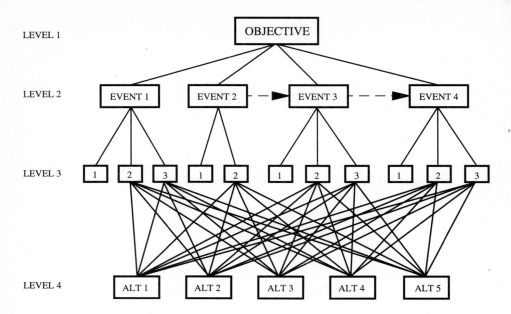

Figure 10-15. A Hierarchical Relationship with Four Levels

10.6.2 DDM Methodology

For the development of the simulation methodology, we define the following notations:

M_i	Main event i
E_{iu}	Subevent u of event i
P_{iu}	Probability that E_{iu} will occur
w_{ab}^{iu}	Likelihood of alternative a over alternative b with respect to E_{iu}
d_{ij}	Overall significance of M_i over M_j
c_{uv}^{ij}	Relative significance of E_{iu} over E_{jv}
E_{iu}'	Subevent u of the event in the ith position of the *rearranged sequence of events*
c_{uv}^{lij}	Relative significance of E_{iu}' over E_{jv}'
x_a^{iu}	Relative weight of alternative a with respect to E_{iu}
S_k	kth scenario
$e_{iu}(S_k)$	Relative weight of E_{iu} calculated with respect to S_k
z_{iu}	Ratio of relative weight of E_{iu} to relative weight of root node, $E_{12}(z_{12} = 1)$
$C(a)$	Composite weight for alternative a

Figure 10–16 shows an example hierarchy for the forecasting of change in project performance. Factors are referred to as main events, factor levels are referred to as subevents, and alternatives are referred to as forecast outcome. We have to assign probabilities to each subevent. Based on the notations above, we have

$$P(E_{iu}) = P_{iu}, \qquad \forall E_{iu} \tag{10.1}$$

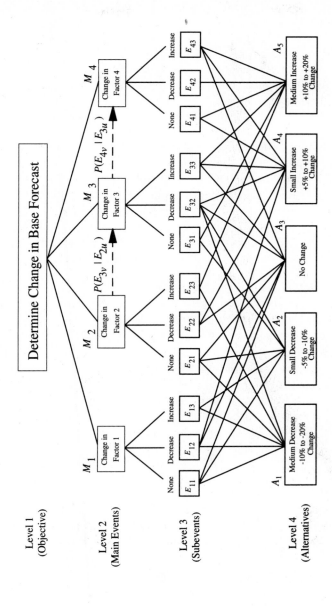

Figure 10-16. DDM Structure for Forecasting Change in Project Performance

where E_{iu} denotes subevent u of event i. The occurrence of an event may be dependent on another event. In Figure 10–15, events 1 and 2 are independent and events 3 and 4 are dependent events. In the example in Figure 10–16, the event change in Factor 3 is dependent on the event change in Factor 2. Information on E_{iu} will affect the likelihood of E_{jv} if M_j is dependent on M_i. Therefore, we incorporate conditional probabilities, $P(E_{jv}|E_{iu})$, for all dependent events i,j and their respective subevents u,v. The next step is pairwise comparison of alternatives (level 4) for each E_{iu}. Let w_{ab}^{iu} denote the importance of alternative a over alternative b with respect to E_{iu}. The conventional weight scale (Saaty 1980) is adopted for w_{ab}^{iu}. The scale was discussed previously in Chapter 9.

The relative importance of subevents of event i over subevents of event j are also needed for the computation of final composite weights for each alternative. Define c_{uv}^{ij} as the relative importance of E_{iu} over E_{jv} in accordance with the weight scale. It is practically impossible for the user to input c_{uv}^{ij} for all pairs of i,j and u,v such that a consistent comparison matrix will be generated for each scenario. It is also practically infeasible to input the comparison matrix for each scenario separately. The DDM procedure requires the user to provide c_{uv}^{ij} for only a subset of i,j and u,v. The remaining comparisons are carried out internally during the simulation. The procedure is as follows.

First, rearrange all the main events in increasing order of the number of subevents. This rearrangement is needed to generate the minimal scenario set. Let E_{iu}' denote subevent u of the event in the ith position of the *rearranged sequence of events*. Let c_{uv}^{lij} denote the comparison weight of E_{iu}' over E_{jv}'. The first subevent of each event is defined as the event not occurring (i.e., no change in the parent event). Hence, c_{uv}^{ij} and c_{uv}^{lij} are defined only for $u \neq 1$ and $v \neq 1$. Let each E_{iu}' be represented by a node. The objective is to define the minimal number of scenarios such that pairwise comparisons of events within each scenario, if consistent, will collectively lead to consistent weight matrices for all scenarios. Although there is more than one way of generating such scenarios, the DDM procedure adopts the heuristic described below.

10.6.3 Minimal Scenario Set

The procedure assumes a definite layout of nodes whereby the nodes (i.e., subevents) belonging to the same event are arranged in rows and the nodes corresponding to the same subevent number are arranged in columns. This is illustrated in Figure 10–17.

STEP 1:   The first subevent of each event is excluded from pairwise comparisons. Thus, the first column of nodes in Figure 10–17 will not be included in the generation of scenarios. In this example, we will start marking the scenarios from column 2 of nodes. Define a scenario, S_k, as the set of subevents such that main events occur at the same level of subevent number. That is, all the subevents making up S_k have the same second subscripts. Mark all columns of S_k having two or more elements. For the network in Figure 10–17, columns 2,

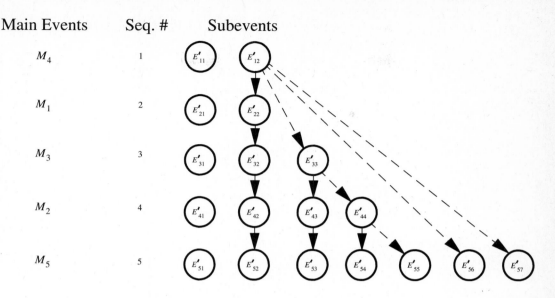

Main Events   Seq. #   Subevents

Figure 10-17.   Generation of DDM Relative Weights

3, and 4 satisfy step 1 requirements. Thus, there are three scenarios:

$$S_1 = \{E'_{12}, E'_{22}, E'_{32}, E'_{42}, E'_{52}\}$$
$$S_2 = \{E'_{33}, E'_{43}, E'_{53}\}$$
$$S_3 = \{E'_{44}, E'_{54}\}$$

The elements of the scenarios are joined by solid arrows in Figure 10–17. Note that the main events do not necessarily appear in numerical order in the first column since they are arranged in increasing order of the number of associated subevents.

STEP 2:   Generate a new scenario. Let the first node in the next unmarked column be the first element of the new scenario. Include the first node in each column which has a lower subevent number subject to the following restrictions:

a. No subevent except E'_{12} can appear in more than one scenario generated in step 2.

b. Not more than one subevent of the same main event can appear in a scenario.

Add E'_{12} to the scenario if it is not already included. Stop if all columns have been marked. Otherwise, repeat step 2. The elements of the resulting scenarios are connected by broken arrows in Figure 10–17. The scenarios are

$$S_4 = \{E'_{12}, E'_{33}, E'_{44}, E'_{55}\}$$
$$S_5 = \{E'_{12}, E'_{56}\}$$
$$S_6 = \{E'_{12}, E'_{57}\}$$

Note that the arcs of the network of Figure 10–17 define a spanning tree rooted at E'_{12}.

By generating a small number of scenarios (defined by the paths of the spanning tree), one can determine the relative weights of E'_{iu} and normalize the weights for each scenario. In summary, once the hierarchy structure is determined, the user needs to input the following:

1. P_{iu} for each subevent u of *independent* event i
2. $P(E_{jv}|E_{iu})$ for each subevent v of *dependent* event j.
3. Comparison matrix $\mathbf{W}_{iu} = [w_{ab}^{iu}]$ of alternatives with respect to E_{iu}
4. Comparison matrix $\mathbf{C}_{s_k} = [c_{uv}^{ij}]$ for each pair of E_{iu} and E_{jv} in S_k with respect to the overall objective

The consistency of each comparison matrix is checked by comparing its maximum eigenvalue to the number of elements in the matrix (Saaty 1980). More specifically, a comparison matrix is consistent if the ratio of $(\lambda_{\max} - n)/(n - 1)$ to the average random index for the same order matrix is less than 10 percent, where $\lambda_{\max}$ denotes the maximum eigenvalue of the comparison matrix and n denotes the size of the matrix. The matrices are modified until the consistency condition is satisfied. The normalized eigenvector of a consistent matrix defines the relative weights of the activities in that matrix. Let x_a^{iu} denote the relative weight of alternative a with respect to the E_{iu}. Let $e_{iu}(S_k)$ be the relative weight of E_{iu} calculated with respect to scenario S_k. The procedure determines $e_{iu}(S_k)$ and x_a^{iu} using and $\mathbf{C}_{s_k}$ and $\mathbf{W}_{iu}$ matrices, respectively. Suppose there exist K independent scenarios (i.e., $S_k, k = 1, 2, \ldots, K$ generated by using the scenario generation procedure). Note that $\bigcup_{k=1}^{K} S_k$ contains all E_{iu} except E_{i1}. Also observe that for the illustrative example, we have

$$\bigcap_{k=1}^{K} S_k = \{E'_{12}, E'_{33}, E'_{44}\}$$

= set of lead subevents in columns having more than one subevent

The relative weight of E_{iu} with respect to a new $S_j \neq S_k, k = 1, 2, \ldots, K$, can be calculated as shown below. Let

$$z_{iu} = \frac{e_{iu}(S_k)}{e_{12}(S_k)}, \qquad \forall E_{iu} \in S_k, \qquad k = 1, 2, \ldots, K \qquad (10.2)$$

where z_{iu} is the ratio of the relative weight of E_{iu} to the relative weight of the root node, E_{12}. Note that $z_{12} = 1$. For a new S_j, we have

$$e_{iu}(S_j) = \frac{z_{iu}}{\displaystyle\sum_{E_{iu} \in S_j} z_{iu}} \qquad (10.3)$$

Once the comparison matrices are filled, new scenarios are generated by using E_{iu}'s, not E'_{iu}'s. This is done internally by the program. First, independent events are randomly generated using the P_{iu} values. Next, conditional probabilities are used to generate dependent events. A scenario is defined by the set of events and subevents which are assumed to have occurred. Then, relative weight, $e_{iu}(S_j)$ of each E_{iu} and the relative weight, x_a^{iu}, of each alternative a are calculated as discussed above. It should be noted that

$$\sum_{a=1}^{N} x_a^{iu} = 1 \qquad\qquad (10.4)$$

with respect to each E_{iu}, where $a = 1, 2, \ldots, N$ alternatives. Finally, the composite weight, $C(a)$, for each alternative a with respect to each scenario j is calculated as

$$C(a) = \sum_i x_a^{iu} e_{iu}(S_j), \qquad \forall E_{iu} \in S_j \qquad\qquad (10.5)$$

The alternative with the highest $C(a)$ value in the current scenario is selected in this specific run of the simulation. The procedure then generates a new set of random numbers and, hence, a new scenario. For each scenario, the most likely alternative is determined. After a sufficient number of repetitions, one can determine the frequency of selection for each alternative. This frequency distribution is presented to the decision maker rather than an estimated single forecast. The hypothetical example below demonstrates the use of DDM software.

10.6.4 DDM Example

Figure 10–18 illustrates a simple hierarchy structure where events 1 and 3 are independent. The occurrence of event 2 depends on the level that event 3 occurs. The first subevent under each event describes nonoccurrence of that event. If all three events do not occur, then no alternative is picked. Main event 1, M_1, contains two subevents, E_{11} and E_{12}. Main event 2, M_2, contains three subevents, $E_{21}, E_{22},$ and E_{23}. Main event 3, M_3, contains two subevents, E_{31} and E_{32}.

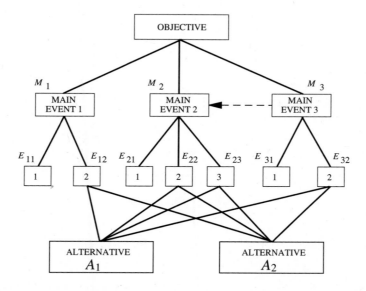

Figure 10-18. AHP Hierarchy for Illustrative Example

Manual computations.   Reordering the events in increasing order of number of subevents, we get the following:

$$E'_{1u} = E_{1u}, \qquad u = 1, 2$$
$$E'_{2u} = E_{3u}, \qquad u = 1, 2$$
$$E'_{3u} = E_{2u}, \qquad u = 1, 2, 3$$

Figure 10–19 shows the spanning tree structure. Table 10–4 displays P_{ij} values for the independent events.

Tables 10–5 and 10–6 display $P(E_{2v}|E_{3v})$ and w^{iu}_{ab} values, respectively. Negative values are used to indicate reciprocal preference in the pairwise comparison matrices. Table 10–6 contains three consistent pairwise comparison matrices. The entries in the table are used later to compute the x^{iu}_a values for the alternatives. Table 10–7 shows the corresponding comparison matrices. For illustrative purposes, comparison matrices are entered for only the two following scenarios:

$$S_1 = \{E'_{12}, E'_{22}, E'_{32}\}$$
$$S_2 = \{E'_{12}, E'_{33}\}$$

That is,

$$S_1 = \{E'_{12}, E'_{32}, E'_{22}\}$$
$$S_2 = \{E'_{12}, E'_{33}\}$$

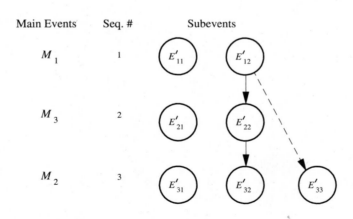

Figure 10-19.   A Spanning Tree of the Example Problem

Table 10–4 P_{iu} for Independent Events

	u	Subevent 1	Subevent 2
Main Event 1	P_{1u}	0.8	0.2
Main Event 3	P_{3u}	0.6	0.4

Table 10-5 Conditional Probabilities for Dependent Events

			E_{2v}	
	Given E_{3u}	E_{21}	E_{22}	E_{23}
Subevent 1 of main event 3	E_{31}	0.6	0.3	0.1
Subevent 2 of main event 3	E_{32}	0.2	0.4	0.4

Table 10-6 Examples of Pairwise Comparisons of Alternatives

w_{ab}^{12}		b		w_{ab}^{22}		b		w_{ab}^{23}		b		w_{ab}^{32}		b	
a	1	2		a	1	2		a	1	2		a	1	2	
1	1	3		1	1	-2		1	1	5		1	1	-7	
2	-3	1		2	2	1		2	-5	1		2	7	1	

Table 10-7 Comparison Matrices for S_1 and S_2

C_{S1}	E_{12}	E_{32}	E_{22}		C_{S2}	E_{12}	E_{23}
E_{12}	1	3	2		E_{12}	1	-2
E_{32}	-3	1	-2		E_{23}	2	1
E_{22}	-2	2	1				

The relative weights of the subevents in scenarios 1 and 2 are calculated from the comparison matrices in Table 10–7:

$$e_{12}(S_1) = 0.53 \qquad e_{12}(S_2) = 0.33$$
$$e_{32}(S_1) = 0.33 \qquad e_{23}(S_2) = 0.67$$
$$e_{22}(S_1) = 0.14$$

Using equation (10.2), the ratios, z_{iu}, are computed for the elements in each scenario:

$$z_{12} = 1, z_{32} = 0.6, z_{22} = 0.26, z_{23} = 2.03$$

To generate a new scenario, a random number, r_1, is generated from a uniform distribution (0,1) for event 1. If $r_1 \leq 0.8$, then, based on P_{11} in Table 10–4, we conclude that E_{11} occurs. Otherwise, E_{12} occurs. Similarly, random number r_2 determines level at which event 3 occurs (E_{3u}). Suppose E_{12} and E_{32} have occurred. Random number r_3 is generated and occurrence of $E_{2u}(u = 1, 2, 3)$ is determined using $P(E_{2u}|E_{32})$. Suppose that $R_3 = 0.8$. Then, using the second row of Table 10–5, we conclude that E_{23} has occurred. The new scenario is now defined by

$$S_3 = \{E_{12}, E_{23}, E_{32}\}$$

Using equation (10.3), e_{iu} for each $E_{iu} \in S_3$ is calculated as

$$v_{12}(S_3) = \frac{z_{12}}{\sum\limits_{E_{iu} \in S_3} z_{iu}} = \frac{1}{(1 + 2.03 + 0.62)} = 0.274$$

$$v_{23}(S_3) = \frac{z_{23}}{\sum\limits_{E_{iu} \in S_3} z_{iu}} = \frac{2.03}{(1 + 2.03 + 0.62)} = 0.556$$

$$v_{32}(S_3) = \frac{z_{32}}{\sum\limits_{E_{iu} \in S_3} z_{iu}} = \frac{0.62}{(1 + 2.03 + 0.62)} = 0.170$$

The relative weights x_a^{iu} for alternatives 1 and 2 are obtained from the comparison matrices based on E_{12}, E_{23}, and E_{32} in Table 10–6.

$$x_1^{12} = 0.750 \quad x_1^{23} = 0.833 \quad x_1^{32} = 0.125$$
$$x_2^{12} = 0.250 \quad x_2^{23} = 0.167 \quad x_2^{32} = 0.875$$

The composite weights for the two alternatives are now calculated based on equation (10.5):

$$C(1) = x_1^{12}e_{12}(S_3) + x_1^{23}v_{23}(S_3) + x_1^{32}v_{32}(S_3)$$
$$= 0.750(0.274) + 0.833(0.556) + 0.125(0.170)$$
$$= 0.690$$
$$C(2) = x_2^{12}v_{12}(S_3) + x_2^{23}v_{23}(S_3) + x_2^{32}v_{32}(S_3)$$
$$= 0.250(0.274) + 0.167(0.556) + 0.875(0.170)$$
$$= 0.310$$

Hence, alternative 1 is forecast for this specific scenario. The above scenario generation and selection of alternatives are repeated n times and the frequency distribution for each alternative is determined. The simulation run in the next section confirms the manual computation shown above.

10.6.5 Sample Simulation Run of DDM

The simulation model is developed as an interactive PC-based program. This facilitates accessibility and ease of execution for users. It is written in C language. Thus, it is very portable on various computer systems. The program will run on any PC compatible computer with an EGA display unit and 640K RAM. The design of the system makes use of drop-down menu structures. The program contains an on-line help from which a user can get clarification on the procedures and parameters involved in using the program. A demo program is also available to introduce users to the general operation of the program. The demo program uses a built-in AHP decision problem for its execution. The program operates in the following sequence:

user input $\Rightarrow$ AHP simulation $\Rightarrow$ distribution of forecast outcomes

The input requirements are the main events, subevents, dependent sets, independent events, conditional probabilities, forecast alternatives, and pair-wise comparison matrices. A dependent set is defined as the collection of two main events that are mutually dependent. DDM simulation methodology is designed to minimize the number of inputs required from the user. Thus, the user is not burdened with data entry when using the simulation program. The check of the consistency ratio for a pair-wise comparison matrix is done internally by the program. If a comparison matrix is not consistent, the user is informed and requested to modify the comparison scales. Depending on available system memory, the program can handle several thousand simulation runs at a time. The output of the system is a frequency histogram of the distribution of the forecast outcomes. The example simulation run below illustrates the operational details of the simulation program.

The AHP decision hierarchy shown in Figure 10–18 is used for this example. For the example, events 1 and 3 are independent. The occurrence of event 2 depends on the level at which event 3 occurs. The first subevent under each event describes nonoccurrence of that event. If all three events do not occur, then no alternative is picked. Main event 1, M_1, contains two subevents, E_{11} and E_{12}. Main event 2, M_2, contains three subevents, $E_{21}, E_{22},$ and E_{23}. Main event 3, M_3, contains two subevents, E_{31} and E_{32}. Since this is a forecasting problem, there is a possible outcome (alternative) defined as "no change." The first alternative in Figure 10–18 (i.e., alternative A) is used to represent the "no change" alternative. Similarly, the first subevent of each main event is selected as the subevent indicating "no change." Thus, the alternatives are compared only with respect to the subevents associated with changes in event levels. Referring to Figure 10–18, the subevents eligible for the pair-wise comparisons are $E_{12}, E_{22}, E_{23},$ and E_{32}.

Figure 10–20 shows the main menu of the simulation program. The options on the menu involve the specification of events and alternatives, entry of probability information, entry of pair-wise comparison matrices, on-line help, running the simulation, data verification, and terminating the program. An option is selected from the menus by pressing the first letter of the name of the desired option.

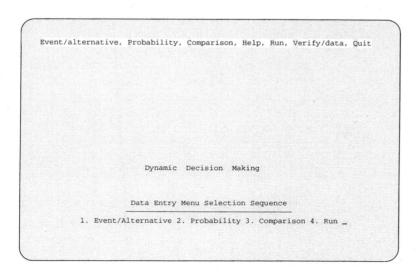

```
Event/alternative, Probability, Comparison, Help, Run, Verify/data, Quit




                        Dynamic  Decision  Making


                    Data Entry Menu Selection Sequence
                    _____
            1. Event/Alternative 2. Probability 3. Comparison 4. Run _
```

Figure 10-20.   DDM Screen 1

Figure 10–21 shows the drop down menu for the first option in the main menu. The specification of events and alternatives must be done before any of the other options are selected from the main menu. Entries required are events and subevents, dependent sets, and alternatives.

Figure 10–22 shows the specification of 3 main events and their respective number of subevents.

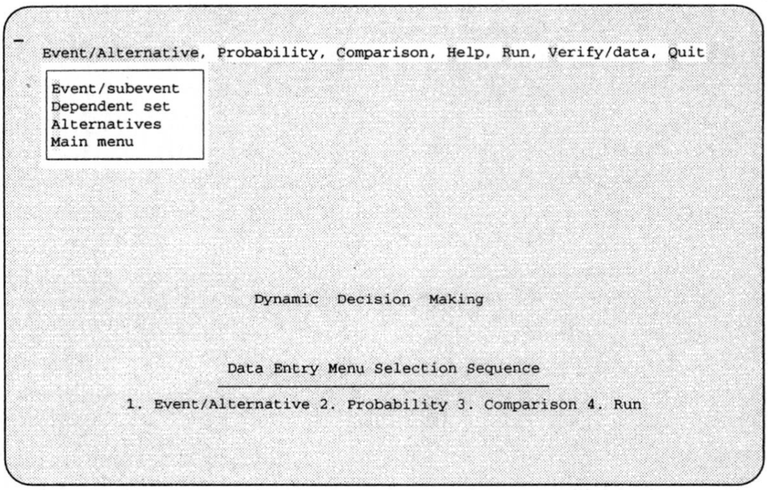

```
 __    Event/Alternative, Probability, Comparison, Help, Run, Verify/data, Quit
        ┌─────────────────┐
        │ Event/subevent  │
        │ Dependent set   │
        │ Alternatives    │
        │ Main menu       │
        └─────────────────┘




                         Dynamic  Decision  Making


                      Data Entry Menu Selection Sequence
                 ───────────────────────────────────────────
                 1. Event/Alternative 2. Probability 3. Comparison 4. Run

```

Figure 10-21.   DDM Screen 2

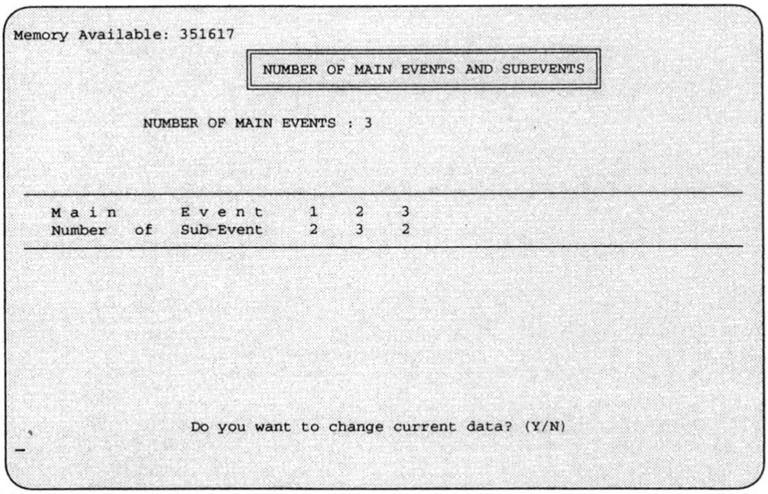

```
 Memory Available: 351617
                          ┌──────────────────────────────────────────┐
                          │ NUMBER OF MAIN EVENTS AND SUBEVENTS       │
                          └──────────────────────────────────────────┘

             NUMBER OF MAIN EVENTS : 3


   ──────────────────────────────────────────────────────────────────────
     M a i n        E v e n t     1    2    3
     Number   of    Sub-Event     2    3    2
   ──────────────────────────────────────────────────────────────────────




                    Do you want to change current data? (Y/N)
 __
```

Figure 10-22.   DDM Screen 3

The number of dependent sets is shown in Figure 10–23. Based on Figure 10–18, there is only one dependent set. Dependent set 1 consists of two main events.

Figure 10–24 shows the sequential entry of the mutually dependent events in each dependent set. Two columns (column 1, column 2) are provided for the specification of the member events in each dependent set. The elements of dependent set 1 are entered as 3 and 2 (i.e., 3 before 2). This indicates that event 2 is dependent on event 3.

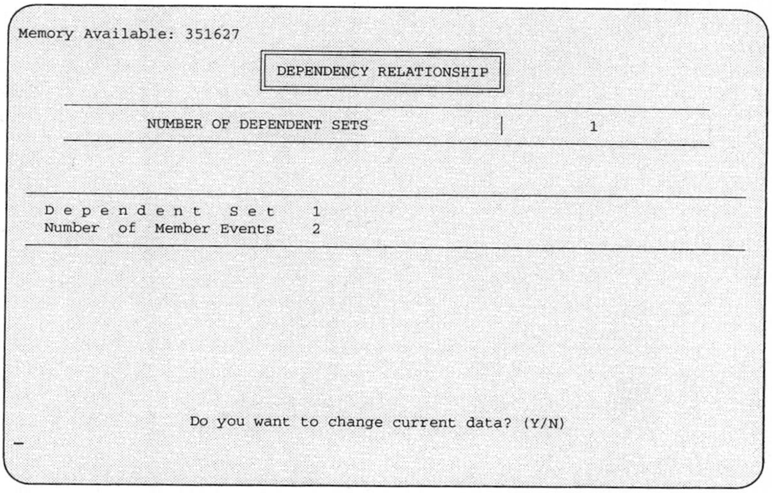

```
Memory Available: 351627

                      ┌─────────────────────────┐
                      │ DEPENDENCY RELATIONSHIP  │
                      └─────────────────────────┘

            NUMBER OF DEPENDENT SETS              │        1


      D e p e n d e n t   S e t     1
      Number  of  Member Events     2




                    Do you want to change current data? (Y/N)
    ─
```

Figure 10-23.   DDM Screen 4

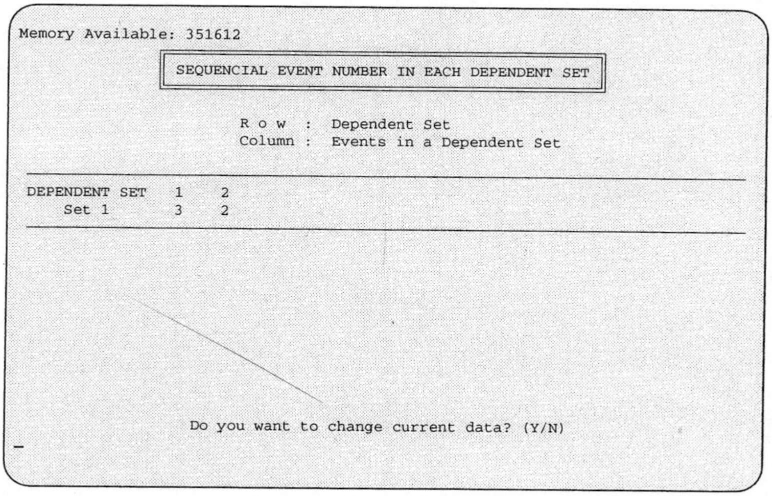

```
Memory Available: 351612

              ┌──────────────────────────────────────────────┐
              │ SEQUENCIAL EVENT NUMBER IN EACH DEPENDENT SET │
              └──────────────────────────────────────────────┘

                      R o w  :  Dependent Set
                      Column :  Events in a Dependent Set

DEPENDENT SET   1    2
     Set 1      3    2




                    Do you want to change current data? (Y/N)
    ─
```

Figure 10-24.   DDM Screen 5

Figure 10–25 indicates that the number of alternatives is 2. As mentioned earlier, alternative 1 is indicated as the alternative representing no change.

The next set of entries involve the probability information for the problem. Figure 10–26 shows options for conditional probability and probability of independent events.

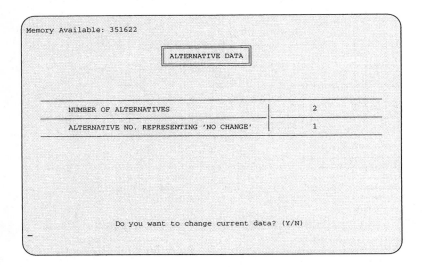

Figure 10-25.   DDM Screen 6

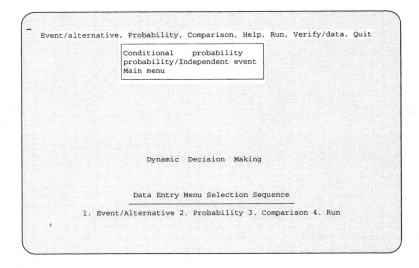

Figure 10-26.   DDM Screen 7

Figure 10–27 is for dependent set 1. The conditional event is event 3 and the dependent event is event 2. The conditional probability information from Table 10–5 is entered as shown on the screen. The three dependent subevents of event 2 are indicated in the columns. The two conditional subevents of event 3 are indicated in the rows. Note that the probabilities entered for each row must sum to 1.

The probabilities of the subevents of the independent events 1 and 3 are entered in Figure 10–28. This probability information is obtained from Table 10–4. Note that dependent event 2 is blocked out of the data entry for this screen. Note also that the sum of the probability entries for each row is 1.

The next set of data entries involves pair-wise comparisons. Figure 10–29 shows that entries are required for comparison matrices for alternatives and subevents.

Figures 10–30 to 10–33 show the comparison of the two alternatives with respect to each of the four subevents E_{12}, E_{22}, E_{23}, and E_{32}. These matrices are presented in Table 10–6. Since each matrix is symmetric, only one-half of the matrix is required to be entered by the user. The program internally completes the matrix. Note that reciprocal pair-wise ratings are entered as negatives on the screen. This is to facilitate ease of data entry. The negative numbers are converted to appropriate reciprocals internally. For example, −5 on the screen is converted to 1/5 when normalizing the matrices.

The pair-wise comparison matrices for the subevents contained in the initial two scenarios, as shown in Table 10–7, are entered in Figures 10–34 and 10–35.

Figure 10–36 shows the next option from the main menu. This option involves running the simulation and printing the histogram.

When the simulation option is selected, Figure 10–37 appears. The desired number of simulation runs is entered on the screen. For this example, we have specified 15,000 simulation runs.

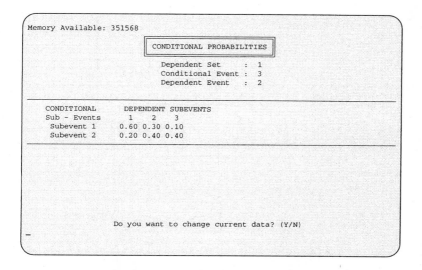

Figure 10-27.   DDM Screen 8

```
Memory Available: 351541

              ┌─────────────────────────────┐
              │  PROBABILITIES OF SUBEVENTS  │
              └─────────────────────────────┘
──────────────────────────────────────────────────────────────────────
    SUB EVENT ->  1    2
      Event 1    0.80 0.20
      Event 2    ██████████
      Event 3    0.60 0.40
──────────────────────────────────────────────────────────────────────




                Do you want to change current data? (Y/N)
 _
```

Figure 10-28.   DDM Screen 9

```
 _
    Event/alternative, Probability, Comparison, Help, Run, Verify/data, Quit
                                 ┌──────────────┐
                                 │ Alternatives │
                                 │ Subevent     │
                                 │ Main menu    │
                                 └──────────────┘




                       Dynamic   Decision   Making


                     ─────────────────────────────────
                      Data Entry Menu Selection Sequence
                1. Event/Alternative 2. Probability 3. Comparison 4. Run
```

Figure 10-29.   DDM Screen 10

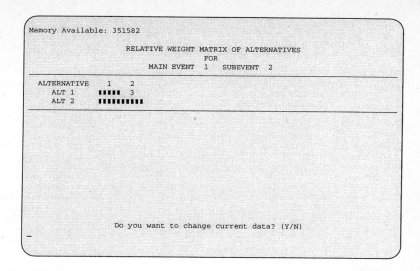

Figure 10-30. DDM Screen 11

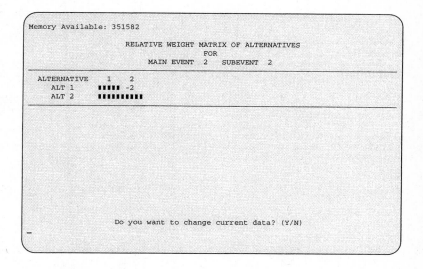

Figure 10-31. DDM Screen 12

461

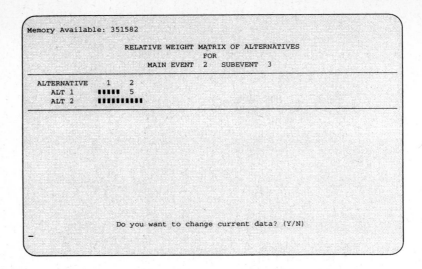

Figure 10-32. DDM Screen 13

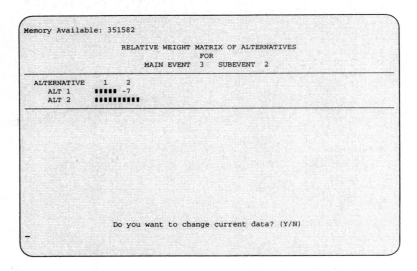

Figure 10-33. DDM Screen 14

462

```
Memory Available: 351582
                   RELATIVE WEIGHT MATRIX OF SUBEVENTS
              Header I-J : The Jth subevent of the Ith event.
_____
SubEvent  1-2  2-3
   1-2    ▪▪▪▪▪ -2
   2-3    ▪▪▪▪▪▪▪▪▪▪
_____




                    Do you want to change current data? (Y/N)
  _
```

Figure 10-34. DDM Screen 15

```
Memory Available: 351572
                   RELATIVE WEIGHT MATRIX OF SUBEVENTS
              Header I-J : The Jth subevent of the Ith event.
_____
SubEvent  1-2  3-2  2-2
   1-2    ▪▪▪▪▪  3    2
   3-2    ▪▪▪▪▪▪▪▪▪▪ -2
   2-2    ▪▪▪▪▪▪▪▪▪▪▪▪▪▪
_____




                    Do you want to change current data? (Y/N)
  _
```

Figure 10-35. DDM Screen 16

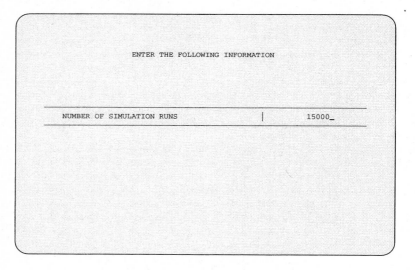

```
 ─ '
     Event/alternative, Probability, Comparison, Help, Run, Verify/data, Quit
                                                         ┌──────────────┐
                                                         │ Simulation   │
                                                         │ Histogram    │
                                                         │ Main menu    │
                                                         └──────────────┘




                              Dynamic   Decision   Making


                          Data Entry Menu Selection Sequence
                   ─────────────────────────────────────────────
                   1. Event/Alternative 2. Probability 3. Comparison 4. Run
```

Figure 10-36. DDM Screen 17

```
                          ENTER THE FOLLOWING INFORMATION




       ────────────────────────────────────────────────────────────────
          NUMBER OF SIMULATION RUNS                  │        15000_
       ────────────────────────────────────────────────────────────────
```

Figure 10-37. DDM Screen 18

464

Figure 10–38 shows the result of the simulation run. Out of the 15,000 forecast outcomes, 9,933 are for alternative 1 while 5,067 are for alternative 2.

Figure 10–39 shows the histogram of the simulation result. Sixty-six percent of the outcomes are for alternative 1 and 34 percent are for alternative 2.

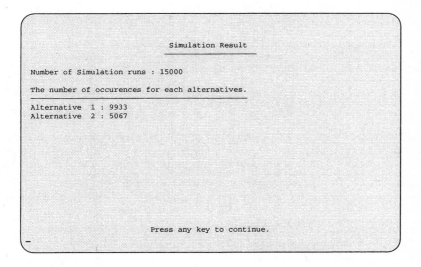

Figure 10-38.   DDM Screen 19

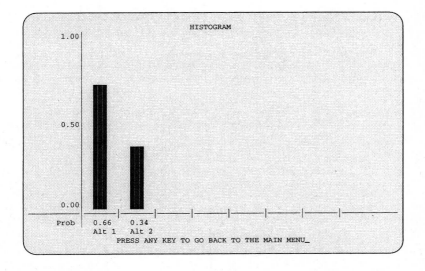

Figure 10-39.   DDM Screen 20

The result of the simulation closely matches the manual computation presented for one specific scenario. It should be recalled that the simulation result is based on 15,000 simulated scenarios. Table 10–8 below shows a comparison of the results of the manual computation and the simulation.

Table 10-8   Comparison of the Results of Manual Computation and Simulation

	Manual	Simulation
Alternative 1	0.69	0.66
Alternative 2	0.31	0.34

10.7 EXPERT SYSTEMS AND PROJECT MANAGEMENT

Expert systems, which form a branch of artificial intelligence (AI), are being used more and more to aid project planning and control. An *expert system* is defined as an interactive computer-based decision tool that uses both facts and heuristics to solve difficult decision problems based on knowledge acquired from an expert (Badiru 1992a). An expert system may be viewed as a computer simulation of a human expert. Badiru and Theodoracatos (1994) discuss the potential use of expert systems for design project management in a manufacturing environment. Some strategies for using AI techniques in project planning and control are discussed by Morad and Vorster (1993), Mango (1992), and Gaarslev (1992). Expert systems are organized in the following three distinct levels:

1. *Knowledge base.* This consists of problem-solving rules, procedures, experiential inferences, and intrinsic data relevant to the problem domain. The general knowledge base can be used to solve a variety of related problems within the same problem domain.
2. *Working memory (data structure).* This refers to task specific data for the problem under consideration. The working memory changes with each problem scenario. This is the most dynamic component of an expert system.
3. *Inference engine.* This is a generic control mechanism that applies the axiomatic knowledge in the knowledge base to the task-specific data to arrive at some solution or conclusion. Efficient search strategies are important in the construction of inference engines.

Just as we can use project management techniques to facilitate successful implementation of expert systems, we can also use expert systems techniques to enhance the implementation of project management functions. One potential use of expert systems to project management is the utilization of state-space representation to monitor and control activities in the project environment. The state-space model can provide up-to-date information about project status before the expert system knowledge base is invoked.

10.7.1 State-Space Project Tracking Model

A state is a set of conditions or values that describe a system at a specified point during processing. The state space is the set of all possible states the system could be in during the problem-solving process. State-space representation solves problems by moving from an initial state in the space to another state, and eventually to a goal state. The movement from state to state is achieved by the means of operators, typically rules or procedures. A goal is a description of an intended state that has not yet been achieved. The process of solving a problem involves finding a sequence of operators that represent a solution path from the initial state to the goal state.

State-space techniques have been used extensively to model continuous system problems in engineering and management decision problems. Using an expert system and state-space modeling, the project management decision process can be greatly enhanced. Thus, while project management techniques are used to manage expert systems projects, expert systems themselves could be used to enhance the techniques of project management.

A state-space model consists of definition state variables that describe the internal state or prevailing configuration of the system being represented. The state variables are related to the system inputs by an equation. One other equation relates both the state variables and system inputs to the outputs of the system. Examples of potential state variables in a project system include product quality level, budget, due date, resource, skill, and productivity level. In the case of a model described by a system of differential equations, the state-space representation is of the form

$$\dot{z} = f(z(t), x(t))$$
$$y(t) = g(z(t), x(t))$$

where f and g are vector-valued functions. In the case of linear systems, the representation is of the form

$$\dot{z} = Az(t) + Bx(t)$$
$$y(t) = Cz(t) + Dx(t)$$

where $z(t), x(t)$, and $y(t)$ are vectors and A, B, C, and D are matrices. The variable y is the output vector while the variable x denotes the inputs. The state vector $z(t)$ is an intermediate vector relating $x(t)$ to $y(t)$.

The state-space representation of a discrete-time linear dynamic system, with respect to a suitable time index, is given by

$$z(t + 1) = Az(t) + Bx(t)$$
$$y(t) = Cz(t) + Dx(t)$$

In generic terms, a project system is transformed from one state to another by a driving function that produces a transitional equation given by

$$\psi = f(x|\theta) + \epsilon$$

where ψ is the subsequent state, x is the state variable, θ is the initial state, and ϵ is the error component. The function f is composed of a given action (or a set of actions) applied to the project structure. Each intermediate state may represent a significant milestone in the project system. Thus, a descriptive state-space model facilitates an analysis of what actions to apply in order to achieve the next desired state or milestone.

10.7.2 The Project Model

Any human endeavor can be defined as a project. A project may be considered as a process of producing a product. The product may be a measurable physical quantity or an intangible conceptual entity. We may consider the product as a single object constructed from a set of subobjects, which are themselves constructed from sub-subobjects. A simple project model is the application of an action to an object which is in a given state or condition.

The application of action constitutes a project activity. The production process involves the planning, coordination, and control of a collection of activities. Project objectives are achieved by state-to-state transformation of successive object abstractions. Figure 10–40 shows the transformation of an object from one state to another through the application of action. The simple representation can be expanded to include other components within the project framework. The hierarchical linking of objects provides a detailed description of the project profile as shown in Figure 10–41.

The project model can further be expanded in accordance with implicit project requirements. These considerations might include grouping object classes, precedence linkings (both technical and procedural), required communication links, and reporting requirements.

The actions to be taken at each state depend on the prevailing conditions. The natures of the subsequent alternate states depend on what actions are actually implemented. Sometimes there are multiple paths that can lead to the desired end result. At other times, there exists only one unique path to the desired objective. In conventional practice, the characteristics of the future states can be recognized only after the fact, thus, making it impossible to develop adaptive plans. In terms of control, deviations are often recognized when it is too late to take effective corrective actions. The occurrence of both events within and outside the project state boundaries can be taken into account in the planning function. These environmental influences are shown in Figure 10–42.

Project state representation. We can describe a project system by M state variables s_i. The composite state of the system at any given time can then be represented

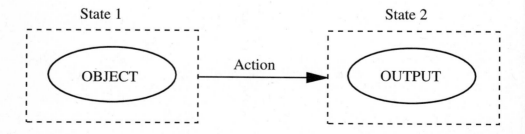

Figure 10-40.   Object Transformation

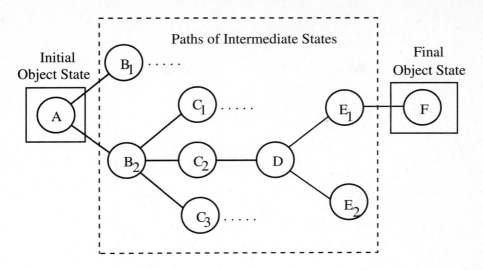

Figure 10-41.   Object State Paths

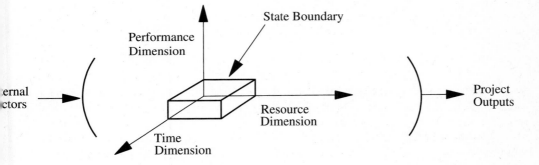

Figure 10-42.   External/Internal Factors Affecting Project Implementation

by an M-component vector **S**.

$$\mathbf{S} = \{s_1, s_2, \ldots, s_M\}$$

The components of the state vector could represent either quantitative or qualitative variables (voltage, investments, energy, color, etc.). We can visualize every state vector as a point in the M-dimensional state space shown in Figure 10–43. The representation is unique since every state vector corresponds to one and only one point in the state space.

Project state transformation.   Suppose we have a set of actions (transformation agents) that we can apply to a project so as to change it from one state to another within the state space. The transformation will change a state vector into another state vector. A transformation, in practical terms, may be a heat treatment, firing a rocket, or a change in management policy. We can let T_k be the kth type of transformation. If T_k

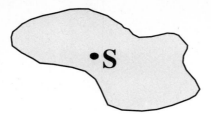

Figure 10-43.   Project State Space

is applied to the project when it is in state $\mathbf{S}$, the new state vector will be $T_k(\mathbf{S})$, which is another point in the state space. The number of transformations (or actions) available for a project may be finite or countably infinite. We can construct trajectories that describe the potential states of a project as we apply successive transformations. Each transformation may be repeated as many times as needed. For convenience, we can use the notation T_i to indicate the ith transformation in the sequence of transformations applied to the project. Given an initial state $\mathbf{S}$, the sequence of state vectors is then given by

$$\mathbf{S}_1 = T_1(\mathbf{S})$$
$$\mathbf{S}_2 = T_2(\mathbf{S}_1)$$
$$\mathbf{S}_3 = T_3(\mathbf{S}_2)$$

$$.$$
$$.$$
$$.$$

$$\mathbf{S}_n = T_n(\mathbf{S}_{n-1})$$

The final state $\mathbf{S}_n$ depends on the initial state $\mathbf{S}$ and the effects of the transformations applied. The sequence is shown graphically within the state space in Figure 10–44.

State performance measurement.   A measure of project performance can be obtained at each state of the transformation trajectories. Thus, we can develop a reward function $r^k(\mathbf{S})$ associated with the kth type transformation on the state vectors (Howard 1971). The reward specifies the magnitude of gain (time, quality, money, revenue, equipment utilization, etc.) to be achieved by applying a given transformation. The difference between a reward and a performance specification may be used as a criterion for determining project control actions. The performance deviation may be defined as

$$\delta = r^k(S) - \rho$$

where ρ is the performance specification.

Project policy development.   Given the number of transformations still available and the current state vector, we can develop a policy, P, to be the rule for determining the next transformation to be applied. The total project reward can then be denoted as

$$r(S|n, P) = r_1(S) + r_2(S_1) + \ldots + r_n(S_{n-1})$$

where n is the number of transformations applied and $r_i(.)$ is the ith reward in the sequence of transformations. We can now visualize a project environment where the starting state vector and the possible actions (transformations) are specified. We have

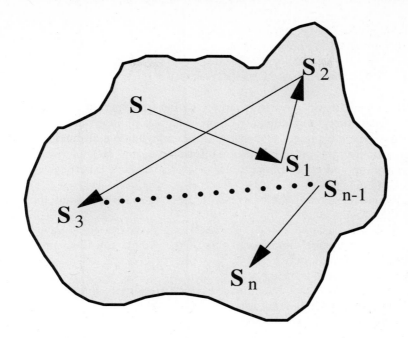

Figure 10-44.   Transformation Trajectories in State Space

to decide what transformations to use in what order so as to maximize the total reward. That is, we must develop the best project plan.

If we let v represent a quantitative measure of the worth of a project plan based on the reward system described above, then the maximum reward is given by

$$v(S|n) = \text{Max}\{r(S|n, P)\}$$

The maximization of the reward function is carried out over all possible project policies that can be obtained with n given transformations.

Probabilistic states.   In many project situations, the results of applying transformations to the system may not be deterministic. Rather, the new state vector, the reward generated, or both may have to be described by random variables. We can define an expected total reward, $Q(S|n)$, as the sum of the individual expected rewards from all possible states. We let S^p be a possible new state vector generated by the probabilistic process and let $P(S^p|S, T^k)$ be the probability that the new state vector will be S^p if the initial state vector is S and the transformation T^k is applied. We can now write a recursive relation for the expected total reward:

$$Q(S|n) = \underset{k}{\text{Max}}\left\{\bar{r}^k(S) + \sum_{s^p} Q(S^p|n-1)P(S^p|S, T^k)\right\}, \qquad \text{for } n = 1, 2, 3, \dots$$

The notation $\bar{r}^k(S)$ is used to designate the expected reward received by applying the kth type transformation to the system when it is described by state vector S. The above procedure allows the possibility that the terminal reward itself may be a random variable. Thus, the state-space model permits a complete analysis of all the ramifications and uncertainties of the project management system.

10.7.3 Expert System Implementation Model

The If-then structure of knowledge representation for expert systems provides a mechanism for evaluating the multiplicity of project states under diversified actions. Expert systems have the advantages of consistency, comprehensive evaluation of all available data, accessibility, infinite retention of information, and lack of bias. An integrated project planning and control system using state space and expert systems is shown in Figure 10–45.

A project manager might interact with the state-space model and the expert system by doing the following:

1. Performing the real-time monitoring of the project and then supplying inputs to the state-space model.
2. Getting feedback from the state model.
3. Consulting the expert system based on the state-space information.
4. Getting recommendations from the expert system.
5. Taking project actions that eventually generate further inputs to the state model.
6. Updating the expert system knowledge base as new project management knowledge is acquired.

The supply of state-space information to both the project manager and the expert system in Figure 10–45 will serve as a control measure. In case the manager does not utilize all available information, the expert system can query him or her for justification.

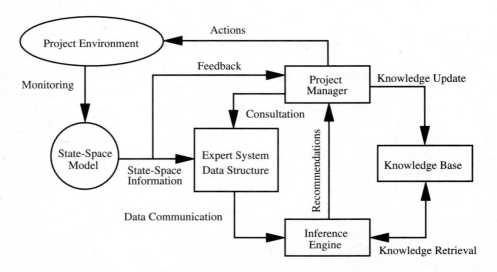

Figure 10-45.   State-Space/Expert System Model

10.7.4 Network Cost Planning Example

The concept of integrating cost into project network analysis provides management with the tools for effective control of project resources. In today's complex organizational cost structures, there are several factors that influence project implementations. The technology of expert systems offers a considerable potential for real-time integration of project costs. Expert systems can facilitate effective and prompt managerial cost control without the need for extensive run-time data collection.

A project network is a representation of a project plan such that the relationships among jobs or activities are clarified and can easily be examined. Network analysis reduces the project evaluation to four stages:

1. Breaking down the project into a set of individual jobs and events and arranging them into a logical network.
2. Estimating the duration and resource requirements of each job.
3. Developing a schedule, and finding which job controls the completion of the project.
4. Reallocating funds or other resources to improve the schedule.

Network cost analysis offers the following advantages:

- It forces a thorough preplanning of project tasks.
- It increases coordination.
- It identifies trouble spots in advance and pinpoints responsibility.
- It focuses management's attention on the most critical activities.
- It indicates cost-effective start and finish times for project activities.
- It enables revisions of project plans without extensive cost changes.
- It suggests where alternative project methods can be used.
- It makes costs tractable.
- It facilitates fair allocation of resources.
- It minimizes cost overruns.

An expert system can serve as the effective mechanism that aids management in answering questions such as

1. What are the actual project costs to date?
   - Single cost estimate
   - Activity direct costs
   - Project indirect costs
   - Three cost estimates
   - Normal and crash activity costs
2. How do the actual costs to date compare with planned costs to date?
3. By how much can the project be expected to overrun the total planned cost?
4. How can limited resources be allocated best?

The first step in network cost planning is constructing a network and performing the basic scheduling which gives the earliest and latest start and finish times for each activity. Then the estimated cost data for each activity can be added to the network and the first cost computation can be made.

As the project progresses, actual expenditures are recorded by activity at specified reporting dates. Any revised estimate of duration or costs of activities are also accounted for at the reporting times. The following computations may be made:

- Summation of all actual costs
- Summation of budgeted costs at this point in time
- Summation of budgeted costs for all activities completed and partial costs of activities partially completed
- Computation of differences in actual costs and planned costs for completed portions of the project
- Computation of a projection on the expected cost of the total project based on the progress made so far

Based on the above cost computations and other prevailing cost circumstances, rules can be developed to serve as guides for courses of managerial actions in diverse cost and project scenarios. These rules can then be incorporated into an expert system to facilitate quick real-time managerial decisions.

Rule-based cost analysis. The cost rules discussed in the previous section may be organized into an expert system knowledge base. For example, the following cost rules may apply to a given project:

Rule 1

If total project budget exceeds $100,000 and expected project duration exceeds one year, then allocate 10 percent of the budget to each 10 percent of expected duration.

This rule establishes the procedure for spreading the project budget evenly over the duration of the project. This rule may be particularly useful for fixed-budget projects that extend over several years. This provides a safeguard against expending a larger portion of the budget early in the project at the expense of later project needs. The rule, of course, assumes that any initial fixed cost (lump sum) requirements of the project have been accounted for. If not, a subrule such as rule 1.1 may be utilized.

Rule 1.1

If fixed cost exceeds 10 percent of total budget and 10 percent of expected duration is less than 1 day, then compute a new budget as total budget minus fixed cost and allocate 10 percent of the new budget to each 10 percent of expected duration.

Rule 2

If project activities are greater than 20, then record half of the budget for each activity at the activity's scheduled start time and record the other half of the budget at the activity's scheduled completion time.

This rule is the common 50/50 rule used in reducing cost variance in projects with a large number of elements. One advantage of the 50/50 rule is that it reduces the necessity of continuously determining the percent complete for each activity in the process of allocating cost.

Rule 3

If current costs are more than 5 percent above initial budget, then review the cost of higher-salaried personnel.

This is an example of a rule that may give management an indication of where to start cost troubleshooting.

Rule 4

If the cost of any given project element exceeds 10 percent of total cost then reexamine the work breakdown structure.

Rules similar to the ones presented above can be developed to fit specific organizational needs. Many practical expert systems have been developed using as few as 20 rules. Some systems designed for more robust applications have been known to contain as many as 1,000 rules. Generally, specific organizational needs and available expertise should determine the contents and size of a knowledge base.

Knowledge base structure.   An example of a cost decision tree is shown in Figure 10–46. The tree can be used to determine the structure of the knowledge base.

The decision tree can be as complex and detailed as there are relevant leads. Each statement of fact consists of three parts, namely,

Attribute: A keyword or phrase chosen to represent the subject of the fact to be determined (e.g., budget).
Value: A description assigned to the attribute (e.g., large).
Predicate: An item used to relate the value to the attribute (e.g., is).

Referring to Figure 10–46, we can generate the following rules:

1. If the project is subcontracted, then cost information is from the subcontractor.
2. If the number of project elements is large, then the 50/50 rule is applicable.
3. If the fixed cost is not zero, then the new budget is computed from procedure for new budget.
4. If the 50/50 rule is applicable and the new budget is greater than $100,000 and the expected duration is longer than one year, then use the 10% allocation procedure.

An external procedure is a computational subroutine used to calculate variable values. The computation procedure for the new budget may be

$$(\text{new budget}) = (\text{budget}) - (\text{fixed cost})$$

The computation procedure for cost allocation may be

$$\text{allocation for 10\% of duration} = 10\% \ (\text{new budget})$$

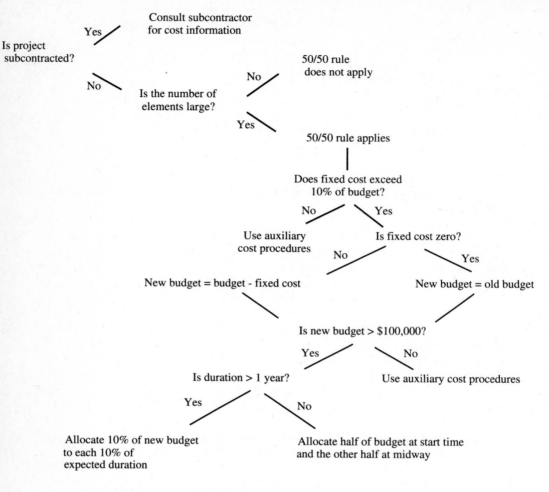

Figure 10-46. Decision Tree for Project Cost Analysis

Prompts are added to the knowledge base to facilitate entering project data during the expert system consultation. The prompts are needed for facts that the expert system cannot implicitly infer. Such prompts may be

1. **Prompt:**   Fixed cost
                       Enter value for fixed cost?
2. **Prompt:**   Budget
                       Enter value for budget?
3. **Prompt:**   Project-type
                       Is the project subcontracted?
4. **Prompt:**   Project size
                       Is the number of project elements large?

With an expert system, valuable cost control recommendations can be obtained in a matter of minutes. This will reduce the need to read voluminous cost guides or waste productive time trying to locate an expert to handle routine cost functions. Expert systems can significantly enhance the functions of project analysts. The systems can handle routine cost analysis and rule-based decision analysis, thus, freeing the human expert to concentrate his or her efforts on critical project problems. Project management has always been a fertile area for the utilization of heuristics. The premise of an expert system is the utilization of heuristics to enhance decision processes. Consequently, both areas are inherently suitable for integration.

An expert system can facilitate the integration and consistent usage of all the relevant information in a complex project environment. Task precedence relaxation assessment and resource-constrained heuristic scheduling are other examples suitable for expert system application. The applicable expert system rules may be expressed in the general forms presented below:

> IF: Logistical conditions are satisfied,
> THEN: Perform the selected scheduling action.
> IF: Condition A is satisfied and
>   condition B is false and
>   evidence C is present or
>   observation D is available,
> THEN: Precedence belongs in class X.
> IF: If precedence belongs to class X,
> THEN: Activate heuristic scheduling procedure Y.

The function of the expert system model is to aid a decision maker in developing a task sequence that fits the needs of concurrent scheduling. Based on user input, the model will determine the type of task precedence, establish precedence relaxation strategy, implement a task scheduling heuristic, match the schedule to resource availability, and present a recommended task sequence to the user. The user can perform what-if analysis by making changes in the input data and conducting sensitivity analysis. The expert system implementation can be achieved in an interactive environment as shown in Figure 10–47. The user will provide task definitions and resource availabilities with appropriate precedence requirements. At each stage, the user is prompted to consider potential points for precedence relaxation. Wherever schedule options exist, they will be presented to the user for consideration and approval. The need for task splitting will be assessed implicitly by the expert system. The user will have the opportunity to make final decisions about task sequence.

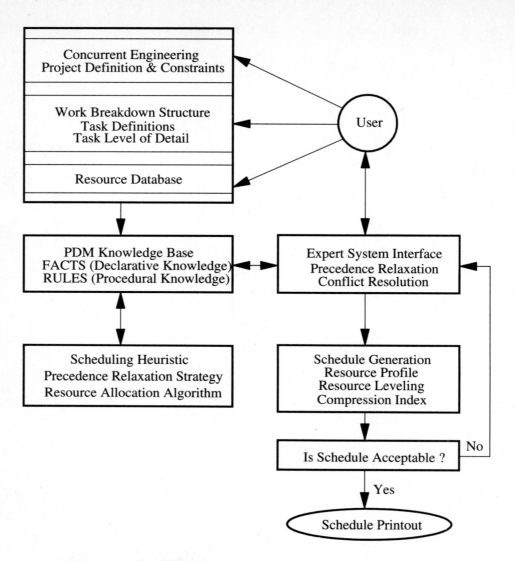

Figure 10-47.   Expert System Decision Support Model for Scheduling

11

Case Study: Reconstruction Project Management at Tinker Air Force Base

This chapter presents a case study[1] that illustrates the application of an integrated project management approach to a real project. The project is a major reconstruction effort that called for the use of innovative management practices, optimization techniques, and computer applications.

11.1 INTRODUCTION

Tinker Air Force Base (TAFB), located in Oklahoma City, Oklahoma, is one of the five overhaul bases in the U.S. Air Force Logistic Command. It is a prime air force rework facility for jet engine parts that support standard engines. The base has the responsibility for the overhaul and repair of six types of jet engines, various aircraft and engine accessories, as well as worldwide management of selected air force military assets. The base is a major economic factor to the state of Oklahoma and a major strategic facility for national defense. Covering more than 5,000 acres, the base has nearly 25,000 military and civilian personnel. Its annual payroll of $676 million is important to the economy of the entire area. TAFB's mission, which includes worldwide support of B-52, B-1 and B-2 bombers as well as a host of other aircraft and missile types, is vital to the nation's defense.

  The base dates back to 1941, when the need for a maintenance and repair depot was recognized and the land was made available by a group of local business and civic leaders. The base was named for Major General Clarence Tinker of Pawhuska, Oklahoma. General Tinker was lost leading a bombing raid on Wake Island early in World War II. During the war and subsequent conflicts, Tinker Air Force Base has

[1]This article is reprinted from the *PMNETwork*, Vol. VII, No. 2, February 1993, with permission of the Project Management Institute, 130 South State Road, Upper Darby, PA 19082, a worldwide organization of advancing the state-of-the-art in project management.

performed its logistics, maintenance and repair jobs with distinction so it was a major concern when a devastating fire in November 1984 threatened to disrupt operations in the propulsion (engine) division in the directorate of maintenance.

11.2 PROJECT BACKGROUND

The fire that necessitated the project occurred in Building 3001, which has the reputation of being the largest single building in Oklahoma. The building houses several key maintenance facilities. The base command quickly organized a team to plan, design, and implement an emergency reconstruction project. The base team was headed by Larry Williams, who was then engineering and planning branch deputy chief, propulsion division, directorate of maintenance, for the base. The base team contracted with a team of faculty and students from the School of Industrial Engineering, University of Oklahoma, on the technical aspects of the reconstruction efforts. The University of Oklahoma team got the contract after some of the national consulting firms that submitted bids for the project claimed that the project could not be done within the required time limit. Some of the firms submitted bids running into millions of dollars. The University of Oklahoma team completed its tasks ahead of schedule at a fraction of what a consulting firm would have cost. A team of Tinker Air Force Base staff was responsible for managing the overall project while the University of Oklahoma (OU) team was responsible for the quantitative and computer analyses required by the project. The analysis by the OU team to aid the transition to the reconstructed layout included developing a simulation program, designing the overhead conveyor system, laying out the plant, and making a routing analysis of the intershop part transfers.

The reconstruction project provided an opportunity to improve on the previous production facilities at TAFB. It resulted in a 50 percent decrease in material handling for long flow items, a 50 percent decrease in organizational transfers for long flow parts, a savings of 30,000 square feet of production floor space, a savings of $3.5 million by the elimination of excess machinery, and $1.8 million in direct labor efficiency. The project has gained international recognition through several publications and by winning the fourth place international award in the 1988 Franz Edelman award presented by the Institute of Management Science (TIMS).

This case study illustrates the project management techniques applied during the reconstruction project. Other aspects of the project have been documented in the literature (see references at the end of the chapter).

11.3 THE NEED FOR A PROJECT MANAGEMENT APPROACH

Project management is the process of managing, allocating, and timing resources to achieve a desired goal in an efficient and expedient manner. The objectives that constitute the desired goal are normally a combination of time, cost, and performance requirements. A project can be simple or very complex. The reconstruction project at Tinker Air Force was an example of a very complex project which required carefully designed interfaces among several teams. The project provided an opportunity to integrate project management concepts with real-world practices.

11.3.1 Project Management Steps

The project management process and steps discussed in Chapter 1 were followed in the project. The life cycle of the project consisted of several functions starting with problem identification, definition, specifications, project formulation, organizing, resource allocation, scheduling, tracking, reporting, control, and ending with project termination. The steps were performed strategically in accordance with the specified project goal. It was necessary to have concurrent implementation of the steps for the several groups involved in the project. Some of the steps are explained below.

Project definition.   The purpose of the project was clarified in this step. A *mission statement* and a *statement of work* (SOW) were the major outputs at this point. The mission statement specified that the base must be back in full and satisfactory operation within the shortest possible time. A condensed sample of the statement of work is presented at the end of this section.

Planning.   Project planning was required to determine how to initiate and execute the objectives of the project. Both the bottom-up and top-down planning strategies were used. The requirements for supervising and delegating authority were considered in the planning step. The critical path method (CPM) was used as the analytical tool for carrying out the project plan. Gantt chart schedules were developed after performing the CPM calculations. A project responsibility matrix was developed to indicate where and how each team would fit into the overall project plan. Each team then developed its own team responsibility chart to provide details of what each individual was responsible for. The project responsibility matrix consisted of columns containing the list of teams and a row containing the list of responsibilities. Cells within the matrix were filled with relationship codes that indicated who was responsible for what. The responsibility chart helped to avoid neglecting crucial responsibility or communication requirements.

Resource allocation.   Project goals and objectives were accomplished through the allocation of resources to functional requirements. Resources for this project were defined in terms of workforce, equipment, and skill requirements. Because of the crucial nature of the project, resource allocation was not a major problem. However, whatever resources were allocated had to be effectively and efficiently utilized.

Scheduling.   The various tasks in the project had to be carefully orchestrated. Task precedence relationships were reviewed in terms of technical, procedural, and resource restrictions. Time lines were developed for specific team schedules. Bottleneck tasks were identified and given special attention.

Project tracking and reporting.   This step involved evaluating whether or not project results conformed to plans and specifications. Frequent project review meetings were conducted to appraise the progress of the project. Reports were generated with the appropriate levels of detail and in simplified form to facilitate quick management review and decisions. Areas of potential conflicts were identified and contingency plans were developed for such areas. When conflicts developed, prompt control actions were taken.

Use of computer simulation.   The reconstruction project management at Tinker Air Force Base called for an integrative use of simulation to enhance the planning, scheduling, and control functions. The output of the simulation of the newly designed production environment provided crucial inputs for better management of the reconstruction project.

Use of optimization techniques.   Optimization techniques were used during the project for the design of the conveyor system, queue sizing in front of each machine, number of machines to place at a workstation, and the routing of parts between shops.

Project phaseout.   The phaseout of a project is as important as its initiation. To prevent the project from dragging on needlessly, strict directives were issued about when the project should end. However, provisions were made for follow-up projects that would further improve the results of the project. These follow-up or spinoff projects were to be managed as totally separate projects but with proper input-output relationships between the sequence of projects. For example, one follow-up project involved transferring the simulation program developed by the OU team to some other air force bases involved in similar maintenance operations.

A sample of the statement of work for the project is presented below. The details in the statement of work are indicative of the complexity and criticality of the reconstruction project.

SAMPLE OF THE STATEMENT OF WORK

SECTION I: SCOPE

This section describes the scope of the project.

1. *Contractor is to provide all labor, equipment, materials, facilities, and transportation necessary to evaluate the alternate production facility layouts for jet engine overhaul provided by the government. Evaluation and simulation will determine the most economical shop arrangement in terms of machine utilization, material handling cost, and floor space occupation that provides sufficient overhaul capability to meet projected requirements. This effort will require the services of simulation and modeling and mechanized material handling (MMHS) and storage devices analysis. The analysis will include all production shops in B3001 assigned to or directly associated with engine overhaul. Specific shop description will be provided by OC-ALC/MAE. Evaluation of the demand and routing of the mechanized material handling system for shop-to-shop material transfer as well as the size, configuration, and location of the mini-stackers for individual shop material staging and queuing are an integral part of the project. This effort is essential to a timely and cost-effective relocation of OC-ALC/MAE production facilities into the fire damaged portion of Building 3001 at Tinker Air Force Base. Site visits will be scheduled as needed to insure contractor understanding of the requirements and evaluate their recommendations.*

2. *While this effort is directly related to recovering from the November 1984 fire at Tinker AFB, it will establish and provide an essential engineering capability for analysis, evaluation, and development of current state-of-the-art theories or findings to overhaul requirements of OC-ALC/MA. This service will provide an essential capability that is not organically available at OC-ALC/MA.*

3. *a. Model must reflect production flow and sequence of operations. Production data will be provided by OC-ALC/MAE for each work control document.*

b. Contractor will provide proposed report formats no later than 10 days after contact award for OC-ALC/MAE review and approval. Report formats will be divided into three areas:

HEADER: Define purpose and application of report

DATA DESCRIPTION: Display/define data elements presented

DATA: Compiled/analyzed/processed raw data that provides specific information for management decisions.

All reports will provide data by individual code elements. Data elements and report formats used in SLAM II simulation software will be the basis for all reports. Selected figures are attached to clarify requirements. Modifications may be required to meet OC-ALC/MAE terminology requirements.

SECTION II: REQUIREMENTS

The deliverables required of the contractor include the application of industrial management expertise, analysis, and evaluation to include application of standard industrial engineering techniques, such as, facility modeling, simulation, operations research, queueing theory, and potential failure analysis. Specific requirements to be fulfilled are

1. Data Security: All production data must be secured from non-AFLC personnel and non-U.S. civilians. Actual capacity projections are considered sensitive, not classified, and sufficient security measures must be taken to insure data protection, and upon completion of this project, all data in the contractor's possession must be destroyed or returned to the government. Certification of material destruction must be submitted to the contracting officer for verification.

2. Training: Instruct MA personnel in the input, analysis, and operation of the capacity planning model and the analysis of mechanized material handling system (MMHS) sizing and routing simulation. A minimum of five MA personnel will require this training.

3. Specific Contractor Reports and Analysis Required:

    a. Reports will reflect shop organizations

    i)   Family groupings of machines/processes

    ii) Individual Modular Repair Centers (MRCs)

    iii) Individual machines/processes

    iv) Total of all production shops

    b. Specific reports for the above areas will include but not limited to:

    1. Output requirements by individual or nested customer order (control numbers) for variable time/production periods.

    2. Process requirements by process code that details the Total Bill of Material (BOM) by Work Control Document (WCD) for each control number for each shop area involved.

    3. Projected operating costs in terms of personnel costs and equipment requirements based on proposed shop layouts, shop resources, and workload mixes.

    4. Reports for individual shops or all shops for various workload projections.

    5. Machinery dedication in terms of operating costs, floor space, and machine utilization, based on sequenced WCD flow and constraints specified by OC-ALC/MAE.

    6. Capability and capacity of each shop, MRC and organization.

    7. Utilization of individual or groups of machines.

    8. Using current operating costs, production outputs and space requirement from the combustion can and gearbox or TF30 assembly organizations as a baseline, provide (±5%) accurate data projections for the following:

        a. Queue time

        b. Flow time and material handling cost within shop/MRC

        c. Flow time and material handling cost between shops/MRCs

9. *Recommendations for sequence and location of equipment within shops/MRCs to reduce operating costs.*

c. *MMHS and Mini-Stacker Analysis: Using data from verified simulation model, contractor will provide the following:*

1. *Economic analysis of planned routing of MMHS.*
2. *Recommended changes to minimize operating and construction costs.*
3. *Number and location of input-output elevators to minimize material handling costs, and other operating costs.*
4. *Identify potential bottlenecks generated by surge situations and cost effective solutions to associated problems.*
5. *Accurate (±5%) prediction of system volume in total and for individual shops for variable time frames.*
6. *Determine if proposed government conveyor design will permit continuous material flow from any shop to any other shop. Recommend corrective actions needed to achieve this requirement.*
7. *Recommend optimum location, size, number of bins, size of bins, and degree of automation for mini-stackers to meet overall shop/MRC requirements.*
8. *Size queuing areas at the following points:*
   c. *Input-Output stations*
   b. *Key machines*
   c. *Mini-stackers*
9. *Develop operating procedures for mini-stackers.*

d. *Software Package, simulation, etc. must operate on VAX 11/780 series computers. Contractor will provide a working model of the facility with support documentation by June 15, 1985. MMHS recommendations are required by July 1, 1985.*

SECTION III: ITEMS TO BE PROVIDED TO CONTRACTOR

The following items will be provided to the contractor by the government.

1. *Process requirements data for each Work Control Document (WCD) on magnetic tape that will contain the following in a standard format:*
   a. *Part description data, WCD number, noun, size, MRC assignment, workload requirements by control number.*
   b. *Individual process requirements in production sequence: The type of process, standard labor, and process flow time.*
2. *Equipment availability by process code and shop/MRC.*
3. *Workload requirements.*
4. *Proposed shop capability and capacity by process code.*

SECTION IV: DURATION

As identified in Section II, capacity reports are due June 15 and MMHS and Mini-Stacker recommendations are due July 1, 1985. Follow-on analysis and additional evaluation will be required throughout the period of relocation or approximately 12–24 months in total.

11.4 ORGANIZING THE PROJECT TEAM

The base reconstruction team was organized quickly with the task to get the maintenance facility back in production status within the minimum amount of time. Using the matrix structure, personnel from various departments within TAFB were assigned to the project team. The team included both civilian and military personnel. Figure 11–1 presents an abbreviated portion of the overall matrix structure. The major groups within the overall project team were

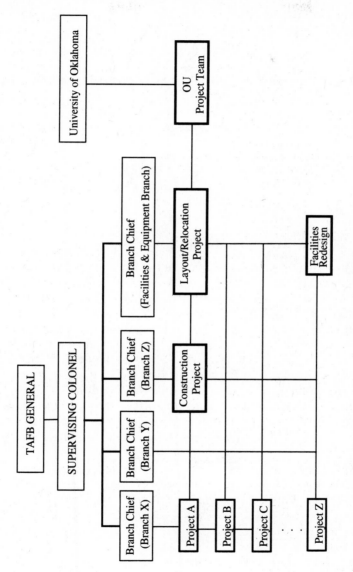

Figure 11-1. Abbreviated Matrix Organizational Structure

- Tinker Task Force management group headed by a colonel
- Tinker project management group headed by Larry Williams
- Tinker facilities department personnel
- Construction subcontractors
- University of Oklahoma research group

The project team was given the authority to implement whatever strategies were necessary to accomplish the reconstruction without disrupting maintenance operations spared by the fire. A temporary project office was created for the project team at the base. During the most critical planning stages, the team was sequestered in the office away from all interruptions. Despite all the reconstruction and relocation activities going on, regular maintenance operations continued to the extent possible and met mission requirements. A control center was established to manage the day-to-day tasks of facility relocation.

11.5 PLANNING WITH CPM

Very careful planning was needed to achieve the reconstruction goals. Machines had to be relocated while new shop layouts were developed. Relocation efforts began on 1 August 1985. The first shops to be reclaimed were the disassembly and cleaning functions. Shops and modular repair centers (MRCs) were time phased to match existing constraints and minimize impact on production requirements. The constraints included space availability, technical service support, machine availability, and impact on production. In order to determine the impact on limited resources and to insure a smooth continuous return to the north end of the building, a critical path method (CPM) chart was established for each individual MRC and the overall project. In order to properly define the required activities, the project was viewed as a component of a larger system. The hierarchy of the system components followed the sequence of *system, program, project, task*, and *activity*. These components are explained next.

System.   A project system consists of interrelated elements organized for the purpose of achieving a common goal. The elements are expected to work synergistically to generate a unified output that is greater than the sum of the individual outputs of the components. For the reconstruction project, the U.S. national defense system was defined as the parent system.

Program.   Program commonly denotes very large and prolonged undertakings. It is a term that is typically applied to project endeavors that span several years. Programs are usually associated with particular systems. In this case study, Tinker Air Force Base's function of maintaining air force jet engines was defined as a program within the national defense system.

Project.   Project is the term generally applied to time-phased efforts of smaller scope and duration than programs. Programs are sometimes viewed as consisting of a set of projects. The reconstruction project was viewed as one project within ongoing TAFB programs.

Task.   A task is a functional element of a project. A project consists of a contiguous collection of tasks that all contribute to the overall project goal. For the reconstruction project, some of the responsibilities assigned to specific groups within the project team were defined as tasks. For example, one of the tasks of the University of Oklahoma research team was to develop a computer simulation program.

Activity.   An activity is defined as a single element of a project. A collection of activities constitutes a task. One activity within the programming task was the collection of input data for simulation.

The above hierarchical definitions were necessary because throughout the project, each function was evaluated in terms of how it would affect other functions at a higher or lower level of the overall project structure. The definitions also helped in developing accurate work breakdown structures (WBS) and clear statements of work (SOW). Activities for each MRC required resources and consumed calendar time, all of which had to be planned, scheduled, and controlled. Each activity had to have an estimate of time required for completion in labor-hours as well as clock hours and days. The level of detail required for each output and format of the output was agreed upon by all the groups involved. Initial time and resource elements were obtained from each branch by 31 May 1985. An abbreviated sample of the requirements format is shown in Table 11–1.

Figure 11–2 presents a portion of the MRC CPM planning guide. The sample network shows the complexity of the project. The larger nodes represent tasks of a bigger scope than activities. The task codes are written as T (for task) plus a task number. The smaller nodes represent activities of a smaller scope than tasks. The activity codes are written as A (for activity) plus an activity number. Expanded and separate CPM networks were drawn for some of the tasks in the sample network. For example, the task node representing the University of Oklahoma project team had its own expanded CPM network. With this network strategy, it was not necessary to draw one huge cluttered network for the overall project. Time Line project management software was used to analyze the CPM networks. The descriptions of some of the task and activity codes in the network are presented as follows:

T1:	Prepare data on Z100	T13:	Input from OU project team
T2:	Conceptual layout	T16:	Develop production schedules
T3:	MRC requirements	A2:	Develop fire safety strategy
T4:	MRC layout	A8:	Utility routing
T5:	Drawing review	A20:	Verify tooling and setup
T6:	Design utilities	A21:	Schedule resources
T7:	Shop drawings	A29:	MRC B.O.M.
T8:	Shop relocation plan	A40:	Modify sequence
T9:	Prepare for shop relocation	A41:	Verify functional operations
T10:	Prepare for machine relocation	A42:	Verify percent complete
T11:	Move machines	A43:	Change drawings
T12:	MRC relocated	A45:	Verify material flow

Table 11-1 Example of Activity Requirements Format for CPM Planning

Engineering planning function	Time (days)	Branch	Format	
			Data input	Data output
Shop requirements				
a. Z100 85 data	Completed	MAE	WCD-process	MRC summary
b. Z100 2000 eng.	3–7	MAE	Control #	Data
c. OU simulation	By June 15	MAE	Z100 mag. tape	MRC simulation
Conceptual layout	NA	MAE	Shop-Z100	Building print
MRC requirements	1–2	MAE	Z100 OU sim.	MRC matrix
a. OU Req.		MAEE		
MRC machine placement				
a. Foundation req.				
1. Size and Specs				
2. Location				
3. Quantity				
b. MMHS Req.				
1. Conveyor I/O consoles				
2. Mini-stackers			OU simulation	
c. Unique MRC req.				
1. Bridge crane				
2. Jib crane				
3. Environ. req.				
d. Bulk utility req.				
Drawing review				
Design utilities				
Shop drawings				
Shop relocation plan				
MRCs relocation				
Move machines to MRCs				
Update shop drawings				
Verify material flow				

To facilitate the project efforts, new organizational and shop layout strategies were developed.

- Maintenance facility was initially sized for one-shift operation. Additional specific processes and equipment were added later for two-shift operations to meet prefire process requirements.
- Modular repair centers were established for parts with similar geometries, part families, and process requirements.
- Machine placements were clustered to reduce material handling and establish work centers.
- General purpose functions such as heat treat, painting, and plating remained unrelocated due to the nature of those processes.

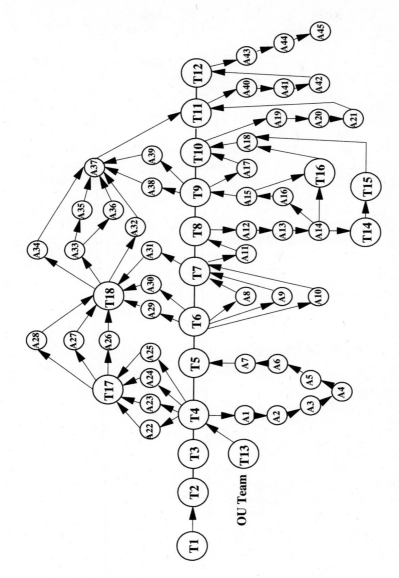

Figure 11-2. Sample of MRC CPM Network

- Each MRC was provided with all equipment and processes that could be economically justified. This significantly reduced MRC to MRC traffic and material flow on the conveyor system.
- The MRC became the single point of production responsibility. Every work control document (WCD) was assigned to a specific MRC.
- The station 57 concept of an independent group checking components prior to delivery to the serviceable stacker was discontinued.
- Known and anticipated productivity enhancements were incorporated into individual MRCs.

The CPM chart helped in establishing the sequence of when each MRC would be created and relocated. However, allowances were made for continuous changes to the initial plan as detailed planning was completed and actual construction proceeded. An abridged example of a relocation plan for an MRC is shown in Table 11–2. Table 11–3 shows an example of a machine relocation plan within an MRC.

With the information contained in Tables 11–2 and 11–3 and other similar memos and notices, it was possible to keep project team members informed of the requirements and due dates. Thus, few scheduling conflicts were experienced. A feedback loop was created whereby each person or team could identify and promptly report potential or real problems.

11.6 COMMUNICATION, COOPERATION, AND COORDINATION

Planning, communication, coordination, and more planning formed the essential basis for the relocation to the damaged portion of Building 3001. Equipment recovery, site preparation, construction, and relocation were carefully coordinated to avoid conflicting schedules. While the military process of command, control, and communication played a role in the initial response to the fire, the project team relied more on the Triple C model of project management (see Chapter 2). Triple C suggests planning and implementing complex projects under a structured approach to communication, cooperation, and coordination. The principle facilitates a systematic approach to the planning, organizing, scheduling, and controlling of a project. The three components of the Triple C model as applied to the project are discussed next.

11.6.1 Communication

Proper communication contributes significantly to better performance in a project environment. The communication function of project management involves making all those concerned become aware of the project requirements and progress. Those who would be affected by the project directly or indirectly, as participants or beneficiaries, were informed as appropriate regarding the following:

Table 11-2 Abridged Example of MRC Relocation Plan

Overall relocation plan: Based on the following Area release dates:
Area 1 (July 28), Area 2 (September 1), Area 3 (August 5), Area 4 (September 1)

Phase I
Relocate to Area I   Start August 1
1. Chemical cleaning
2. Disassembly
3. Fork lifts
4. Blast cleaning
5. Combustion cans

Phase II
MMHS Construction   Start August 1
Constructed by area - TBD

Phase III
NDI construction   Start August 1
1. Case NDI
2. Nozzle and Cans NDI
3. Gear box - bearing housing NDI
4. Turbine compressor rotor
5. General

Phase IV
Continued MRC relocation   Start August 1
6. Nozzles
7. TDR
8. J79
.
.
.

Phase V
Modifications to existing shops
1. Blades
2. QVC
3. Plating
.   .
.   .
.   .

Table 11-3 Example of Machine Relocation Plan

MAEE-1
Pacer Response—Relocation of Machines to the Case Shop

MA-1
1. In order to meet the April 5, 1986 due date, the machines for the Case Shop Part "A" will have to be relocated in accordance with the schedule specified below.
2. Pressure tanks, 1 each 48-inch Dia and 1 each 60-inch will have to be fabricated with controllers, pumps and filters prior to start of move: Part "A." Three VTLs must be installed when received from DIPEC.
3. Project monitor for MAEE-1 is *(individual's name and telephone)*. Please inform us of all problems in meeting this requirement.
4. Parts B, C, D, and E will follow in 7-day intervals.

Date	Machine	Ser. #	OC/AF #	Present location	New location
3-3-86	Engine lathe	79347	OC2715	CK Area	D93, 10,S
	Welder		OC1360	Y14, W/S 10	B91, 10,N
	Jig borer	M16951	405201	DIPEC	C93
	.	.	.	.	.
	.	.	.	.	.
3-4-86	Radial drill	79347	OC2559	Y58, 15,W	B93, 10,S
	.	.	.	.	.
	.	.	.	.	.
3-5-86	VTL 42"	27112	OC2096	S63	D-E91, 35,N
	.	.	.	.	.
	.	.	.	.	.

- Scope of the project
- Need for the project
- Organization of the project
- Expected impact of the project
- Individual in charge of the project
- Time frame for implementing the project
- Level of personnel participation required

The communication channel was kept open throughout the reconstruction project. In addition to in-house communication, external sources were also consulted as appropriate.

11.6.2 Cooperation

The cooperation of the project personnel was explicitly sought. Merely being informed of the project requirements was not enough assurance of full cooperation. The personnel were convinced of the merits and the urgency of the project. Factors that are typically of concern in cooperation problems were addressed by the project coordinators. These factors included resource requirements, relocation requirements, revised priorities, and so on. Each project team member was made aware of the following:

- The cooperative efforts needed
- The implication of lack of cooperation
- The criticality of cooperation to the project
- The time frame involved for personnel participation
- The organizational impact of cooperation

11.6.3 Coordination

After successfully initiating the communication and cooperation functions, the efforts of the project team were directed at coordination. Coordination facilitates harmonious organization of project efforts. The development of a responsibility chart was very helpful at this stage. A responsibility chart showing individuals, teams, or functional departments and assigned tasks and responsibilities was developed. The responsibility chart helped to avoid overlooking critical communication requirements as well as commitments. It helped resolve questions such as

- Who was to do what?
- Who was responsible for which results?
- What personnel interfaces were involved?
- Who was to inform whom of what?
- Whose approval was needed for what?
- What support was needed from whom for what functions?

11.7 TRAINING AND PROJECT TRANSFER

An important component of the reconstruction project was a training program conducted by the University of Oklahoma research team for Tinker Air Force Base personnel. The training was conducted over several weeks and it involved how to run the simulation programs developed for production planning in the reconstructed shops. The simulation model, called Tinker integrated planning and simulation (TIPS), was written using discrete even orientation in SLAM II and it contains over 1,750 lines of customized FORTRAN code (see Foote et al. 1988; Ravindran et al. 1989; Leemis et al. 1990 for further details on TIPS). A separate training program was conducted for managers. For Tinker technical staff, the training covered the technical aspects of the simulation model. The training for managers covered how to use the outputs of the simulation to make decisions. For this purpose, customized and simplified output formats were incorporated into the simulation model. After the two training programs, the project of the University of Oklahoma team was essentially transferred to the Tinker staff. The team, however, continued to provide technical support. Modified or customized versions of the simulation model were later transferred to some air force bases with functions similar to those of Tinker Air Force Base.

11.8 CONCLUSION

The integrated project management approach employed on the reconstruction project was the major factor in the quick recovery from the disastrous fire. The project team started the project in January 1985, approved the organizational concept in February 1985, and developed the industrial process code concept and started data collection in late February. The first detailed shop layout was required to be completed in July 1985. Data required to meet material and scheduling lead times was needed by 15 June 1985. All simulation runs had to be completed by September 1985 to finalize shop resource allocations and to allow for design lead times. The outputs of the simulation model were needed to allocate personnel, machines, and floor space to the various shops. All of these requirements were met on time and within budget. The role played by project management on this reconstruction project is a good example that should benefit other organizations that face crisis management problems. The case study also serves as a good motivating example for the use of an integrated project management approach in complex project environments.

11.9 REFERENCES FOR CASE STUDY

BADIRU, A. B., B. L. FOOTE, L. LEEMIS, A. RAVINDRAN, and L. WILLIAMS. 1993. "Recovering from a Crisis at Tinker Air Force Base." *PM Network* 7, no. 2 (February): 10–23.

FOOTE, B., A. RAVINDRAN, A. B. BADIRU; L. LEEMIS; and L. WILLIAMS. 1988. "Simulation and Network Analysis Pay Off in Conveyor System Design." *Industrial Engineering* 20, no. 6 (June): 48–53.

LEEMIS, L., A. B. BADIRU, B. L. FOOTE, and A. RAVINDRAN. 1990. "Job Shop Configuration Optimization at Tinker Air Force Base." *Simulation* 54, no. 6 (June): 287–290.

RAVINDRAN, A.; B. FOOTE; A. BADIRU; and L. LEEMIS. 1988. "Mechanized Material Handling Systems Design & Routing." *Computers & Industrial Engineering* 14, no. 3: 251–270.

———, and L. WILLIAMS. 1989. "An Application of Simulation and Network Analysis to Capacity Planning and Material Handling Systems at Tinker Air Force Base." *TIMS Interfaces* 19, no. 1 (January–February): 102–115.

A

Standard Normal Table

z	.00	.01	.02	.03	.04	.05	.06	.07	.08	.09
.0	.5000	.5040	.5080	.5120	.5160	.5199	.5239	.5279	.5319	.5359
.1	.5398	.5438	.5478	.5517	.5557	.5596	.5636	.5675	.5714	.5753
.2	.5793	.5832	.5871	.5910	.5948	.5987	.6026	.6064	.6103	.6141
.3	.6179	.6217	.6255	.6293	.6331	.6368	.6406	.6443	.6480	.6517
.4	.6554	.6591	.6628	.6664	.6700	.6736	.6772	.6808	.6844	.6879
.5	.6915	.6950	.6985	.7019	.7054	.7088	.7123	.7157	.7190	.7224
.6	.7257	.7291	.7324	.7357	.7389	.7422	.7454	.7486	.7517	.7549
.7	.7580	.7611	.7642	.7673	.7704	.7734	.7764	.7794	.7823	.7852
.8	.7881	.7910	.7939	.7967	.7995	.8023	.8051	.8078	.8106	.8133
.9	.8159	.8186	.8212	.8238	.8264	.8289	.8315	.8340	.8365	.8389
1.0	.8413	.8438	.8461	.8485	.8508	.8531	.8554	.8577	.8599	.8621
1.1	.8643	.8665	.8686	.8708	.8729	.8749	.8770	.8790	.8810	.8830
1.2	.8849	.8869	.8888	.8907	.8925	.8944	.8962	.8980	.8997	.9015
1.3	.9032	.9049	.9066	.9082	.9099	.9115	.9131	.9147	.9162	.9177
1.4	.9192	.9207	.9222	.9236	.9251	.9265	.9279	.9292	.9306	.9319
1.5	.9332	.9345	.9357	.9370	.9382	.9394	.9406	.9418	.9429	.9441
1.6	.9452	.9463	.9474	.9484	.9495	.9505	.9515	.9525	.9535	.9545
1.7	.9554	.9564	.9573	.9582	.9591	.9599	.9608	.9616	.9625	.9633
1.8	.9641	.9649	.9656	.9664	.9671	.9678	.9686	.9693	.9699	.9706
1.9	.9713	.9719	.9726	.9732	.9738	.9744	.9750	.9756	.9761	.9767
2.0	.9772	.9778	.9783	.9788	.9793	.9798	.9803	.9808	.9812	.9817
2.1	.9821	.9826	.9830	.9834	.9838	.9842	.9846	.9850	.9854	.9857
2.2	.9861	.9864	.9868	.9871	.9875	.9878	.9881	.9884	.9887	.9890
2.3	.9893	.9896	.9898	.9901	.9904	.9906	.9909	.9911	.9913	.9916
2.4	.9918	.9920	.9922	.9925	.9927	.9929	.9931	.9932	.9934	.9936
2.5	.9938	.9940	.9941	.9943	.9945	.9946	.9948	.9949	.9951	.9952
2.6	.9953	.9955	.9956	.9957	.9959	.9960	.9961	.9962	.9963	.9964
2.7	.9965	.9966	.9967	.9968	.9969	.9970	.9971	.9972	.9973	.9974
2.8	.9974	.9975	.9976	.9977	.9977	.9978	.9979	.9979	.9980	.9981
2.9	.9981	.9982	.9982	.9983	.9984	.9984	.9985	.9985	.9986	.9986
3.0	.9987	.9987	.9987	.9988	.9988	.9989	.9989	.9989	.9990	.9990
3.1	.9990	.9991	.9991	.9991	.9992	.9992	.9992	.9992	.9993	.9993
3.2	.9993	.9993	.9994	.9994	.9994	.9994	.9994	.9995	.9995	.9995
3.3	.9995	.9995	.9995	.9996	.9996	.9996	.9996	.9996	.9996	.9997
3.4	.9997	.9997	.9997	.9997	.9997	.9997	.9997	.9997	.9997	.9998

B

LINDO Model and Output

```
Script started on Tue Aug 17 14:33:39 1993
% lindo
 LINDO (23 DEC 87 CHICAGO)
 : retr
 FILE NAME=
proje
 : look
 ROW:
all

  MIN      T10
  SUBJECT TO
     2)    X22 + X23 + X24 + X25 + X26 + X27 + X28 + X29 + X210
   =    1
     3)    X32 + X33 + X34 + X35 + X36 + X37 =    1
     4)    X42 + X45 + X43 + X44 + X46 + X47 =    1
     5)    X52 + X53 + X54 + X55 + X56 + X57 + X58 + X59 + X510
   + X511 =    1
     6)    X67 + X68 + X69 + X610 + X611 + X612 + X613 + X614
   =    1
     7)    X75 + X76 + X77 + X78 + X79 + X710 =    1
     8)    X84 + X85 + X86 + X87 + X88 + X89 =    1
     9)    X97 + X98 + X99 + X910 + X911 + X912 =    1
    10)    X108 + X109 + X1010 + X1011 + X1012 + X1013 =    1
    11) - 2 X22 - 3 X23 - 4 X24 - 5 X25 - 6 X26 - 7 X27 - 8 X28
   - 9 X29 - 10 X210 + 6 X66 + 7 X67 + 8 X68 + 9 X69 + 10 X610
   + 11 X611 + 12 X612 + 13 X613 + 14 X614 >=    4
    12) - 2 X32 - 3 X33 - 4 X34 - 5 X35 - 6 X36 - 7 X37 + 5 X75
   + 6 X76 + 7 X77 + 8 X78 + 9 X79 + 10 X710 >=    3
    13) - 2 X42 - 5 X45 - 3 X43 - 4 X44 - 6 X46 - 7 X47 + 4 X84
   + 5 X85 + 6 X86 + 7 X87 + 8 X88 + 9 X89 >=    2
    14) - 2 X52 - 3 X53 - 4 X54 - 5 X55 - 6 X56 - 7 X57 - 8 X58
   - 9 X59 - 10 X510 - 11 X511 + 8 X108 + 9 X109 + 10 X1010
   + 11 X1011 + 12 X1012 + 13 X1013 >=    2
    15) - 5 X75 - 6 X76 - 7 X77 - 8 X78 - 9 X79 - 10 X710
   + 7 X97 + 8 X98 + 9 X99 + 10 X910 + 11 X911 + 12 X912
   >=    2
    16) - 4 X84 - 5 X85 - 6 X86 - 7 X87 - 8 X88 - 9 X89 + 7 X97
   + 8 X98 + 9 X99 + 10 X910 + 11 X911 + 12 X912 >=    3
    17) - 7 X97 - 8 X98 - 9 X99 - 10 X910 - 11 X911 - 12 X912
   + 8 X108 + 9 X109 + 10 X1010 + 11 X1011 + 12 X1012
   + 13 X1013 >=    1
    18) - 6 X66 - 7 X67 - 8 X68 - 9 X69 - 10 X610 - 11 X611
   - 12 X612 - 13 X613 - 14 X614 + T10 >=    1
    19) - 8 X108 - 9 X109 - 10 X1010 - 11 X1011 - 12 X1012
   - 13 X1013 + T10 >=    2
    20)    X22 + 4 X32 + 3 X42 + 5 X52 <=    10
    21)    X22 + X23 + 4 X32 + 4 X33 + 3 X42 + 3 X43 + 5 X52
   + 5 X53 <=    10
    22)    X22 + X23 + X24 + 4 X32 + 4 X33 + 4 X34 + 3 X43
   + 3 X44 + 5 X53 + 5 X54 + 6 X84 <=    10
    23)    X22 + X23 + X24 + X25 + 4 X33 + 4 X34 + 4 X35 + 3 X45
   + 3 X44 + 5 X54 + 5 X55 + 2 X75 + 6 X84 + 6 X85 <=    10
    24)    X23 + X24 + X25 + X26 + 4 X34 + 4 X35 + 4 X36 + 3 X45
   + 3 X46 + 5 X55 + 5 X56 + 8 X66 + 2 X75 + 2 X76 + 6 X84
   + 6 X85 + 6 X86 <=    10
    25)    X24 + X25 + X26 + X27 + 4 X35 + 4 X36 + 4 X37 + 3 X46
   + 3 X47 + 5 X56 + 5 X57 + 8 X66 + 8 X67 + 2 X76 + 2 X77
   + 6 X85 + 6 X86 + 6 X87 + 9 X97 <=    10
    26)    X25 + X26 + X27 + X28 + 4 X36 + 4 X37 + 3 X47 + 5 X57
   + 5 X58 + 8 X67 + 8 X68 + 2 X77 + 2 X78 + 6 X86 + 6 X87
   + 6 X88 + 9 X98 + 3 X108 <=    10
    27)    X26 + X27 + X28 + X29 + 4 X37 + 5 X58 + 5 X59 + 8 X68
   + 8 X69 + 2 X78 + 2 X79 + 6 X87 + 6 X88 + 6 X89 + 9 X99
   + 3 X108 + 3 X109 <=    10
    28)    X27 + X28 + X29 + X210 + 5 X59 + 5 X510 + 8 X69
```

498

```
       + 8 X610 + 2 X79 + 2 X710 + 6 X88 + 6 X89 + 9 X910 + 3 X108
       + 3 X109 + 3 X1010 <=    10
   29)    X28 + X29 + X210 + 5 X510 + 5 X511 + 8 X610 + 8 X611
   + 2 X710 + 6 X89 + 9 X911 + 3 X109 + 3 X1010 + 3 X1011
   <=     10
   30)    X29 + X210 + 5 X511 + 8 X611 + 8 X612 + 9 X912
   + 3 X1010 + 3 X1011 + 3 X1012 <=    10
   31)    X210 + 8 X612 + 8 X613 + 3 X1011 + 3 X1012 + 3 X1013
   <=    10
   32)    8 X613 + 8 X614 + 3 X1012 + 3 X1013 <=    10
   33)    8 X614 + 3 X1013 <=    10
END
INTE        X22
INTE        X23
INTE        X24
INTE        X25
INTE        X26
INTE        X27
INTE        X28
INTE        X29
INTE        X210
INTE        X32
INTE        X33
INTE        X34
INTE        X35
INTE        X36
INTE        X37
INTE        X42
INTE        X45
INTE        X43
INTE        X44
INTE        X46
INTE        X47
INTE        X52
INTE        X53
INTE        X54
INTE        X55
INTE        X56
INTE        X57
INTE        X58
INTE        X59
INTE        X510
INTE        X511
INTE        X66
INTE        X67
INTE        X68
INTE        X69
INTE        X610
INTE        X611
INTE        X612
INTE        X613
INTE        X614
INTE        X75
INTE        X76
INTE        X77
INTE        X78
INTE        X79
INTE        X710
INTE        X84
INTE        X85
INTE        X86
INTE        X87
INTE        X88
INTE        X89
INTE        X97
INTE        X98
```

```
INTE      X99
INTE      X910
INTE      X911
INTE      X912
INTE      X108
INTE      X109
INTE      X1010
INTE      X1011
INTE      X1012
INTE      X1013
```

OBJECTIVE FUNCTION VALUE

1)   14.0000000

VARIABLE	VALUE	REDUCED COST
X22	0.000000	0.000000
X23	1.000000	0.000000
X24	0.000000	0.000000
X25	0.000000	0.000000
X26	0.000000	0.000000
X27	0.000000	0.000000
X28	0.000000	0.000000
X29	0.000000	0.000000
X210	0.000000	0.000000
X32	1.000000	0.000000
X33	0.000000	0.000000
X34	0.000000	0.000000
X35	0.000000	0.000000
X36	0.000000	0.000000
X37	0.000000	0.000000
X42	1.000000	0.000000
X45	0.000000	0.000000
X43	0.000000	0.000000
X44	0.000000	0.000000
X46	0.000000	0.000000
X47	0.000000	0.000000
X52	0.000000	0.000000
X53	0.000000	0.000000
X54	1.000000	0.000000
X55	0.000000	0.000000
X56	0.000000	0.000000
X57	0.000000	0.000000
X58	0.000000	0.000000
X59	0.000000	0.000000
X510	0.000000	0.000000
X511	0.000000	0.000000
X66	0.000000	6.000000
X67	0.000000	7.000000
X68	0.000000	8.000000
X69	0.000000	9.000000
X610	0.000000	10.000000
X611	0.000000	11.000000
X612	0.000000	12.000000
X613	1.000000	13.000000
X614	0.000000	14.000000
X75	1.000000	0.000000
X76	0.000000	0.000000
X77	0.000000	0.000000
X78	0.000000	0.000000
X79	0.000000	0.000000
X710	0.000000	0.000000
X84	0.000000	0.000000
X85	0.000000	0.000000
X86	1.000000	0.000000
X87	0.000000	0.000000
X88	0.000000	0.000000
X89	0.000000	0.000000
X97	0.000000	0.000000
X98	0.000000	0.000000
X99	1.000000	0.000000
X910	0.000000	0.000000
X911	0.000000	0.000000
X912	0.000000	0.000000
X108	0.000000	0.000000
X109	0.000000	0.000000

```
           X1010           1.000000                0.000000
           X1011           0.000000                0.000000
           X1012           0.000000                0.000000
           X1013           0.000000                0.000000
             T10          14.000000                0.000000


           ROW    SLACK OR SURPLUS          DUAL PRICES
            2)           0.000000                0.000000
            3)           0.000000                0.000000
            4)           0.000000                0.000000
            5)           0.000000                0.000000
            6)           0.000000                0.000000
            7)           0.000000                0.000000
            8)           0.000000                0.000000
            9)           0.000000                0.000000
           10)           0.000000                0.000000
           11)           6.000000                0.000000
           12)           0.000000                0.000000
           13)           2.000000                0.000000
           14)           4.000000                0.000000
           15)           2.000000                0.000000
           16)           0.000000                0.000000
           17)           0.000000                0.000000
           18)           0.000000               -1.000000
           19)           2.000000                0.000000
           20)           3.000000                0.000000
           21)           2.000000                0.000000
           22)           0.000000                0.000000
           23)           2.000000                0.000000
           24)           1.000000                0.000000
           25)           4.000000                0.000000
           26)           4.000000                0.000000
           27)           1.000000                0.000000
           28)           7.000000                0.000000
           29)           7.000000                0.000000
           30)           7.000000                0.000000
           31)           2.000000                0.000000
           32)           2.000000                0.000000
           33)          10.000000                0.000000

   NO. ITERATIONS=    35429
   BRANCHES= 1609 DETERM.=   1.000E  0
    : quit
 % ^D
 script done on Tue Aug 17 14:37:02 1993
```

Practitioner's Toolbox:
Conversion Factors

NUMBER PREFIXES

Prefix	SI symbol	Multiplication factors	Interpretation
tera	T	$1\ 000\ 000\ 000\ 000 = 10^{12}$	trillionfold
giga	G	$1\ 000\ 000\ 000 = 10^{9}$	billionfold
mega	M	$1\ 000\ 000 = 10^{6}$	millionfold
kilo	k	$1\ 000 = 10^{3}$	thousandfold
hecto	h	$100 = 10^{2}$	hundredfold
deca	da	$10 = 10^{1}$	tenfold
deci	d	$0.1 = 10^{-1}$	tenth part
centi	c	$0.01 = 10^{-2}$	hundredth part
milli	m	$0.001 = 10^{-3}$	thousandth part
micro	μ	$0.000\ 001 = 10^{-6}$	millionth part
nano	n	$0.000\ 000\ 001 = 10^{-9}$	billionth part
pico	p	$0.000\ 000\ 000\ 001 = 10^{-12}$	trillionth part
femto	f	$0.000\ 000\ 000\ 000\ 001 = 10^{-15}$	quadrillionth part
atto	a	$0.000\ 000\ 000\ 000\ 000\ 001 = 10^{-18}$	quintillionth part

UNITS OF MEASURE

Acre: An area of 43,560 square feet.

Agate: 1/14 inch (used in printing for measuring column length).

Ampere: Unit of electric current.

Astronomical (A.U.): 93,000,000 miles; the average distance of the earth from the sun (used in astronomy).

Bale: A large bundle of goods. In United States, approximate weight of a bale of cotton is 500 pounds. Weight of a bale may vary from country to country.

Board Foot: 144 cubic inches (12 by 12 by 1 used for lumber).

Bolt: 40 yards (used for measuring cloth).

Btu: British thermal unit; amount of heat needed to increase the temperature of one pound of water by one degree Fahrenheit (252 calories).

Carat: 200 milligrams or 3.086 troy; used for weighing precious stones (originally the weight of a seed of the carob tree in the Mediterranean region). *See also Karat.*

Chain: 66 feet; used in surveying (one mile = 80 chains).

Cubit: 18 inches (derived from distance between elbow and tip of middle finger).

Decibel: Unit of relative loudness.

Freight Ton: 40 cubic feet of merchandise (used for cargo freight).

Gross: 12 dozen (144).

Hertz: Unit of measurement of electromagnetic wave frequencies (measures cycles per second).

Hogshead: 2 liquid barrels or 14,653 cubic inches.

Horsepower: The power needed to lift 33,000 pounds a distance of one foot in one minute (about $1\frac{1}{2}$ times the power an average horse can exert); used for measuring power of mechanical engines.

Karat: A measure of the purity of gold. It indicates how many parts out of 24 are pure. 18 karat gold is $\frac{3}{4}$ pure gold.

Knot: Rate of speed of 1 nautical mile per hour; used for measuring speed of ships (not distance).

League: Approximately 3 miles.

Light-year: 5,880,000,000,000 miles; distance traveled by light in one year at the rate of 186,281.7 miles per second; used for measurement of interstellar space.

Magnum: Two-quart bottle; used for measuring wine.

Ohm: Unit of electrical resistance.

Parsec: Approximately 3.26 light-years of 19.2 trillion miles; used for measuring interstellar distances.

Pi (π): 3.14159265+; the ratio of the circumference of a circle to its diameter.

Pica: $\frac{1}{6}$ inch or 12 points; used in printing for measuring column width.

Pipe: 2 hogsheads; used for measuring wine and other liquids.

Point: 0.013837 (approximately $\frac{1}{72}$) inch or $\frac{1}{12}$ pica; used in printing for measuring type size.

Quintal: 100,000 grams or 220.46 pounds avoirdupois.

Quire: 24 or 25 sheets; used for measuring paper (20 quires is one ream).

Ream: 480 or 500 sheets; used for measuring paper.

Roentgen: Dosage unit of radiation exposure produced by X-rays.

Score: 20 units.

Span: 9 inches or 22.86 cm; derived from the distance between the end of the thumb and the end of the little finger when both are outstretched.

Square: 100 square feet; used in building.

Stone: 14 pounds avoirdupois in Great Britain.

Therm: 100,000 Btu's.

Township: U.S. land measurement of almost 36 square miles; used in surveying.

Tun: 252 gallons (sometimes larger); used for measuring wine and other liquids.

Watt: Unit of power.

AREA

Multiply	by	to obtain
acres	43,560	sq feet
	4,047	sq meters
	4,840	sq yards
	0.405	hectare
sq cm	0.155	sq inches
sq feet	144	sq inches
	0.09290	sq meters
	0.1111	sq yards
sq inches	645.16	sq millimeters
sq kilometers	0.3861	sq miles
sq meters	10.764	sq feet
	1.196	sq yards
sq miles	640	acres
	2.590	sq kilometers

VOLUME

Multiply	by	to obtain
acre-foot	1233.5	cubic meters
cubic cm	0.06102	cubic inches
cubic feet	1728	cubic inches
	7.480	gallons (U.S.)
	0.02832	cubic meters
	0.03704	cubic yards
liter	1.057	liquid quarts
	0.908	dry quart
	61.024	cubic inches
barrel of oil	40	gallons
gallons (US)	231	cubic inches
	3.7854	liters
	4	quarts
	0.833	British gallons
	128	U.S. fluid ounces
quarts (U.S.)	0.9463	liters

MASS

Multiply	by	to obtain
carat	0.200	cubic grams
grams	0.03527	ounces
kilograms	2.2046	pounds
ounces	28.350	grams
pound	16	ounces
	453.6	grams
stone (UK)	6.35	kilograms
	14	pounds
ton (net)	907.2	kilograms
	2000	pounds
	0.893	gross ton
	0.907	metric ton
ton (gross)	2240	pounds
	1.12	net tons
	1.016	metric tons
tonne (metric)	2,204.623	pounds
	0.984	gross ton
	1000	kilograms

TEMPERATURE

Conversion formulas

Celsius to kelvin	$K = C + 273.15$
Celsius to Fahrenheit	$F = (9/5)C + 32$
Fahrenheit to Celsius	$C = (5/9)(F - 32)$
Fahrenheit to kelvin	$K = (5/9)(F + 459.67)$
Fahrenheit to Rankin	$R = F + 459.67$
Rankin to kelvin	$K = (5/9)R$

ENERGY, HEAT, POWER

Multiply	by	to obtain
Btu	1055.9	joules
	0.2520	kg-calories
watt-hour	3600	joules
	3.409	Btu
HP (electric)	746	watts
Btu/second	1055.9	watts
watt-second	1.00	joules

VELOCITY

Multiply	by	to obtain
feet/minute	5.080	mm/second
feet/second	0.3048	meters/second
inches/second	0.0254	meters/second
km/hour	0.6214	miles/hour
meters/second	3.2808	feet/second
	2.237	miles/hour
miles/hour	88.0	feet/minute
	0.44704	meters/second
	1.6093	km/hour
	0.8684	knots
knot	1.151	miles/hour

PRESSURE

Multiply	by	to obtain
atmospheres	1.01325	bars
	33.90	feet of water
	29.92	inches of mercury
	760.0	mm of mercury
bar	75.01	cm of mercury
	14.50	pounds/sq inch
dyne/sq cm	0.1	N/sq meter
newtons/sq cm	1.450	pounds/sq inch
pounds/sq inch	0.06805	atmospheres
	2.036	inches of mercury
	27.708	inches of water
	68.948	millibars
	51.72	mm of mercury

LENGTH

Multiply	by	to obtain
angstrom	10^{-10}	meters
feet	0.30480	meters
	12	inches
inches	25.40	millimeters
	0.02540	meters
	0.08333	feet
kilometers	3280.8	feet
	0.6214	miles
	1094	yards
meters	39.370	inches
	3.2808	feet
	1.094	yards
miles	5280	feet
	1.6093	kilometers
	0.8694	nautical miles
millimeters	0.03937	inches
nautical miles	6076	feet
	1.852	kilometers
yards	0.9144	meters
	3	feet
	36	inches

CONSTANTS

speed of light	$2.997,925 \times 10^{10}$ cm/sec
	983.6×10^{6} ft/sec
	186,284 miles/sec
velocity of sound	340.3 meters/sec
	1116 ft/sec
gravity (acceleration)	9.80665 m/sec square
	32.174 ft/sec square
	386.089 inches/sec square

D

Researcher's Toolbox:
Useful Mathematical Expressions

$$\sum_{n=0}^{\infty} \frac{x^n}{n!} = e^x$$

$$\sum_{n=0}^{\infty} \frac{x^n}{n} = \ln\left(\frac{1}{1-x}\right)$$

$$\sum_{n=0}^{k} x^n = \frac{x^{k+1} - 1}{x - 1}, \qquad x \neq 1$$

$$\sum_{n=1}^{k} x^n = \frac{x - x^{k+1}}{1 - x}, \qquad x \neq 1$$

$$\sum_{n=2}^{k} x^n = \frac{x^2 - x^{k+1}}{1 - x}, \qquad x \neq 1$$

$$\sum_{n=0}^{\infty} p^n = \frac{1}{1 - p}, \qquad \text{if } |p| < 1$$

$$\sum_{n=1}^{\infty} (1 - p)^n - \frac{1 - p}{p}, \qquad \text{if } |p| < 1$$

$$\sum_{n=0}^{\infty} nx^n = \frac{x}{(1 - x)^2}, \qquad x \neq 1$$

$$\sum_{n=0}^{\infty} n^2 x^n = \frac{2x^2}{(1 - x)^3} + \frac{x}{(1 - x)^2}, \qquad x \neq 1$$

$$\sum_{n=0}^{\infty} n^3 x^n = \frac{6x^3}{(1 - x)^4} + \frac{6x^2}{(1 - x)^3} + \frac{x}{(1 - x)^2}, \qquad x \neq 1$$

$$\sum_{n=0}^{M} nx^n = \frac{x[1 - (M + 1)x^M + Mx^{M+1}]}{(1 - x)^2}, \qquad x \neq 1$$

$$\sum_{x=0}^{\infty} \binom{r + x - 1}{x} u^x = (1 - u)^{-r}, \qquad \text{if } |u| < 1$$

$$\sum_{k=1}^{\infty} (-1)^{k+1} \frac{1}{k} = 1 - \frac{1}{2} + \frac{1}{3} - \frac{1}{4} + \frac{1}{5} - \frac{1}{6} + \ldots = \ln 2$$

$$\sum_{k=1}^{\infty} (-1)^{k+1} \frac{1}{(2k - 1)} = 1 - \frac{1}{3} + \frac{1}{5} - \frac{1}{7} + \frac{1}{9} - \ldots = \frac{\pi}{4}$$

$$\sum_{k=0}^{\infty} (-1)^k x^k = \frac{1}{1 + x}, \qquad -1 < x < 1$$

$$\sum_{k=1}^{n} (-1)^k \binom{n}{k} = 1, \qquad \text{for } n \geq 2$$

$$\sum_{k=0}^{n} \binom{n}{k}^2 = \binom{2n}{n}$$

$$\sum_{k=1}^{n} k = 1 + 2 + 3 + \ldots + n = \frac{n(n + 1)}{2}$$

$$\sum_{k=1}^{n} (2k) = 2 + 4 + 6 + \ldots + 2n = n(n - 1)$$

$$\sum_{k=1}^{n} (2k - 1) = 1 + 3 + 5 + \ldots + (2n - 1) = n^2$$

$$\sum_{k=0}^{\infty} (a + kd)r^k = a + (a + d)r + (a + 2d)r^2 + \ldots + = \frac{a}{1 - r} + \frac{rd}{(1 - r)^2}$$

$$\sum_{k=1}^{n} k^2 = 1 + 4 + 9 + \ldots + n^2 = \frac{n(n + 1)(2n + 1)}{6}$$

$$\sum_{k=1}^{n} k^3 = 1 + 8 + 27 + \ldots + n^3 = \frac{n^2(n + 1)^2}{4} = \left[\frac{n(n + 1)^2}{2}\right] = \left[\sum_{k=1}^{n} k\right]^2$$

$$\sum_{x=1}^{\infty} \frac{1}{x} = 1 + \frac{1}{2} + \frac{1}{3} + \ldots \text{(does not converge)}$$

$$\sum_{m=0}^{k} ma^m = \frac{a}{(1 - a)^2}[1 - (k + 1)a^k + ka^{k+1}] = \sum_{m=1}^{k} ma^m$$

$$\sum_{k=0}^{n} (1) = n$$

$$\sum_{k=0}^{n} \binom{n}{k} = 2^n$$

$$(a + b)^n = \sum_{k=0}^{n} \binom{n}{k} a^k b^{n-k}$$

$$\prod_{n=1}^{\infty} a_n = e^{\left(\sum_{n=1}^{\infty} \ln a_n \right)}$$

$$\ln \left(\prod_{n=1}^{\infty} a_n \right) = \sum_{n=1}^{\infty} \ln a_n$$

$$\ln(x) = \sum_{k=1}^{\infty} \frac{1}{k} \left(\frac{x-1}{x} \right)^k, \qquad x \geq \frac{1}{2}$$

$$\lim_{h \to \infty} (1 + h)^{1/h} = e$$

$$\lim_{n \to \infty} \left(1 - \frac{x}{n} \right)^n = e^{-x}$$

$$\lim_{n \to \infty} \sum_{k=0}^{n} \frac{e^{-n} n^r}{K!} = \frac{1}{2}$$

$$\lim_{k \to \infty} \left(\frac{x^k}{k!} \right) = 0$$

$$|x + y| \leq |x| + |y|$$

$$|x - y| \geq |x| - |y|$$

$$\ln(1 + x) = \sum_{k=1}^{\infty} (-1)^{k+1} \left(\frac{x^k}{k} \right), \qquad \text{if } -1 < x \leq 1$$

$$\Gamma \left(\frac{1}{2} \right) = \sqrt{\pi}$$

$$\Gamma(\alpha + 1) = \alpha \Gamma(\alpha)$$

$$\Gamma \left(\frac{n}{2} \right) = \frac{\sqrt{\pi}(n-1)!}{2^{n-1} \left(\frac{n-1}{2} \right)!}, \qquad n \text{ odd}$$

$$\int_0^{\infty} e^{-x} x^{n-1} \, dx = \Gamma(n)$$

$$\binom{n}{2} = \frac{1}{2}(n^2 - n) = \sum_{k=1}^{n-1} K$$

$$\binom{n+1}{2} = \binom{n}{2} + n$$

$$2.4.6.8 \ldots 2n = \prod_{k=1}^{n} 2k = 2^n n!$$

$$1.3.5.7 \ldots (2n - 1) = \frac{(2n - 1)!}{2^{2n-2}(2n - 2)!} = \frac{2n - 1}{2^{2n-2}}$$

Derivation of closed form expression for $\sum_{k=1}^{n} kx^k$:

$$\sum_{k=1}^{n} kx^k = x \sum_{k=1}^{n} kx^{k-1}$$

$$= x \sum_{k=1}^{n} \frac{d}{dx}[x^k]$$

$$= x \frac{d}{dx}\left[\sum_{k=1}^{n} x^k\right]$$

$$= x \frac{d}{dx}\left[\frac{x(1-x^n)}{1-x}\right]$$

$$= x \left[\frac{(1-(n+1)x^n)(1-x)-x(1-x^n)(-1)}{(1-x)^2}\right]$$

$$= \frac{x[1-(n+1)x^n+nx^{n+1}]}{(1-x)^2}, \qquad x \neq 1$$

E

Project Management Software

Software	Vendor	Platforms	Target
ABT Project Workbench Standard	Applied Business Technologies, Inc. New York, NY	IBM PC/XT/AT	Small- to medium-size projects
Action-Network w/Project Query Language	Information Research Corp. Charlottesville, VA	IBM PC/XT/AT	Project tracking, budget mgmt., multiproject control
Action-Tracker	Information Research Corp. Charlottesville, VA	IBM PC/XT/AT	Project tracking, budget mgmt.
AcuVision	Systonetics Fullerton, CA	IBM MVS,VM	Evaluating, scheduling, tracking
Advanced Pro-PATH 6	SoftCorp, Inc. Clearwater, FL	IBM PC/XT/AT, PS/2	Uses CPM for planning; cost/resource mgmt.
AMS Time Machine	Diversified Information Services, Inc. Studio City, CA	IBM PC/XT/AT, HP 150	Scheduling, resource management, CPM
APECS/800	Automatic Data Processing Ann Arbor, MI	DEC VAX, Micro VAX	Project estimating and control

Software	Vendor	Platforms	Target
Artemis	Metier Management Systems Houston, TX	IBM MVS,VMHP	CPM project mgmt. data base
Artemis Project	Metier Management Systems Houston, TX	IBM PC/XT/AT,PS/2	What-if analysis, resource leveling
CA-Estimacs	Computer Associates Garden City, NY	IBM PC/MS-DOS	Estimates for mainframe application dev.
CA-Planmacs	Computer Associates Garden City, NY	IBM PC/XT/AT	Planning tool, resource mgmt.
CA-SuperProject Expert	Computer Associates Garden City, NY	IBM PC/XT/AT,PS/2	PERT, Gantt charts, CPM, probability analysis
CA-Tellapla	Computer Associates Garden City, NY	IBM MVS, VM, DEC VAX/VMS, Apollo, Sun	CPM, Gantts, planning analysis
Capital Project Management System	Data Design Associates Sunnyvale, CA	IBM OS/MVS, DOS/VSE	Tracks large expenditure projects against budget
Easytrak	Cullinet Software Westwood, MA	IBM mainframe	Cost and resource mgmt. system
Estiplan	AGS Mgmt. Systems King of Prussia, PA	IBM PC/XT/AT	Software project mgmt. and control
Harvard Project Manager	Software Publishing Corp. Mountain View, CA	IBM PC/XT/AT, PS/2	WBS, project control/mgmt.; PERT/Gantt charts
InstaPlan	InstaPlan Corp. Mill Valley, CA	IBM PC/XT/AT, PS/2	Top-down, outline-oriented
MAPPS	Mitchell Management Systems Westborough, MA	IBM PC, VM, HP150, Wang PC, DEC VAX, DG	Time, cost, resource planning and control
MI-Projec	Matterhorn, Inc. Minneapolis, MN	IBM DOS/VSE,OS/MVS, VS1	Project planning control

Software	Vendor	Platforms	Target
MicroMan II	Poc-It Management Systems Santa Monica, CA	IBM PC/XT/AT, PS/2	Tracks, manages IS activity
Micro Planner	Micro Planning Software San Francisco, CA	IBM PC/XT/AT	Planning, resource management
Micro Planner Plus	Micro Planning Software San Francisco, CA	Apple Macintosh	Planning, resource management
Micro Trak	SofTrak Systems Salt Lake City, UT	IBM PC/XT/AT, PS/2, Unix, xenix,VMS	Project scheduling, control resource management using CPM
Microsoft Project	Microsoft Corp. Redmond, WA	IBM PC/XT/AT, MS-Windows	Resource planning, cost analysis, Gantt charts
Multitrak	Multisystems Cambridge, MA	IBM MVS, DOS, CICS, IBM PC	Resource allocation
N1100	Nichols & Comp. Culver City, CA	IBM PC, HP150, Wang PC	Critical path system; project mgmt.; control
N5500	Nichols & Comp. Culver City, CA	IBM OS/MVS, DOS, Unisys, DEC, Wang, DG, HP, Honeywell, Prime	Critical path system; network up to 3,000 activities
OpenPlan	Welcom Software Technology Houston, TX	IBM PC/XT/AT, PS/2	Critical path analysis, resource scheduling
Opera	Welcom Software Technology Houston, TX	IBM PC/XT/AT, PS/2	Risk analysis extension
PAC II	AGS Mgmt. Systems King of Prussia, PA	IBM DOS/MVS, DOS/VSE, VS/1, DEC VAX/VMS, Wang VS	Resource allocation, cost analysis, reporting
PAC III	AGS Mgmt. Systems King of Prussia, PA	IBM OS/MVS, VM, DOS/VSE; S/38; DEC VAX/VMS	Integrated networking, scheduling

Software	Vendor	Platforms	Target
PAC Micro	AGS Mgmt. Systems King of Prussia, PA	IBM PC/XT/AT	Multiple-project scheduling, CPM
Parade	Primavera Systems Bala Cynwyd, PA	IBM PC/XT/AT, PS/2	Performance measurement, earned value analysis
Pertmaster Advanced	Pertmaster International Santa Clara, CA	IBM PC/XT/AT, PS/2	Plan, schedule, manage projects with interrelated tasks
PEVA	Engineering Mgmt. Consultants Troy, MI	IBM PC/XT/AT	Project earned value
PICOM	K & H Professional Management services Wayne, PA	IBM OS,VM	Project cost information management
Planner	Productivity Solutions Waltham, MA	DEC VAX/VMS	CPM, resource
Plantrac	ComputerLine, Inc. Pembroke, MA	IBM PC/XT/AT	Planning, tracking tool
PlotTrak	SofTrak Systems Salt Lake City, UT	IBM PC/XT/AT, PS/2,Unix,	For microtracking, graphics, Gantt charts and network diagrams
PMS-11	North America Mica San Diego, CA	IBM PC/XT/AT	CPM, PERT
PMS-80 Advanced	Pinnell Engineering Portland, OR	IBM PC/XT/AT	CPM, PERT, resource, cost
PREMIS	K & H Professional Management services Wayne, PA	IBM OS, VM	Time analysis, resource mgmt.
Pride PMS	M. Bryce & Associates Palm Harbor, FL	DEC VAX/VMS, IBM OS/MVS	Project mgmt. system for CASE
Primavera Project Planner	Primavera Systems Bala Cynwyd, PA	IBM PC/XT/AT, PS/2	Cost analysis and control, activity coding
Pro Tracs	Applied Microsystems Roswell, CA	IBM PC/XT/AT	Action/item tracking

Software	Vendor	Platforms	Target
ProjectAlert	CRI, Inc. Santa Clara, CA	DEC VAX, HP 3000, Apollo	PERT and Gantt, scheduling monitoring
PROJECT-MAN AGER	Manager Software Products Lexington, MA	IBM MVS, VS1, DOS/VSE	Resource management, budget control
Project OUTLOOK	Strategic Software Planning Corp. Cambridge, MA	IBM PC/XT/AT, PS/2, MS-Windows	Interactive, builds networks on screen with mouse
Project Scheduler	Scitor Corp. Foster City, CA	IBM PC/XT/AT, PS/2, Wang PC, HP150	Handles multiple projects
Project/2	Project Software & Development, Inc. Cambridge, MA	IBM MVS,VM,DEC VAX/VMS	For complex projects; schedule and cost mgmt.
Project Workbench Advanced	Applied Business Technologies, Inc. New York, NY	IBM PC/XT/AT, Wang PC	Complex projects; Gantt/resource screens
Prothos	New Technology Association Evansville, IN	IBM PC/XT/AT, DEC VAX, Unix Sys V	planning, resource, PERT
Quick-Plan II	Mitchell Management Systems Westborough, MA	IBM PC/XT/AT	Planning, resource scheduling
Quick Schedule	Power Up San Mateo, CA	IBM PC	Scheduling
Qwiknet Profesional	Project Software & Development, Inc. Cambridge, MA	IBM PC/XT/AT, PS/2, DEC VAX/VMS	CPM, multiproject scheduling
SAS System	SAS Institute Cary, NC	IBM PC/XT/AT, PS/2, OS, DOS/VSE, DEC VAX/VMS, Prime	Scheduling, control, and analysis
Scheduling & Control	Softext Publishing Corp. New York, NY	IBM PC/AT	Uses CPM
Skyline	Applitech Software Cambridge, MA	IBM PC/XT/AT, PS/2	Project outliner, uses CPM

Software	Vendor	Platforms	Target
SSP's PROMIS	Strategic Software Planning Corp. Cambridge, MA	IBM PC/XT/AT, PS/2	Subnetworking for large projects
Synergy	Bechtel Software Acton, MA	IBM PC/XT, DEC VAX/VMS	Budgeting, control, performance mgmt.
Task Monitor	Monitor Software Los Altos, CA	IBM PC	CPM, PERT Gantt chart
Time Line	Symantec Corp. Novato, CA	IBM PC/XT/AT, PS/2	Project planning, resource allocation, cost tracking
Topdown Project Planner	Ajida Technologies Santa Rosa, CA	IBM PC/XT/AT PS/2	Top-down approach, WBS
TRAK	The Bridge, Inc. Millbrae, CA	IBM OS/MVS,VS/1, DOS/VSE CICS,TSO	Calculates, reports time totals, cost
TRAKKER	Dekker, Ltd. San Bernardino, CA	DEC/VAX, MS DOS, LANS, UNIX	Handles earned value, scheduling, cost, resources
VAX Software Project Manager	Digital Equipment Corp. Maynard, MA	DEC VAX/VMS	Software development, project mgmt.
ViewPoint	Computer Aided Management, Inc. Petaluma, CA	IBM PC/XT/AT, PS/2	Top-down planning, scheduling
VISION	Systonetics Fullerton, CA	DEC VAX/VMS, Prime	Scheduling, resource allocation, performance measurement
VISIONmicro	Systonetics Fullerton, CA	IBM PC/XT/AT	Mouse-driven interface; project mgmt. and graphics
Vue	National Info Systems Cupertino, CA	IBM PC/XT/AT, DEC VAX/VMS, Unix, HP 3000, Honeywell	Scheduling—uses CPM
Who/What/When	Chronos Software San Francisco, CA	IBM PC/XT/AT	Personal project, time manager

Bibliography

ABERNATHY, W. J., and K. WAYNE. 1974. "Limits of the Learning Curve." *Harvard Business Review* 52: 109–119.

ADRIAN, J. J. 1973. *Quantitative Methods in Construction Management.* New York: Elsevier Publishing Co.

AFIESIMAMA, B. T. 1987. "Aggregate Manpower Requirements for Strategic Project Planning." *Computers and Industrial Engineering* 12, no. 4: 249–62.

AKAO, Y., and T. ASAKA, eds. 1990. *Quality Function Deployment.* Cambridge, MA: Productivity, Inc.

ALCHIAN, A. 1963. "Reliability of Progress Curves in Airframe Production." *Econometrica* 31, no. 4: 679–93.

ALLMENDINGER, G. 1985. "Management Goals for Manufacturing Technology." *Manufacturing Engineering* (November): 83–84.

AMRINE, H. T., J. A. RITCHEY, and C. L. MOODIE. 1987. *Manufacturing Organization and Management.* 5th ed. Englewood Cliffs, NJ: Prentice Hall.

ANTHONISSE, J. M., K. M. VAN HEE, and J. K. LENSTRA. 1988. "Resource-Constrained Project Scheduling: An International Exercise in DSS Development." *Decision Support Systems* 4, no. 2: 249–57.

AQUILANO, N. J., and D. E. SMITH. 1980. "A Formal Set of Algorithms for Project Scheduling with Critical Path Method—Material Requirements Planning." *Journal of Operations Management* 1, no. 2: 57–67.

ARCHIBALD, R. D. 1976. *Managing High-Technology Programs and Projects.* New York: John Wiley.

ARINZE, B. and F. Y. PARTOVI. 1992. "A Knowledge-Based Decision Support System for Project Management." *Computers & Operations Research* 19, no. 5: 321–34.

ASHER, H. 1956. "Cost-Quantity Relationships in the Airframe Industry." *Report No. R-291* (July 1). Santa Monica, CA: The Rand Corporation.

ASSAD, A. A., and E. A. WASIL. 1986. "Project Management Using A Microcomputer." *Computers and Operations Research* 13, no. 2/3: 231–60.

ASSAD, M. G., and G. P. J. PELSER. 1983. "Project Management: A Goal-Directed Approach." *Project Management Quarterly* (June): 49–58.

BADIRU, A. B. 1987. "Communication, Cooperation, Coordination: The Triple C of Project Management." In *Proceedings of 1987 IIE Spring Conference*, Norcross, GA: Institute of Industrial Engineers. 401–04.

———. 1988a. *Project Management in Manufacturing and High Technology Operations*. New York: John Wiley.

———. 1988b. "Cost-Integrated Network Planning Using Expert Systems." *Project Management Journal* 19, no. 2 (April): 59–62.

———. 1988c. "Graphic Evaluation of Amortization Schedules." *Industrial Engineering* 20, no. 9 (September): 18–22.

———. 1988d. "Towards the Standardization of Performance Measures for Project Scheduling Heuristics." *IEEE Transactions on Engineering Management* 35, no. 2 (May): 82–89.

———. 1990. "A Management Guide to Automation Cost Justification." *Industrial Engineering* 22, no. 2 (February): 26–30.

———. 1991a. *Project Management Tools for Engineering and Management Professionals*. Norcross, GA: Industrial Engineering & Management Press.

———. 1991b. "A Simulation Approach to PERT Network Analysis." *Simulation* 57, no. 4 (October): 245–55.

———. 1991c. "Manufacturing Cost Estimation: A Multivariate Learning Curve Approach." *Journal of Manufacturing Systems* 10, no. 6: 431–41.

———. 1991d. "STARC 2.0: An Improved PERT Network Simulation Tool." *Computers and Industrial Engineering* 20, no. 3: 389–400.

———. 1992a. *Expert Systems Applications in Engineering and Manufacturing*. Englewood Cliffs, NJ: Prentice Hall.

———. 1992b. "Computational Survey of Univariate and Multivariate Learning Curve Models." *IEEE Transactions on Engineering Management* 39, no. 2 (May): 176–88.

———. 1993. *Quantitative Models for Project Planning, Scheduling, and Control*. Westport, CT: Quorum Books.

———. 1994. "Multi-Factor Learning and Forgetting Models for Productivity and Performance Analysis." *International Journal of Human Factors in Manufacturing*. 4, no. 1, 37–54.

———, B. L. FOOTE, L. LEEMIS, A. RAVINDRAN, and L. WILLIAMS. 1993. "Recovering from a Crisis at Tinker Air Force Base." *PM Network* 7, no. 2 (February): 10–23.

———, B. FOOTE, and J. W. PETERS. 1993. "A Case Study of Economic Justification of Expert Systems in Maintenance Operations." *Engineering Economist* 38, no. 2: 99–117.

———, P. SIMIN PULAT, and M. KANG. 1993. "DDM: Decision Support System for Hierarchical Dynamic Decision Making." *Decision Support Systems* 10: 1–18.

———, and D. RUSSELL. 1987. "Minimum Annual Project Revenue Requirement Analysis." *Computers and Industrial Engineering* 13, nos. 1–4, 366–70.

———, and V. E. THEODORACATOS. 1994. "Analytical and Integrative Expert System Model for Design Project Management." *Journal of Design and Manufacturing*. 4, no. 2 (in press).

————, and G. E. WHITEHOUSE. 1989. *Computer Tools, Models, and Techniques for Project Management*. Blue Ridge Summit, PA: TAB Books.

BAHOUTH, S. B. 1981. *Project Management: Why, Who, and How*. Unpublished Masters thesis, Vanderbilt University.

————. 1987. "Case Study of Multinational Project Organization." Project Management Seminar, School of Industrial Engineering, University of Oklahoma.

BAKER, B. 1990. "Real-time PC-based project tracking system." Class report, Engineering Project Management, School of Industrial Engineering, University of Oklahoma.

BAKER, N. R., and W. H. POUND. 1964. "Project Selection: Where We Stand." *IEEE Transactions on Engineering Management* EM-11: 124–34.

BALLOU, P. O. 1985. "Decision-Making Environment of a Program Office." *Program Manager: The Journal of the Defense Systems Management College* (September–October): 36–39.

BALOFF, N. 1971. "Extension of the Learning Curve: Some Empirical Results." *Operations Research Quarterly* 22, no. 4: 329–340.

BANERJEE, B. P. 1965. "Single Facility Sequencing with Random Execution Times." *Operations Research* 13 (May-June): 358–64.

BARAN, R. H. 1988. "A Case Study in the Allocation of Research Funds." *Systems and Decisions* 1, no. 1: 1–11.

BARBER, T. J., J. T. BOARDMAN, and N. BROWN. 1990. "Practical Evaluation of an Intelligent Knowledge-Based Project Control System." *IEE Proceedings* 137, part A, no. 1 (January): 35–51.

BATSON, R. G. 1987. "Critical Path Acceleration and Simulation in Aircraft Technology Planning." *IEEE Transactions of Engineering Management* EM34, no. 4: 244–51.

BAUER, C. S., and A. B. BADIRU. 1985. "Stochastic Rate of Return Analysis on a Microcomputer." *Softcover Software*. Norcross, GA: Industrial Engineering & Management Press. 127–35.

BAZARAA, M. S., J. J. JARVIS, and H. D. SHERALI. 1990. *Linear Programming and Network Flows* (2nd ed.). New York: John Wiley.

BAZARAA, M. S., and J. J. JARVIS. 1977. *Linear Programming and Network Flows*. New York: John Wiley.

BEDWORTH, D. D. 1973. *Industrial Systems: Planning, Analysis and Control*. New York: Ronald Press.

————, and J. E. BAILEY. 1982. *Integrated Production Control Systems: Management, Analysis, Design*. New York: John Wiley.

BEIMBORN, E. A., and W. A. GARVEY. 1972. "The Blob Chart." *Industrial Engineering* (December): 4, no. 12: 17–19.

BELKAOUI, A. 1986. *The Learning Curve*. Westport, CT: Quorum Books.

BELL, C. E. 1989. "Maintaining Project Networks in Automated Artificial Intelligence Planning." *Management Science* 35, no. 10: 1192–1214.

BELLMORE, M., and G. L. NEMHAUSER. 1968. "The Travelling Salesman Problem: A Survey." *Operations Research* 16: 538–58.

BEMIS, J. C. 1981. "A Model for Examining the Cost Implications of Production Rate." *Concepts: The Journal of Defense Systems Acquisition Management* 4, no. 2: 84–94.

BENNIS, W., and B. NANUS. 1985. *Leaders: The Strategies For Taking Charge*. New York: Harper & Row.

BENT, J. A., and A. THUMANN. 1989. *Project Management for Engineering and Construction*. Lilburn, GA: Fairmont Press.

BERGEN, S. A. 1990. *RandD Management: Managing Projects and New Products*. Cambridge, MA: Basil Blackwell.

BERMAN, E. B. 1964. "Resource Allocation in a PERT Network Under Continuous Activity Time-Cost Functions." *Management Science* 10, no. 4 (July): 734–45.

BERSOFF, E. H. 1984. "Elements of Software Configuration Management." *IEEE Transactions on Software Engineering* SE-10, no. 1 (January): 79–87.

BEY, R. B., R. H. DOERSCH, and J. H. PATTERSON. 1981. "The Net Present Value Criterion: Its Impact on Project Scheduling." *Project Management Quarterly* 12, no. 2 (June): 223–33.

BIDANDA, B. 1989. "Techniques to Assess Project Feasibility." *Project Management Journal* 20, no. 2 (June): 5–10.

BLANCHARD, F. L. 1990. *Engineering Project Management*. New York: Marcel Dekker.

BOCTOR, F. F. 1990. "Some Efficient Multi-Heuristic Procedures for Resource-Constrained Project Scheduling." *European Journal of Operational Research* 49: 3–13.

BOROVITS, I. 1984. *Management of Computer Operations*. Englewood Cliffs, NJ: Prentice Hall.

BOULDING, K. E. 1956. "General Systems Theory: The Skeleton of Science." *Management Science* 2, no. 3 (April): 197–208.

BOX, G. E. P., and G. M. JENKINS. 1976. *Time Series Analysis: Forecasting and Control*. San Francisco: Holden-Day.

BOZER, Y., SCHORN, E. C., and SHARP, G. P. 1990. "Geometric Approaches to Solve the Chebyshev Traveling Salesman Problem." *IIE Transactions* 22, no. 3: 238–54.

BRAND, J. D., W. L. MEYER, and L. R. SHAFFER. 1964. "The Resource Scheduling Method for Construction." *Civil Engineering Studies Report*, no. 5, University of Illinois.

BRIGHT, D. S. 1985. *Gearing Up for the Fast Lane: New Tools For Management in a High-Tech World*. New York: Random House.

BROADWELL, M. M., and R. S. HOUSE. 1986. *Supervising Technical & Professional People*. New York: John Wiley.

BROOKS, G. H., and C. R. WHITE. 1965. "An Algorithm for Finding Optimal or Near Optimal Solutions to the Production Scheduling Problem." *Journal of Industrial Engineering* (January-February): 34–40.

BROWN, D. E., and A. R. SPILLANE. 1989. "A Knowledge-Based Design Aid for Superheaters Employing Pseudo-Random Search." *Journal of Operational Research Society* 40, no. 6: 539–50.

BROWNE, J., J. HARHEN, and J. SHIVNAN. 1988. *Production Management Systems: A CIM Perspective*. Reading, MA: Addison-Wesley.

BUFFA, E. S., and J. S. DYER. 1977. *Management Science and Operations Research: Model Formulation and Solution Methods*. New York: John Wiley.

BURGESS, A. R., and J. B. KILLEBREW. 1962. "Variation in Activity Level on a Cyclic Arrow Diagram." *Journal of Industrial Engineering* 13, no. 2 (March-April).

BUSSEY, L. E., and T. G. ESCHENBACH. 1992. *The Economic Analysis of Industrial Projects*. 2d ed. Englewood Cliffs, NJ: Prentice Hall.

CAMM, J. D., J. R. EVANS, and N. K. WOMER. 1987. "The Unit Learning Curve Approximation of Total Cost." *Computers and Industrial Engineering* 12, no. 3: 205–13.

CANADA, J. R., and W. G. SULLIVAN. 1989. *Economic and Multiattribute Evaluation of Advanced Manufacturing Systems*. Englewood Cliffs, NJ: Prentice Hall.

CARLSON, E. W., and W. G. CARLSON. 1987. "A Case Study in the Application of Microcomputer Technology in the Construction Industry." *Computers and Industrial Engineering* 13, nos. 1–4: 156–159.

CARR, C. 1988. "Using Expert Job Aids: A Primer." *Educational Technology* 29, no. 6: 18–22.

CARR, G. W. 1946. "Peacetime Cost Estimating Requires New Learning Curves." *Aviation* 45 (April): 76–77.

CARRAWAY, R. L., and R. L. SCHMIDT. 1991. "An Improved Discrete Dynamic Programming Algorithm for Allocating Resources among Interdependent Projects." *Management Science* 37, no. 9: 1195–1200.

CARROLL, J. M. 1987. *Simulation Using Personal Computers.* Englewood Cliffs, NJ: Prentice Hall.

CHAE, K. C., and S. KIM. 1990. "Estimating the Mean and Variance of PERT Activity Time Using Likelihood-Ratio of the Mode and the Midpoint." *IIE Transactions* 22, no. 3 (September): 198–202.

CHAPMAN, C. B. 1970. "The Optimal Allocation of Resources to a Variable Timetable." *Operational Research Quarterly* 21, no. 1 (March): 81–90.

———, D. F. COOPER, and M. J. PAGE. 1987. *Management for Engineers.* New York: John Wiley.

CHARNES, A., W. W. COOPER, and G. L. THOMPSON. 1964. "Critical Path Analyses Via Chance Constrained and Stochastic Programming." *Operations Research* 12 (May-June): 460–70.

CHASE, R. B., and N. J. AQUILANO. 1977. *Production Operations Management*, rev. ed. New York: Richard D. Irwin.

CHIANG, M. K., and S. A. ZENIOS. 1989. "On the Use of Expert Systems in Network Optimization: With an Application to Matrix Balancing." *Annals of Operations Research* 20: 111–40.

CHRISTENSEN, D. S. 1993. "The Estimate at Completion Problem: A Review of Three Studies." *Project Management Journal* 24, no. 1 (March): 37–42.

CHRISTIAN, P. H. 1993. "Project Success or Project Failure: It's Up to You." *Industrial Management* (March/April) 36, no. 2 : 8–9.

CLARK, C. E. 1961. "The Optimum Allocation of Resources Among the Activities of a Network." *Journal of Industrial Engineering* 12 (January-February): 11–17.

CLELAND, D. I. 1989. "Strategic Issues in Project Management." *Project Management Journal* 20, no. 1 (March): 31–39.

———. 1990. *Project Management: Strategic Design and Implementation.* New York: TAB Professional & Reference Books.

———, and K. M. BURSIC. 1992. *Strategic Technology Management.* New York: American Management Association.

———, and W. R. KING, eds. 1983a. *Project Management Handbook.* New York: Van Nostrand Reinhold.

———, and W. R. KING. 1983b. *Systems Analysis and Project Management* 3d ed. New York: McGraw-Hill.

———, and D. F. KOCAOGLU. 1981. *Engineering Management.* New York: McGraw-Hill.

CLEWS, G., and R. LEONARD. 1985. *Technology and Production.* Oxford, England: Philip Allan Publishers Limited.

COHEN, W. A. 1986. *High-Tech Management.* New York: John Wiley.

COLLIER, M., ed. 1986. *Microcomputer Software for Information Management: Case Studies.* Brookfield, VT: Gover Publishing.

CONLEY, P. 1970. "Experience Curves as a Planning Tool." *IEEE Spectrum* 7, no. 6, 63–68.

COOPER, D. F. 1976. "Heuristics for Scheduling Resource-Constrained Projects: An Experimental Investigation." *Management Science* 22 (July): 1186–94.

COOPER, D. and C. CHAPMAN. 1987. *Risk Analysis for Large Projects: Models, Methods, and Cases*. New York: John Wiley.

COOPER, J. D., and M. J. FISHER. 1979. *Software Quality Management*. New York: Petrocelli Books.

COOPER, K. G. 1993a. "The Rework Cycle: Benchmarks for the Project Manager." *Project Management Journal* 24, no. 1 (March): 17–22.

———. 1993b. "The Rework Cycle: How it Really Works and Reworks." *PM Network* 7, no. 2 (February): 25–28.

———. 1993c. "The Rework Cycle: Why Projects are Mismanaged." *PM Network* 7, no. 2 (February): 5–7.

COOPER, L. and D. STEINBERG. 1974. *Methods and Applications of Linear Programming*. Philadelphia: W. B. Saunders.

COX, L. W., and J. S. GANSLER. 1981. "Evaluating the Impact of Quantity, Rate, and Competition." *Concepts: The Journal of Defense Systems Acquisition Management* 4, no. 4: 29–53.

COYLE, R. G. 1984. "A Systems Approach to the Management of a Hospital for Short-Term Patients." *Socio-Econ Planning Science* 18, no. 4: 219–26.

CRANDALL, K. C. 1973. "Project Planning with Precedence Lead/Lag Factors." *Project Management Quarterly* 4, no. 3 (September): 18–27.

CROSBY, P. B. 1979. *Quality Is Free*. New York: McGraw-Hill.

CUNNINGHAM, M. 1984. *Powerplay: What Really Happened at Bendix*. New York: Simon & Schuster.

DALLENBACK, H., and J. GEORGE. 1978. *Introduction to Operations Research Techniques*. Boston: Allyn & Bacon.

DALY, E. B. 1979. "Organizing for Successful Software Development." *Datamation* 25, no. 10 (December): 106–20.

DANTZIG, G. B. 1963. *Linear Programming and Extensions*. Princeton, NJ: Princeton University Press.

DAR-EL, E. M., and Y. TUR. 1977. "A Multi-Resource Project Scheduling Algorithm." *AIIE Transactions* 9 (March): 44–52.

DAVIES, C., A. DEMB, and R. ESPEJO. 1979. *Organization for Project Management*. New York: John Wiley.

DAVIES, E. M. 1974. "An Experimental Investigation of Resource Allocation in Multiactivity Projects." *Operational Research Quarterly* 24, no. 4: 587–91.

———. 1966. "Resource Allocation in Project Network Models: A Survey." *Journal of Industrial Engineering* (April): 177–88.

———. 1975. "Project Network Summary Measures and Constrained-Resource Scheduling." *AIIE Transactions* 7, no. 2 (June): 132–42.

———, and G. E. HEIDORN. 1971. "An Algorithm for Optimal Project Scheduling under Multiple Resource Constraints." *Management Science* 17, no. 8: B803–B814.

DAVIS, E. W., ed. 1983. *Project Management: Techniques, Applications, and Managerial Issues*. 2d ed. Norcross, GA: Industrial Engineering & Management Press.

———, and J. W. PATTERSON. 1975. "A Comparison of Heuristic and Optimum Solutions in Resource-Constrained Project Scheduling." *Management Science* 21 (April): 944–55.

DAY-COPELAND, L. 1988. "Project Management Software Market Shows Indications of Growth." *PC Week* (August 22): 5, no. 32: 102.

DECKRO, R. F., J. E. HEBERT, and W. A. VERDINI. 1992. "Project Scheduling with Work Packages." *Omega* 20, no. 2: 169–82.

DEGARMO, E. P., W. G. SULLIVAN, and J. A. BONTADELLI. 1988. *Engineering Economy.* 8th ed. New York: Macmillan.

DEJONG, J. R. 1957. "The Effects of Increasing Skill on Cycle Time and Its Consequences for Time Standards." *Ergonomics* 1, no. 1 (November): 51–60.

DEMING, W. E. 1986. *Out of the Crisis.* Cambridge, MA: MIT Center for Advanced Engineering Study.

DERAKHSHANI, S. 1983. "Factors Affecting Success in International Transfers of Technology: A Synthesis, and a Test of a New Contingency Model." *Developing Economies* 21: 27–45.

DHILLON, B. S. 1987. *Engineering Management: Concepts, Procedures and Models.* Lancaster, PA: Technomic Publishing.

DINSMORE, P. C. 1990. *Human Factors In Project Management,* rev. ed. New York: American Management Association.

DOD. 1967. *Performance Measurement for Selected Acquisitions.* DOD Instruction no. 7000.2. Washington, DC: U.S. Department of Defense. Govt. Printing Office.

DOE. 1981. *Cost and Schedule Control Systems: Criteria for Contract Performance Measurement: Work Breakdown Structure Guide.* Washington, DC: U.S. Department of Energy, Office of Project and Facilities Management.

DOERSCH, R. H., and J. H. PATTERSON. 1977. "Scheduling a Project to Maximize Its Present Value: A Zero-One Programming Approach." *Management Science* 23, no. 8: 882–89.

DONALDSON, W. A. 1965. "The Estimation of the Mean and Variance of A 'PERT' Activity Time." *Operations Research* 13: 383–87.

DOUGHERTY, E. R. 1990. *Probability & Statistics for the Engineering, Computing, and Physical Sciences.* Englewood Cliffs, NJ: Prentice Hall.

DREXL, A. 1991. "Scheduling of Project Networks by Job Assignment." *Management Science* 37, no. 12: 1590–1602.

DREYFUS, S. E. 1969. "An Appraisal of Some Shortest Path Algorithms." *Operations Research* 17: 395–412.

DRIGANI, F. 1989. *Computerized Project Control.* New York: Marcel Dekker.

DRUCKER, P. F. 1985. *Innovation and Entrepreneurship.* New York: Harper & Row.

DUMBLETON, J. H. 1986. *Management of High-Technology Research and Development.* New York: Elsevier.

DUNCAN, A. J. 1965. *Quality Control and Industrial Statistics.* 3d ed. Homewood, IL: Richard D. Irwin.

EAST, E. W., and J. G. KIRBY. 1989. *A Guide to Computerized Project Scheduling.* New York: Van Nostrand Reinhold.

ECK, R. 1976. *Operations Research for Business.* Belmont CA: Wadsworth.

EGLESE, R. W. 1990. "Simulated Annealing: A Tool For Operational Research." *European Journal of Operational Research* 46: 271–81.

EIN-DOR, P., and C. R. JONES. 1985. *Information Systems Management: Analytical Tools and Techniques.* New York: Elsevier Science Publishing.

ELMAGHRABY, S. E. 1977. *Activity Networks: Project Planning and Control by Network Models.* New York: John Wiley.

————. 1990. "Project Bidding Under Deterministic and Probabilistic Activity Durations." *European Journal of Operational Research* 49: 14–34.

————, and W. S. HERROELEN. 1980. "On The Measurement of Complexity in Activity Networks." *European Journal of Operations Research* 5: 223–34.

————. 1990. "The Scheduling of Activities to Maximize the Net Present Value of Projects." *European Journal of Operational Research* 49: 35–49.

ELMAGHRABY, S. E., and P. S. PULAT. 1979. "Optimal Project Compression with Due-Dated Events." *Naval Research Logistics Quarterly* 26, no. 2: 331–48.

ELSAYED, E. A. 1982. "Algorithms for Project Scheduling with Resource Constraints." *International Journal of Production Research* 20, no. 1: 95–103.

FABRYCKY, W. J., and B. S. BLANCHARD. 1991. *Life-Cycle Cost and Economic Analysis.* Englewood Cliffs, NJ: Prentice Hall.

FARID, F., and R. KANGARI. 1991. "A Knowledge-Based System for Selecting Project Management Microsoftware Packages." *Project Management Journal* 22, no. 3: 55–61.

FARNUM, N. R., and L. W. STANTON. 1987. "Some Results Concerning the Estimation of Beta Distribution Parameters in PERT." *Journal of Operational Research Society* 38: 287–90.

FATHI, E. T., and C. V. W. ARMSTRONG. 1985. *Microprocessor Software Project Management.* New York: Marcel Dekkar.

FENDLEY, L. G. 1968. "Toward the Development of a Complete Multiproject Scheduling System." *Journal of Industrial Engineering* 19 (October): 505–15.

FILLEY, R. D. 1986. "1986 Project Management Software Buyer's Guide." *Industrial Engineering* 18, no.1 (January): 51–63.

FISHER, M. L. 1980. "Worst-Case Analysis of Heuristic Algorithms." *Management Science* 26, no. 1: 1–17.

FLEISCHER, G. A. 1984. *Engineering Economy: Capital Allocation Theory.* Monterey, CA: Brooks/Cole Engineering Division.

FLYNN, R. R. 1987. *An Introduction to Information Science.* New York: Marcel Dekker.

FOOTE, B., A. RAVINDRAN, A. B. BADIRU, L. LEEMIS, and L. WILLIAMS. 1988. "Simulation and Network Analysis Pay Off in Conveyor System Design." *Industrial Engineering* 20, no. 6 (June): 48–53.

FORMAN, E. H., T. L. SAATY, M. A. SELLY, and R. WALDOM. 1983. *Expert Choice.* McLean, VA: Decision Support Software, Inc.

FRAME, J. D. 1987. *Managing Projects in Organizations.* San Francisco: Jossey-Bass.

GAARSLEV, A. 1992. "Application of Artificial Intelligence in Project Management." *Proceedings of 1992 Project Management Institute Annual Seminar/Symposium*, Pittsburgh, PA: Webster, NC: Project Management Institute. 38–43.

GALBRAITH J. R. 1971. "Matrix Organization Design." *Business Horizons* 14, no. 2 (February): 29–40.

GALLAGHER, C. 1987. "A Note on PERT Assumptions." *Management Science* 33, no. 10: 1360.

GALLIMORE, J. M. 1983. "Planning To Automate Your Factory." *Production Engineering* 30, no. 5 (May): 50–52.

GESSNER, R. A. 1984. *Manufacturing Information Systems: Implementation Planning.* New York: John Wiley.

GHARAJEDAGHI, J. 1986. *A Prologue to National Development Planning.* New York: Greenwood Press.

GIBSON, J. E. 1990. *Modern Management of the High-Technology Enterprise.* Englewood Cliffs, NJ: Prentice Hall.

GIDO, J. 1985. *Project Management Software Directory.* New York: Industrial Press.

GILBREATH, R. D. 1986. *Winning At Project Management: What Works, What Fails, and Why.* New York: John Wiley.

GILYUTIN, I. 1993. "Managing Resources for a Construction Project in a Mixed (PC-Mainframe) Environment." *Project Management Journal* 24, no. 2 (June): 34–40.

GLASSER, A. 1982. *Research and Development Management.* Englewood Cliffs, NJ: Prentice Hall.

GLOVER, J. H. 1966. "Manufacturing Progress Functions: An Alternative Model and Its Comparison with Existing Functions." *International Journal of Production Research* 4, no. 4: 279–300.

GOLDBERGER, A. S. 1968. "The Interpretation and Estimation of Cobb-Douglas Functions." *Econometrica* 35, no. 3–4: 464–72.

GOLDEN, B., BODIN, L., DOYLE, T., and STEWART, W. JR. 1977. "Approximate Traveling Salesman Algorithms." *Operational Research* 28: 694–711.

GOLDEN, B. L., E. A. WASIL, and P. T. HARKER, eds. 1989. *The Analytic Hierarchy Process: Applications and Studies.* New York: Springer-Verlag.

GOLDSTEIN, I. R. 1990. "Information—The Driving Force." *IE News: Computer & Information Systems* 24, no. 1 (Winter): 1–2.

GOLENKO-GINZBURG, D. 1988. "On the Distribution of Activity-Time in PERT." *Journal of Operational Research Society* 39: 767–71.

———. 1989a. "A New Approach to the Activity-Time Distribution in PERT." *Journal of Operational Research Society* 40, no. 4: 389–93.

———. 1989b. "PERT Assumptions Revisited." *OMEGA* 17, no. 4: 393–96.

GONEN, T. 1990. *Engineering Economy for Engineering Managers.* New York: John Wiley.

GOODMAN, D. 1974. "A Goal Programming Approach to Aggregate Planning of Production and Work Force." *Management Science* 20: 1569–75.

GORDON, G., and I. PRESSMAN. 1978. *Quantitative Decision-Making for Business.* Englewood Cliffs, NJ: Prentice Hall.

GORDON, M. M. 1985. *The Iacocca Management Technique.* New York: Dodd, Mead.

GORDON, T. J., and O. HELMER. 1964. "Report on a Long-Range Forecasting Study." New York: Rand Corporation, Paper 2982 (September).

GORENSTEIN, S. 1972. "An Algorithm for Project (Job) Sequencing with Resource Constraints." *Operations Research* 20 (July-August): 835–50.

GRAHAM, R. J. 1985. *Project Management: Combining Technical and Behavioral Approaches for Effective Implementation.* New York: Van Nostrand Reinhold.

GRANT, E. L., W. G. IRESON, and R. S. LEAVENWORTH. 1982. *Principles of Engineering Economy.* 7th ed. New York: John Wiley.

GREINER, L. E., and V. E. SCHEIN. 1981. "The Paradox of Managing a Project-Oriented Matrix: Establishing Coherence Within Chaos." *Sloan Management Review* (Winter). 23, no. 1, 25–34.

GRUBBS, F. E. 1962. "Attempts to Validate Certain PERT Statistics or 'Picking on PERT.'" *Operations Research* 10: 912–15.

GUENTHNER, F., H. LETHAMANN, and W. SCHONFELD. 1986. "A Theory for the Representation of Knowledge." *IBM Journal of Research and Development* 30, no. 11 (January): 39–56.

GULEZIAN, R. C. 1979. *Statistics for Decision Making.* Philadelphia: W. B. Saunders.

GULLEDGE, T. R., JR., and L. A. LITTERAL, eds. 1989. *Cost Analysis Applications of Economics and Operations Research*. New York: Springer-Verlag.

GUPTA, S. K., J. KYPARISIS, and C. IP. 1992. "Project Selection and Sequencing to Maximize Net Present Value of the Total Return." *Management Science* 38, no. 5 (May): 751–52.

GUTIERREZ, G. J., and P. KOUVELIS. 1991. "Parkinson's Law and Its Implications for Project Management." *Management Science* 37, no. 8: 990–1001.

HAJEK, V. G. 1965. *Project Engineering: Profitable Technical Program Management*. New York: McGraw-Hill.

———. 1977. *Management of Engineering Projects*. New York: McGraw-Hill.

HALL, D. L., and A. NAUDA. 1990. "An Interactive Approach for Selecting R & D Projects." *IEEE Transactions on Engineering Management* 37, no. 2: 126–33.

HAMMILL, J., L. C. T. SALLABANK, and P. B. KELLY. 1990. "Management of Large Projects." *IEE Proceedings* 137, part A, no. 1 (January): 55–64.

HANNAN, E. L. 1978. "The Application of Goal Programming Techniques to the CPM Problem." *Socio-Economic Planning Science* 12: 267–70.

HANRAHAN, J. D. 1983. *Government by Contract*. New York: W. W. Norton.

HARHALAKIS, G. 1990. "Special Features of Precedence Network Charts." *European Journal of Operational Research* 49: 50–59.

HARRISON, F. L. 1985. *Advanced Project Management*. 2d ed. New York: John Wiley.

HARVEY, J. B. 1974. "The Abilene Paradox: The Management of Agreement." *Organizational Dynamics* 3, no. 1: 63–80.

HATRY, H. P. 1983. *A Review of Private Approaches for Delivery of Public Services*. Washington DC: Urban Institute Press.

HEALY, T. L. 1962. "Activity Subdivision and PERT Probability Statements." *Operations Research* 9, no. 3: 341–48.

HERZBERG, F. 1960. *Work and the Nature of Man*. Cleveland: World Publishing.

———. 1968. "One More Time: How Do You Motivate Employees?" *Harvard Business Review* 45, no. 1: 53–62.

HESS, S. W. 1962. "A Dynamic Programming Approach to RandD Budgeting and Project Selection." *IEEE Transactions on Engineering Management*, EM-9, no. 4: 170–99.

HICKS, P. E. 1977. *Introduction to Industrial Engineering and Management Science*. New York: McGraw-Hill.

HILLIER, F. S., and G. J. LIEBERMAN. 1974. *Operations Research*, 2d ed. San Francisco: Holden-Day.

HINDELANG, T. J., and J. F. MUTH. 1979. "A Dynamic Programming Algorithm for Decision CPM Networks." *Operations Research* 27, no. 2: 225–41.

HINES, W. W., and D. C. MONTGOMERY. 1980. *Probability and Statistics in Engineering and Management Science*. 2d ed. New York: John Wiley.

HIRCHMANN, W. B. 1964. "Learning Curve." *Chemical Engineering* 71, no. 7: 95–100.

HIRSHLEIFER, J. 1965. "Investment Decision under Uncertainty: Choice-Theoretic Approaches." *Quarterly Journal of Economics* 79, no. 4: 509–36.

HOELSCHER, H. E. 1987. "Managing in an Unmanageable World." *Engineering Management International* 4: 151–54.

HOFFMAN, T. R. 1967. *Production Management and Manufacturing Systems*. Belmont, CA: Wadsworth.

HOGG, R. V., and J. LEDOLTER. 1992. *Applied Statistics for Engineers and Physical Scientists.* New York: Macmillan.

HOLLOWAY, C. A., R. T. NELSON, and V. SURAPHONGSCHAI. 1979. "Comparison of a Multi-Pass Heuristic Decomposition Procedure with Other Resource-Constrained Project Scheduling Procedures." *Management Science* 25 (September): 862–72.

HOUSE, R. S. 1988. *The Human Side of Project Management.* Reading, MA: Addison-Wesley.

HOWARD, R. 1971. *Dynamic Probabilistic Systems.* New York: John Wiley. Pp. 949–55.

HOWELL, S. D. 1980. "Learning Curves for New Products." *Industrial Marketing Management* 9, no. 2: 97–99.

HRIBAR, J. P. 1985. "Development of An Engineering Manager." *Journal of Management In Engineering* 1: 36–41.

HUMPHREY, W. S. 1987. *Managing For Innovation: Leading Technical People.* Englewood Cliffs, NJ: Prentice Hall.

HUMPHREYS, K. K., ED. 1984. *Project and Cost Engineer's Handbook.* 2d ed. New York: Marcel Dekker.

HUYLER, G., and K. CROSBY. 1993. "The Best Investment a Project Manager Can Make: Improve Meetings." *PM Network* 7, no. 6 (June): 33–35.

IGNIZIO, J. P. 1976. *Goal Programming and Extensions.* Lexington, MA: D. C. Heath.

IZUCHUKWU, J. I. 1990. "Shortening the Critical Path." *Mechanical Engineering* (February): no. 2 59–60. 112.

JELEN, F. C., and J. H. BLACK. 1983. *Cost and Optimization Engineering.* New York: McGraw-Hill.

JENNETT, E. 1973. "Guidelines for Successful Project Management." *Chemical Engineering* (July): 70–82. 80, no. 7.

JEWELL, W. S. 1984. "A Generalized Framework for Learning Curve Reliability Growth Models." *Operations Research* 32, no. 3 (May-June): 547–58.

JOHNSON, D. S. 1990. "Local Optimization and the Traveling Salesman Problem." *Proceedings Seventeenth Colloquium on Automata, Languages and Programming.* New York: Springer-Verlag, 446–61.

JOHNSON, G. A., and C. D. SCHOU. 1990. "Expediting Projects in PERT with Stochastic Time Estimates." *Project Management Journal* 21, no. 2: 29–34.

JOHNSON, J. R. 1977. "Advanced Project Control." *Journal of Systems Management* 28, no. 5 (May): 24–27.

JOHNSON, L. A., and D. C. MONTGOMERY. 1974. *Operations Research in Production, Scheduling, and Inventory Control.* New York: John Wiley.

JOHNSON, R. A., F. E. KAST, and J. A. ROSENZWEIG. 1967. *The Theory and Management of Systems.* 2d ed. New York: McGraw-Hill.

JOHNSON, R., and P. WINN. 1976. *Quantitative Methods for Management.* Boston: Houghton Mifflin.

JOHNSON, R. V. 1992. "Resource Constrained Scheduling Capabilities of Commercial Project Management Software." *Project Management Journal* 22, no. 4 (December): 39–43.

JOHNSON, T. J. R. 1967. *An Algorithm for the Resource-Constrained Project Scheduling Problem.* Ph.D. thesis, MIT.

JURAN, J. M. 1988. *Juran on Planning for Quality.* New York: Free Press.

———, and F. M. GRYNA, JR. 1980. *Quality Planning and Analysis.* 2d ed. New York: McGraw-Hill.

KAIMANN, R. A. 1974. "Coefficient of Network Complexity." *Management Science* 21, no. 2: 172–77.

———. 1975. "Coefficient of Network Complexity: Erratum." *Management Science* 21, no. 10: 1211–12.

KALU, T. U. 1993. "A Framework for the Management of Projects in Complex Organizations." *IEEE Transactions on Engineering Management* 40, no. 2 (May): 175–80.

KANE, J. S. 1964. "Origin of CPM and PERT." In H. L. Wattel, ed., *The Dissemination of New Business Techniques: Network Scheduling and Control Systems (CPM/PERT)*, pp. 50–54. Hofstra University Yearbook of Business, Series 1, vol. 2. Hempstead, NY: Hofstra University.

KANNEWURF, A. S. 1980. "How to Present Your Proposals to Management." *Achieving Success in Manufacturing Management*, Charles F. Hoitash, ed. Dearborn, MI: Society of Manufacturing Engineers.

KAPLAN R. S. 1984. "Yesterday's Accounting Undermines Production." *Harvard Business Review* (July-August): 62, no. 4: 95–101.

KARGER, D. W., and F. H. BAYHA. 1977. *Engineered Work Measurement*. 3d ed. New York: Industrial Press.

KARNI, R. 1973. "Heuristic Resource Analysis." *Management Information* 2, no. 4 (April): 57–70.

KASEVICH, L. S. 1986. *Harvard Project Manager/Total Project Manager: Controlling Your Resources*. Blue Ridge Summit, PA: TAB Books.

KEEN, P. G. W. 1981. "Information Systems and Organizational Change." *Communications of the ACM* 24, no. 1 (January): 24–33.

KEENEY, R. L., and H. RAIFFA. 1976. *Decisions with Multiple Objectives: Preferences and Value Tradeoffs*. New York: John Wiley.

KELLER, R. T. 1985. "Project Group Performance in Research and Development Organizations." *Academy of Management Proceedings, San Diego*. New York: Academy of Management: 315–18.

———, and R. R. CHINTA. 1990. "International Technology Transfer: Strategies for Success." *The Executive* 4, no. 2 (May): 4–11.

KELLEY, A. J., ed. 1982. *New Dimensions of Project Management*. Lexington, MA: Lexington Books.

KELLEY, J. E. 1961. "Critical Path Planning and Scheduling: Mathematical Basis." *Operations Research* 9, no. 3: 296–320.

———, and M. R. WALKER. 1959. "Critical Path Planning and Scheduling." *Proceedings of the Eastern Joint Computer Conference, New York*. Boston: Eastern Joint Computer Conference. 160–73.

KENDRICK, J. W. 1961. *Productivity Trends in the United States*. Princeton, NJ: Princeton University Press.

KERRIDGE, A. E., and C. H. VERVALIN, eds. 1986. *Engineering and Construction Project Management*. Houston, TX: Gulf Publishing.

KERZNER, H. 1986. *Project Management Operating Guidelines: Directives, Procedures, and Forms*. New York: Van Nostrand Reinhold.

———. 1989. *Project Management: A Systems Approach to Planning, Scheduling, and Controlling*. 3rd ed. New York: Van Nostrand Reinhold.

KEZSBOM, D. S., D. L. SCHILLING, and K. A. EDWARD. 1989. *Dynamic Project Management: A Practical Guide for Managers and Engineers.* New York: John Wiley.

KHARBANDA, O. P., and E. A. STALLWORTHY. 1986. *Management Disasters and How to Prevent Them.* London, England: Gower Publishing Company.

KHORRAMSHAHGOL, R., H. AZANI, and Y. GOUSTY. 1988. "An Integrated Approach to Project Evaluation and Selection." *IEEE Transactions on Engineering Management* 35, no. 4: 265–70.

KIM, C. 1976. *Quantitative Analysis for Managerial Decisions.* Reading, MA: Addison-Wesley.

KIMMONS, R. L. 1990. *Project Management Basics: A Step-By-Step Approach.* New York: Marcel Dekker.

———, and J. H. LOWEREE. 1989. *Project Management: A Reference for Professionals.* New York: Marcel Dekker.

KING, R. S., and B. JULSTROM. 1982. *Applied Statistics Using the Computer.* Sherman Oaks, CA: Alfred Publishing Co.

KINTNER, S. S. 1986. "Using Harvard Total Project Manager for Public Facilities Construction and Administration." In Harvey A. Levine, *Project Management Using Microcomputer*, 334–46. Berkeley, CA: McGraw-Hill.

KNECHT, G. R. 1974. "Costing, Technological Growth and Generalized Learning Curves." *Operations Research Quarterly* 25, no. 3 (September): 487–91.

KNIGHT, F. 1921. *Risk, Uncertainty and Profit.* Boston: Houghton Mifflin.

KNUTSON, J., and I. BITZ. 1991. *Project Management: How to Plan and Manage Successful Projects.* New York: American Management Association.

KOCHHAR, A. K. 1979. *Development of Computer-Based Production Systems.* New York: John Wiley.

KOENIG, M. H. 1978. "Management Guide to Resource Scheduling." *Journal of Systems Management* 29 (January): 24–29.

KOONTZ, H., and C. O'DONNEL. 1959. *Principles of Management.* 2d ed. New York: McGraw-Hill.

KOOPMANS, T. C., ed. 1951. *Activity Analysis of Production and Allocation.* New York: John Wiley.

KOPCSO, D. P. 1983. "Learning Curves and Lot Sizing for Independent and Dependent Demand." *Journal of Operations Management* 4, no. 1 (November): 73–83.

KORNBLUTH, J. S. 1973. "A Survey of Goal Programming." *Omega* (April): 193–206.

KOSTNER, J., and C. STRBIAK. 1993. "3-D Leadership: The Key to Inspired Performance." *PM Network* (August): 7, no. 8, 50–52.

KRAKOW, I. H. 1985. *Project Management with the IBM PC.* Bowie, MD: Brady Communications Co.

KUHN, H. W. 1955. "The Hungarian Method for the Assignment Problem." *Naval Research Logistics Quarterly* 2, no. 1: 83–97.

———. 1956. "Variants of the Hungarian Method for the Assignment Problem." *Naval Research Logistics Quarterly* 3, no. 2: 253–58.

KUME, H. 1987. *Statistical Methods for Quality Improvement.* White Plains, NY: Quality Resources.

KURTULUS, I. S., and E. W. DAVIS. 1982. "Multi-Project Scheduling: Categorization of Heuristic Rules Performace." *Management Science* 28, no. 2 (February): 161–72.

LAPIN, L. 1976. *Quantitative Management for Business Decisions.* New York: Harcourt Brace Jovanovich.

————. 1978. *Statistics for Modern Business Decisions.* 2d ed. New York: Harcourt Brace Jovanovich.

————. 1991. *QuickQuant+: Decision Making Software.* Pleasanton, CA: Alamo Publishing Co.

LAWLER, E. 1976. *Combinatorial Optimization: Networks and Matroids.* New York: Holt, Rinehart and Winston.

LAWRENCE, K. D., and S. H. ZANAKIS. 1984. *Production Planning and Scheduling: Mathematical Programming Applications.* Norcross, GA: Industrial Engineering and Management Press.

LEE, S. M. 1972. *Goal Programming for Decision Analysis.* Philadelphia: Auerbach Publishers.

————. 1979. *Goal Programming Methods for Multiple Objective Integer Programs.* OR Monograph Series no. 2. Norcross, GA: Institute of Industrial Engineers.

————, and E. R. CLAYTON. 1972. "A Goal Programming Model for Academic Resource Allocation." *Management Science* 18, no. 8: B395–B408.

————, and H. B. EOM. 1989. "A Multi-Criteria Approach to Formulating International Project-Financing Strategies." *Journal of Operational Research Society* 40, no. 6: 519–28.

LEEMIS, L., A. B. BADIRU, B. L. FOOTE, and A. RAVINDRAN. 1990. "Job Shop Configuration Optimization at Tinker Air Force Base." *Simulation* 54, no. 6 (June): 287–90.

LEVIN, R., and C. KIRKPATRICK. 1978. *Quantitative Approaches to Management.* 4th ed. New York: McGraw-Hill.

LEVINE, H. A. 1986. *Project Management Using Microcomputers.* New York: Osborne McGraw-Hill.

LEVY, F. K. 1965. "Adaptation in the Production Process." *Management Science* 11, no. 6 (April): B136–B154.

LIAO, W. M. 1979. "Effects of Learning on Resource Allocation Decisions." *Decision Sciences* 10: 116–25.

LIBERATORE, M. J. 1987. "An Extension of the Analytic Hierarchy Process for Industrial RandD Project Selection and Resource Allocation." *IEEE Transactions on Engineering Management* EM34, no. 1: 12–18.

LIEPINS, G. E., and V. R. R. UPPULURI, eds. 1990. *Data Quality Control.* New York: Marcel Dekker.

LIGHTHALL, F. F. 1991. "Launching the Space Shuttle Challenger: Disciplinary Deficiencies in the Analysis of Engineering Data." *IEEE Transactions on Engineering Management* 38, no. 1: 63–74.

LILLRANK, P., and N. KANO. 1989. *Continuous Improvement: Quality Control Circles in Japanese Industry.* Ann Arbor: Center for Japanese Studies, University of Michigan.

LIN, S. 1965. "Computer Solutions of the Traveling Salesman Problem." *Bell Systems Technical Journal* 44, no. 5: 2245–69.

————, and KERNIGHAN, B. W. 1973. "An Effective Heuristic Algorithm for the Traveling Salesman Problem." *Operational Research* 2: 498–516.

LITTLE, J. D., G. MARTY, D. W. SWEENEY, and C. KAREL. 1963. "An Algorithm for the Travelling Salesman Problem." *Operations Research* 11, no. 6: 972–89.

LITTLEFIELD, T. K., JR. and P. H. RANDOLPH. 1987. "An Answer to Sasieni's Question on PERT Times." *Management Science* 33: no. 10, 1357–59.

LOCHNER, R. H., and J. E. MATAR. 1990. *Designing for Quality.* White Plains, NY: Quality Resources.

LOVE, S. F. 1989. *Achieving Problem Free Project Management.* New York: John Wiley.

MacCRIMMON, K. R., and C. A. RYAVEC. 1964. "An Analytical Study of the PERT Assumptions." *Operations Research* 12: 16–21.

MACHINA, M. J. 1987. "Decision-Making in the Presence of Risk." *Science* 236 (May): 537–43.

MACKIE, D. 1984. *Engineering Management of Capital Projects: A Practical Guide.* Toronto: McGraw-Hill Ryerson Ltd.

MADANSKY, A. 1988. *Prescriptions for Working Statisticians.* New York: Springer-Verlag.

MADU, C. N. 1988. "An Economic Decision Model for Technology Transfer." *Engineering Management International* 5, no. 1: 53–62.

MAGERS, C. S. 1987. "Managing Software Development in Microprocessor Projects." *IEEE Computer* 11, no. 6 (June): 34–42.

MAIN, J. 1985. "When Public Services Go Private." *Fortune* (May 27): 3–4.

MALCOMB, D. G., J. H. ROSEBOOM, C. E. CLARK, and W. FAZAR. 1959. "Application of a Technique for Research and Development Program Evaluation." *Operations Research* 7, no. 5: 646–99.

MALSTROM, E. M. 1981. *What Every Engineer Should Know About Manufacturing Cost Estimating.* New York: Marcel Dekkar.

MANGO, A. 1992. "Expert System Concepts for Project Planning." *Proceedings of Project Management Institute Annual Seminar/Symposium, Pittsburgh, PA*: Webster, NC: Project Management Institute, 93–100.

MARSHALL, G., T. J. BARBER, and J. T. BOARDMAN. 1987. "Methodology for Modelling a Project Management Control Environment." *IEEE Proceeding-D: Control Theory and Applications* 134, part D, no. 4 (July): 278–85.

MASLOW, A. H. 1943. "A Theory of Human Motivation," *Psychological Review* 1: 370–96.

———. 1954. *Motivation and Personality.* New York: Harper & Brothers.

McBRIDE, W. J., and C. W. McCLELLAND. 1967. "PERT and the Beta Distribution." *IEEE Transactions on Engineering Management* EM-14: 166–69.

McCAHON, C. S. 1993. "Using PERT as an Approximation of Fuzzy Project-Network Analysis." *IEEE Transactions on Engineering Management* 40, no. 2 (May): 146–53.

McDAVID, J. C. 1985. "The Canadian Experience with Privatizing Residential Solid Waste Collection Services." *Public Administration Review* (September/October): 15–20.

McENTEE, G. W. 1987. "The Case against Privatization." *American City & County* (January). 3–5.

McFADDEN, F. R., and J. A. HOFFER. 1985. *Data Base Management.* Menlo Park, CA: Benjamin/Cummings.

McFARLANE, D. 1993. "Enterprise-Wide Project Management." *Industrial Engineering* 25, no. 6 (June): 44.

McGILL, F. 1986. *Factory Automation Case Books.* Walker-Davis Publication, Inc.

McGREGOR, D. 1960. *The Human Side of Enterprise.* New York: McGraw-Hill.

McINTYRE, E. V. 1977. "Cost-Volume-Profit Analysis Adjusted for Learning." *Management Science* 24, no. 2: 149–60.

MECKLER, G. 1987. "Systems Integration—A State-of-the-Art Report." *Consulting-Specifying Engineer* (August): 44–51.

MELCHER, B. H., and H. KERZNER. 1988. *Strategic Planning Development and Implementation.* Blue Ridge Summit, PA: TAB Professional and Reference Books.

MELCHER, R. 1967. "Roles and Relationships: Clarifying the Manager's Job." *Personnel* (May-June): 8–13.

MEREDITH, J. R., ed. 1986. *Justifying New Manufacturing Technology*. Norcross, GA: Industrial Engineering & Management Press.

———, and T. E. GIBBS. 1984. *The Management of Operations*. 2d ed. New York: John Wiley.

———, and N. C. SURESH. 1986. "Justification Techniques for Advanced Manufacturing Technologies." *International Journal of Production Research* 24, no. 5 (September–October): 1043–58.

———, and S. L. MANTEL, JR. 1989. *Project Management: A Managerial Approach*. 2d ed. New York: John Wiley.

MERINO, D. N. 1989. "Developing Economic and Noneconomic Models Incentives to Select Among Technical Alternatives." *Engineering Economist* 34, no. 4 (Summer): 275–90.

METZ, H. J., and J. A. GINGRICH. 1986. "Managing Complexity in Mature Industries." *Manufacturing Issues*. New York: Booz-Allen & Hamilton: 1–8.

METZGER, P. W. 1987. *Managing Programming People: A Personal View*. Englewood Cliffs, NJ: Prentice Hall.

MICHAELS, J. V., and W. P. WOOD. 1989. *Design to Cost*. New York: John Wiley.

MIDDLETON, C. J. 1967. "How to Set Up A Project Organization." *Harvard Business Review* 45, no. 2 (March/April): 73–82.

MILLER, D. M., and J. W. SCHMIDT. 1984. *Industrial Engineering and Operations Research*. New York: John Wiley.

MILLER, I. and J. E. FREUND. 1977. *Probability and Statistics for Engineers*. 2d ed. Englewood Cliffs, NJ: Prentice Hall.

MILLER, J. A. 1986. *From Idea To Profit: Managing Advanced Manufacturing Technology*. New York: Van Nostrand Reinhold.

MINTZBERG, H. 1973. *The Nature of Managerial Work*. New York: Harper & Row.

MIZUNO, S., ed. 1988. *Management for Quality Improvement*. White Plains, NY: Quality Resources.

MODER, J. J., C. R. PHILLIPS, and E. W. DAVIS. 1983. *Project Management with CPM, PERT and Precedence Diagramming*. 3d ed. New York: Van Nostrand Reinhold.

MODER, J. J., and E. G. RODGERS. 1968. "Judgment Estimates of the Moments of PERT Type Distributions." *Management Science* 15: B76–B83.

MOFFAT, D. W. 1987. *Handbook of Manufacturing and Production Management Formulas, Charts, and Tables*. Englewood Cliffs, NJ: Prentice Hall.

MOHRING, R. H. 1984. "Minimizing Costs of Resource Requirements in Project Networks Subject to Fixed Completion Time." *Operations Research* 32, no. 1: 89–120.

MONDEN, Y., R. SHIBAKAWA, S. TAKAYANAGI, AND T. NAGAO. 1985. *Innovations in Management: The Japanese Corporation*. Norcross, GA: Industrial Engineering & Management Press.

MOODIE, C. L., and H. H. YOUNG. 1965. "A Heuristic Method of Assembly Line Balancing for Assumptions of Constant or Variable Work Element Times." *Journal of Industrial Engineering* 26, no. 1: 23–29.

MOORE, F. G. 1964. *A Management Sourcebook*. New York: Harper & Row.

———. 1973. *Production Management*. 6th ed. Homewood, IL: Richard D. Irwin.

MOORE, F. T. 1983. "Technological Change and Industrial Development, Issues and Opportunities." World Bank Paper no. 613, Washington, DC: World Bank.

MOORE, L. J., B. W. TAYLOR III, E. R. CLAYTON, and S. M. LEE. 1978. "Analysis of a Multi-Criteria Project Crashing Model." *AIIE Transactions* 10, no. 2 (June): 163–69.

MORAD, A. A., and M. C. VORSTER. 1993. "Network-Based Versus AI-Based Techniques in Project Planning." *Project Management Journal* 24, no. 1 (March): 23–30.

MUELLER, F. W. 1986. *Integrated Cost and Schedule Control for Construction Projects.* New York: Van Nostrand Reinhold.

MUKHERJEE, S. P., and A. K. CHATTOPADHYAY. 1989. "A Stochastic Analysis of a Staffing Problem." *Journal of Operational Research Society* 40, no. 5: 489–94.

MURDICK, R. G., and J. E. ROSS. 1975. *Information Systems For Modern Management.* 2d ed. Englewood Cliffs, NJ: Prentice Hall.

MURPHY, K. J. 1983. *Macroproject Development In The Third World: An Analysis of Transnational Partnerships.* Boulder, CO: Westview Press.

MUSTAFA, M. A. 1989. "An Integrated Hierarchical Programming Approach for Industrial Planning." *Computers & Industrial Engineering* 16, no. 4: 525–34.

———, and J. F. AL-BAHAR. 1989. "Project Risk Assessment Using the Analytic Hierarchy Process." *IEEE Transactions on Engineering Management* 38, no. 1: 46–52.

———, and E. LILE MURPHREE. 1989. "A Multicriteria Decision Support Approach for Project Compression." *Project Management Journal* 20, no. 2 (June): 29–34.

NAISBITT, J., and P. ABURDENE. 1985. *Re-Inventing The Corporation.* New York: Warner Books.

NANDA, R. 1979. "Using Learning Curves in Integration of Production Resources." *Proceedings of IIE Fall Conference: Atlanta, GA.* Norcross, GA: Institute of Industrial Engineering. 376–80.

NAVON, R. 1990. "Financial Planning in a Project-Oriented Industry." *Project Management Journal* 21, no. 1: 43–48.

NELSON, C. A. 1986. "A Scoring Model for Flexible Manufacturing Systems Project Selection." *European Journal of Operations Research* 24, no. 3: 346–59.

NELSON, C. R. 1973. *Applied Times Series Analysis for Mangerial Forecasting.* San Francisco: Holden-Day.

NEWMAN, W. H., E. K. WARREN, and A. R. MCGILL. 1987. *The Process of Management: Strategy, Action, Results.* Englewood Cliffs, NJ: Prentice Hall.

NEWNAN, D. G. 1991. *Engineering Economic Analysis.* 4th ed. San Jose, CA: Engineering Press, Inc.

NIEBEL, B. W. 1976. *Motion and Time Study.* 6th ed. Homewood, IL: Richard D. Irwin.

NIXON, R. M. 1990. *In the Arena of Victory, Defeat, and Renewal.* New York: Simon & Schuster.

NOORI, H. 1990. *Managing the Dynamics of New Technology: Issues in Manufacturing Management.* Englewood Cliffs, NJ: Prentice Hall.

———, and R. W. RADFORD. 1990. *Readings and Cases in the Management of New Technology: An Operations Perspective.* Englewood Cliffs, NJ: Prentice Hall.

O'BRIEN, J. J. 1974. "The Project Manager: Not Just a Firefighter." *S.A.M. Advanced Management Journal* 39 (January): 52–56.

———. 1984. *CPM in Construction Management.* 3d ed. New York: McGraw-Hill.

O'NEAL, K. R. 1987. "Project Management Microcomputer Software Buyer's Guide." *Industrial Engineering* 19, no. 1 (January): 53–63.

OBRADOVITCH, M. M., and S. E. STEPHANOU. 1990. *Project Management: Risks and Productivity.* Malibu, CA: Daniel Spencer Publishers.

ODIORNE, G. S. 1965. *Management by Objectives: A System of Management Leadership.* New York: Pitman.

ORAL, M., O. KETTANI, and P. LANG. 1991. "A Methodology for Collective Evaluation and Selection of Industrial R&D Projects." *Management Science* 37, no. 7: 871–85.

ORCZYK, J. J., and L. CHANG. 1991. "Parametric Regression Model for Project Scheduling." *Project Management Journal* 22, no. 4: 41–47.

OSTWALD, P. F. 1974. *Cost Estimating for Engineering and Management*. Englewood Cliffs, NJ: Prentice Hall.

OUCHI, W. G. 1981. *Theory Z: How American Business Can Meet the Japanese Challenge*. Reading, MA: Addison-Wesley.

OZDEN, M. 1987. "A Dynamic Planning Technique for Continuous Activities Under Multiple Resource Constraints." *Management Science* 33, no. 10: 1333–47.

PANDIA, R. M. 1989. "Transfer of Technology: Techniques for Chemical and Pharmaceutical Projects." *Project Management Journal* 20, no. 3 (September): 39–45.

PARK, C. S., and G. P. SHARP-BETTE. 1990. *Advanced Engineering Economics*. New York: John Wiley.

————, and Y. K. SON. 1988. "An Economic Evaluation Model for Advanced Manufacturing Systems." *The Engineering Economist* 34, no. 1 (Fall): 1–26.

PARKINSON, C. N. 1957. *Parkinson's Law*. Boston: Houghton Mifflin.

PASCALE, R. T., and A. G. ATHOS. 1982. *The Art of Japanese Management*. New York: Penguin.

PATTERSON, J. H. 1976. "Project Scheduling: The Effects of Problem Structure on Heuristic Performance." *Naval Research Logistics Quarterly* 23, no. 1: 95–123.

————. 1984. "Comparison of Exact Approaches for Solving Multiconstrained Resource Project Scheduling." *Management Science* 30, no. 7: 854–67.

————, and W. D. HUBER. 1974. "A Horizon-varying, Zero-one Approach to Project Scheduling." *Management Science* 20, no. 6: 990–98.

————, and G. W. ROTH. 1976. "Scheduling a Project under Multiple Resource Constraints: A Zero-One Programming Approach." *AIIE Transactions* 8, no. 4 (December): 449–55.

————, F. B. TALBOT, R. SLOWINSKI, and J. WEGLARZ. 1990. "Computational Experience with a Backtracking Algorithm for Solving a General Class of Precedence and Resource-Constrained Scheduling Problems." *European Journal of Operational Research* 49: 68–79.

PEGELS, C. C. 1969. "On Startup or Learning Curves: An Expanded View." *AIIE Transactions* 1, no. 3 (September): 216–22.

————. 1976. "Start Up or Learning Curves—Some New Approaches." *Decision Sciences* 7, no. 4 (October): 705–13.

PETER, L. J., and R. HULL. 1969. *The Peter Principle—Why Things Always Go Wrong*. New York: William Morrow.

PETERS, T. J., and R. H. WATERMAN JR. 1982. *In Search of Excellence: Lessons from America's Best-Run Companies*. New York: Harper & Row.

PETERSON, R., and E. A. SILVER. 1979. *Decision Systems for Inventory Management and Production Planning*. New York: John Wiley.

PETERSON, R. O. 1987. *Managing The Systems Development Function*. New York: Van Nostrand.

PHADKE, M. S. 1989. *Quality Engineering Using Robust Design*. Englewood Cliffs, NJ: Prentice Hall.

PHILLIPS, D. T., and A. GARCIA-DIAZ. 1981. *Fundamentals of Network Analysis*. Englewood Cliffs, NJ: Prentice Hall.

————, and G. L. HOGG. 1976. "Stochastic Network Analysis with Resource Constraints, Cost Parameters and Queueing Capabilities Using GERTS Methodologies." *Computers and Industrial Engineering* 1, no. 1: 13–25.

PIENAR, A., ET AL. 1986. "An Evaluation Model for Quantifying System Value." *IIE Transactions* 18, no. 1 (March): 10–15.

PINCHOT, G. III. 1985. *Intrapreneuring*. New York: Harper & Row.

PITTS, C. E. 1990. "For Project Managers: An Inquiry into the Delicate Art and Science of Influencing Others." *Project Management Journal* 21, no. 1: 21–23, 42.

PMI (Project Management Institute). 1987. *The Project Management Body of Knowledge*. Drexel Hill, PA: Project Management Institute: 4–1.

PRITSKER, A. A. 1986. *Introduction To Simulation and SLAM II*. 3d ed. New York: John Wiley.

PRITSKER, A. A. B., C. E. SIGAL, AND R. D. J. HAMMERSFAHR. 1989. *SLAM II Network Models for Decision Support*. Englewood Cliffs, NJ: Prentice Hall.

PRITSKER, A. B., L. J. WALTERS, and P. M. WOLFE. 1969. "Multi-Project Scheduling with Limited Resources: A Zero-One Programming Approach." *Management Science* 16, no. 1 (September): 93–108.

PULAT, P. S., and A. B. BADIRU. 1990. "Optimization of Oil Industry Investment Yield." Consulting Working Paper, School of Industrial Engineering, University of Oklahoma.

QUINN, J. B. 1985. "Managing Innovation: Controlled Chaos." *Harvard Business Review* 63, no. 3 (May-June): 73–84.

RAVINDRAN, A., B. L. FOOTE, A. B. BADIRU, L. M. LEEMIS, AND L. WILLIAMS. 1989. "An Application of Simulation and Network Analysis to Capacity Planning and Material Handling Systems at Tinker Air Force Base." *TIMS Interfaces* 19, no. 1 (January-February): 102–15.

————, D. T. PHILLIPS, and J. J. SOLBERG. 1987. *Operations Research: Principles and Practice*. 2d ed. New York: John Wiley.

REIFER, D. J. 1981. "The Nature of Software Management: A Primer." In *Software Management*, IEEE Computer Society Tutorial, Los Alamitos, CA: IEEE Computer Society, 9–12.

REINFELD, N. V., and W. R. VOGEL. 1958. *Mathematical Programming*. Englewood Cliffs, NJ: Prentice Hall.

RENDER, B., and R. M. STAIR, JR. 1988. *Quantitative Analysis for Management*. 3d ed. Needham, MA: Allyn & Bacon.

RENNINGER, J. P. 1979. *Multinational Cooperation for Development in West Africa*. New York: Pergamon Press.

RHODES, D., M. WRIGHT, and M. JARRETT. 1984. *Computers, Information & Manufacturing Systems*. New York: Praeger Publishers.

RICHARDSON, W. J. 1978. "Use of Learning Curves to Set Goals and Monitor Progress in Cost Reduction Programs." *Proceedings of IIE Spring Conference*: Norcross, GA: Institute of Industrial Engineering: 235–39.

RIGGS, H. E. 1983. *Managing High-Technology Companies*. Malibu, CA: Lifetime Learning Publications.

RIGGS, J. L., and T. M. WEST. 1986. *Engineering Economics*. 3d ed. New York: McGraw-Hill.

ROBERSON, B. F., and R. O. WEIJO. 1988. "Using Market Research to Convert Federal Technology Into Marketable Products." *Journal of Technology Transfer* 13, no. 1 (Fall): 27–33.

ROCHE, W. J., and N. L. MACKINNON. 1970. "Motivating People With Meaningful Work." *Harvard Business Review* 48, no. 3 (May-June).

ROGERS, T. 1993. "Project Management: Emerging As a Requisite for Success." *Industrial Engineering* 25, no. 6 (June): 42–43.

ROMAN, D. D. 1986. *Managing Projects: A Systems Approach.* New York: Elsevier Science.

RONEN, D. 1992. "Allocation of Trips to Trucks Operating from a Single Terminal." *Computers and Operations Research* 19, no. 5: 445–51.

ROSANDER, A. C. 1989. *The Quest for Quality in Services.* White Plains, NY: Quality Resources.

ROSARIO-BRAID, F. 1983. *Communication Strategies for Productivity Improvement.* Rev. ed. White Plains, NY: Quality Resources.

ROSENAU, M. D., JR. 1981. *Successful Project Management.* Belmont, CA: Lifetime Learning Publications.

———. 1984. *Project Management for Engineers.* Belmont, CA: Lifetime Learning Publications.

ROSENKRATZ, D. J., R. E. STEARNS, and P. M. LEWIS. 1977. "An Analysis of Several Heuristics for the Traveling Salesman Problem." *SIAM Journal of Computing* 6, no. 3: 563–81.

ROTHKOPF, M. H. 1966. "Scheduling with Random Service Times." *Management Science* 12, no. 9: 707–13.

RUEFLI, T. 1971. "A Generalized Goal Decomposition Model." *Management Science* 17 (April): B505–B518.

RUSKIN, A. M., and W. E. ESTES. 1982. *What Every Engineer Should Know About Project Management.* New York: Marcel Dekker.

———. 1992. "Project Risk Management." *PM Network* (April): 30–37.

RUSSELL, E. J. 1969. "Extension of Dantzig's Algorithm to Finding an Initial Near-Optimal Basis for the Transportation Problem." *Operations Research* 17: 187–91.

RUSSELL, R. A. 1986. "A Comparison of Heuristics for Scheduling Projects with Cash Flows and Resource Restrictions." *Management Science* 32, no. 10 (October): 1291–1300.

SAATY, T. L. 1977. "A Scaling Method for Priorities in Hierarchical Structures." *Journal of Mathematical Psychology* 15 (June): 235–81.

———. 1980. *The Analytic Hierarchy Process.* New York: McGraw-Hill.

SADLER, P. 1971. "Designing An Organizational Structure." *Management International Review* 11, no. 6: 19–33.

SAMARAS, T. T., and K. YENSUANG. 1979. *Computerized Project Management Techniques for Manufacturing and Construction Industries.* Englewood Cliffs, NJ: Prentice Hall.

SASIENI, M. W. 1986. "A Note on PERT Times." *Management Science* 32, no. 12: 1652–53.

SATHI, A., T. E. MORTON, and S. F. ROTH. 1986. "Callisto: An Intelligent Project Management System." *AI Magazine* (Winter): 7, no 5, 34–52.

SAUNDERS, R. J., ET AL. 1983. *Telecommunications and Economic Development.* World Bank publication. Baltimore, MD: Johns Hopkins University Press.

SAVAS, E. S. 1979. "Public vs. Private Refuse Collection: A Critical Review of the Evidence." *The Journal of Urban Analysis* 6: 25–34.

———. 1981. "Intracity Competition between Public and Private Service Delivery." *Public Administration Review* 41, no 1 (January/February), 8–15.

———. 1982. *Privatizing the Public Sector.* NJ: Chatham House Publishers.

SCHEER, A. W. 1989. *Enterprise-Wide Data Modelling: Information Systems in Industry.* New York: Springer-Verlag.

SCHEINBERG, M. V. 1992. "Planning of Portfolio of Projects." *Project Management Journal* 23, no. 2 (June): 31–36.

SCHNEIDER, K. C., and C. R. BYERS. 1979. *Quantitative Management: A Self-Teaching Guide.* New York: John Wiley.

SCHONBERGER, R. J. 1981. "Why Projects Are Always Late: A Rationale Based on Manual Simulation of a PERT/CPM Network." *Interfaces* 11, no. 5 (October): 66–70.

SCHRAGE, L. 1986. *LINDO: Linear Interactive and Discrete Optimizer.* 3d ed. Palo Alto: Scientific Press.

———. 1991. *LINDO, User's Manual, Release 5.0.* Palo Alto: Scientific Press.

SCHULTZ, R. L., D. P. SLEVIN, and J. K. PINTO. 1987. "Strategy and Tactics in a Process Model of Project Implementation." *Interfaces* 17, no. 2 (May-June): 135–143.

SCHUYLER, J. R. 1993a. "Decision Analysis in Projects; Expected Value: The Cornerstone." *PM Network* 7, no. 1 (January): 27–31.

SCHUYLER, J. R. 1993b. "Decision Analysis in Projects: Decision Trees." *PM Network* 7, no. 7 (July): 31–34.

SCHUYLER, J. R. 1993c. "Decision Analysis in Projects: Value of Information," *PM Network*, 7, no. 10, (October 1993): 19–23.

SCHWARTZ, S. L., and I. VERTINSKY. 1977. "Multi-Attribute Investment Decisions: A Study of RandD Project Selection." *Management Science* 24: 285–301.

SEMPREVIVO, P. C. 1982. *Systems Analysis: Definition, Process, and Design.* 2d ed. Chicago: Science Research Associates, Inc.

SHAIKEN, H. 1985. "The Automated Factory: The View from the Shop Floor." *Technology Review* (January): 17–27.

SHARAD, D. 1986. "Management by Projects, An Ideological Breakthrough." *Project Management Journal* 17, no. 1 (March): 61–63.

SHARDA, R. 1992. "Linear Programming Software for Personal Computers: Survey." *OR/MS Today* 19, no. 3 (June): 44–60.

SHTUB, A. 1991. "Scheduling of Programs with Repetitive Projects." *Project Management Journal* 22, no. 4: 49–53.

———, J. F. BARD, and S. GLOBERSON. 1994. *Project Management: Engineering, Technology, and Implementation.* Englewood Cliffs, NJ: Prentice Hall.

SILVER, E. A., and H. C. MEAL. 1973. "A Heuristic for Selecting Lot Size Requirements for the Case of a Deterministic Time-Varying Demand Rate and Discrete Opportunities for Replenishment." *Production and Inventory Management* 14, no. 2: 64–74.

———, and R. PETERSON. 1985. *Decision Systems for Inventory Management and Production Planning.* 2d ed. New York: John Wiley.

SILVERMAN, M. 1987. *The Art of Managing Technical Projects.* Englewood Cliffs, NJ: Prentice Hall.

———. 1988. *Project Management: A Short Course for Professionals.* 2d ed. New York: John Wiley.

SIMON, H. A. 1957. *Administrative Behaviour: A Study of Decision Making Processes in Administrative Organization.* 2d ed. New York: Macmillan.

———. 1977. *The New Science of Management Decision.* Rev. ed. Englewood Cliffs, NJ: Prentice Hall.

SKINNER, W. 1985. *Manufacturing: The Formidable Competitive Weapon.* New York: John Wiley.

SLOWINSKI, R. 1980. "Two Approaches to Problems of Resource Allocation among Project Activities: A Comparative Study." *Journal of the Operational Research Society* 31 (August): 711–23.

SMILOR, R. W., and D. V. GIBSON. 1991. "Technology Transfer in Multi-Organizational Environments: The Case of R&D Consortia." *IEEE Transactions on Engineering Management* 38, no. 1 (February): 3–13.

SMITH, D. E. 1977. *Quantitative Business Analysis.* New York: John Wiley.

SMITH, J. 1989. *Learning Curve for Cost Control.* Norcross, GA: Industrial Engineering and Management Press.

SMITH-DANIELS, D. E., and N. J. AQUILANO. 1980. "Constrained Resource Project Scheduling Subject to Material Constraints." *Journal of Operations Management* 4, no. 4: 369–88.

SMUNT, T. L. 1986. "A Comparison of Learning Curve Analysis and Moving Average Ratio Analysis for Detailed Operational Planning." *Decision Sciences* 17, no. 4: 475–95.

SOMASUNDARAM, S., and A. B. BADIRU. 1992. "Project Management for Successful Implementation of Continuous Quality Improvement." *International Journal of Project Management* 10, no. 2 (May): 89–101.

SPINNER, M. 1981. *Elements of Project Management: Plan, Schedule, and Control.* Englewood Cliffs, NJ: Prentice Hall.

SPRAGUE, J. C., and J. D. WHITTAKER. 1986. *Economic Analysis for Engineers and Managers.* Englewood Cliffs, NJ: Prentice Hall.

STARK, R. M., and R. H. MAYER, JR. 1983. *Quantitative Construction Management: Uses of Linear Optimization.* New York: John Wiley.

STEFANI, R. T. 1983. "Optimal Control of a Developing Nation's Economy." *IEEE Transactions on Systems, Man, and Cybernetics* SMC-13, no. 6 (November/December): 1076–89.

STEINER, G. A., and W. G. RYAN. 1968. *Industrial Project Management.* New York: Macmillan.

STEINER, H. M. 1988. *Basic Engineering Economy.* Glen Echo, MD: Books Associates.

———. 1992. *Engineering Economic Principles.* New York: McGraw-Hill.

STEPHANOU, S. E. 1981. *Management: Technology, Innovation, & Engineering.* Malibu, CA: Daniel Spencer.

———, and M. M. OBRADOVITCH. 1985. *Project Management: System Developments and Productivity.* Malibu, CA: Daniel Spencer Publishers.

STEPMAN, K. 1986. *1986 Buyer's Guide to Project Management Software.* Milwaukee, WI: New Issues, Inc.

STEVENS, G. T., JR. 1979. *Economic and Financial Analysis of Capital Investments.* New York: John Wiley.

STEWART, D. L. 1993. "Management Meetings Could Be New Career Field." *Norman Transcript* (April 25): 10A.

STINSON, J. P., E. W. DAVIS, and B. M. KHUMAWALA. 1978. "Multiple Resource-Constrained Scheduling Using Branch and Bound." *AIIE Transactions* 10, no. 3 (September): 252–59.

STUCKENBRUCK, L. C. 1979. "The Matrix Organization." *Project Management Quarterly* 5, no. 3 (September): 21–33.

———, ed. 1981. *The Implementation of Project Management: The Professional's Handbook.* Reading, MA: Addison-Wesley.

SULE, D. R. 1978. "The Effect of Alternate Periods of Learning and Forgetting on Economic Manufacturing Quantity." *AIIE Transactions* 10, no. 3: 338–43.

SULLIVAN, W. G. 1986. "Models IE's Can Use to Include Strategic Nonmonetary Factors in Automation Decisions." *Industrial Engineering* 18, no. 3 (March): 42–50.

SUMMERS, E. L. 1974. *An Introduction To Accounting For Decision Making and Control.* Homewood, IL: Richard D. Irwin.

SWALM, R. O., and J. L. LOPEZ-LEAUTAUD. 1984. *Engineering Economic Analysis: A Future Wealth Approach.* New York: John Wiley.

TAGUCHI, G. 1986. *Introduction to Quality Engineering: Designing Quality into Products and Processes.* White Plains, NY: Quality Resources.

TAHA, H. A. 1982. *Operations Research: An Introduction.* 3d ed. New York: Macmillan.

TALBOT, F. B. 1982. "Resource-Constrained Project Scheduling with Time-Resource Tradeoffs: The Nonpreemptive Case." *Management Science* 28, no. 10 (October): 1197–1210.

———, and J. H. PATTERSON. 1978. "An Efficient Integer Programming Algorithm with Network Cuts for Solving Resource-Constrained Scheduling Problems." *Management Science* 28, no. 11 (July): 1163–74.

TAVARES, V. 1990. "A Multi-Stage Non-Deterministic Model for Project Scheduling Under Resources Constraints." *European Journal of Operational Research* 49: 92–101.

TAYLOR, F. W. 1911. *Scientific Management.* New York: Harper & Row.

TEES, D. W., and S. E. WILKES, JR. 1982. "The Private Connection: A Texas City Official's Guide for Contracting with the Private Sector." Arlington, TX: Institute of Urban Studies, University of Texas.

TEPLITZ, C. J. 1991. *The Learning Curve Deskbook.* New York: Quorum Books.

THAYER, R. H., A. PYSTER, and R. C. WOOD. 1980. "The Challenge of Software Engineering Project Management." *IEEE Computer* 14, no. 8 (August): 51–59.

THESEN, A. 1976. "Heuristic Scheduling of Activities Under Resource and Precedence Restrictions." *Management Science* 23, no. 4 (December): 412–22.

———. 1978. *Computer Methods in Operations Research.* New York: Academic Press.

THIERAUF, R. J. 1978. *An Introductory Approach to Operations Research.* New York: John Wiley.

THUESEN, H. G., W. J. FABRYCKY, and G. J. THUESEN. 1977. *Engineering Economy.* 5th ed. Englewood Cliffs, NJ: Prentice Hall.

TINGLEY, K. M., and J. S. LIEBMAN. 1984. "A Goal Programming Example in Public Health Resource Allocation." *Management Science* 30, no. 3 (March): 279–89.

TOELLE, R. A., and J. WITHERSPOON. 1990. "From 'Managing the Critical Path' to 'Managing Critical Activities.'" *Project Management Journal* 21, no. 4: 33–36.

TOWILL, D. R., and U. KALOO. 1978. "Productivity Drift in Extended Learning Curves." *Omega* 6, no. 4: 295–304.

TRAYLOR, R. C., R. C. STINSON, J. L. MADSEN, R. S. BELL, and K. R. BROWN. 1984. "Project Management under Uncertainty." *Project Management Journal* 15, no. 1 (March): 66–75.

TROUTT, M. D. 1989. "On the Generality of the PERT Average Time Formula." *Decision Sciences* 20 no. 3 (Summer): 410–12.

TROXLER, J. W., and L. BLANK. 1989. "A Comprehensive Methodology for Manufacturing System Evaluation and Comparison." *Journal of Manufacturing Systems* 8, no. 3: 176–83.

TSUBAKITANI, S., and R. F. DECKRO. 1990. "A Heuristic for Multi-Project Scheduling with Limited Resources in the Housing Industry." *European Journal of Operational Research* 49, no. 1: 80–91.

TURNER, W. C., J. H. MIZE, and K. E. CASE. 1987. *Introduction to Industrial and Systems Engineering.* 2d ed. Englewood Cliffs, NJ: Prentice Hall.

TURNER, W. S. III. 1980. *Project Auditing Methodology.* Amsterdam: North-Holland.

TUSHMAN, M. L., and W. L. MOORE, eds. 1988. *Readings in the Management of Innovation.* 2d ed. Cambridge, MA: Ballinger.

ULLMANN, J. E., D. A. CHRISTMAN, and B. HOLTJE. 1986. *Handbook of Engineering Management.* New York: John Wiley.

VAN SLYKE, R. M. 1963. "Monte Carlo Methods and the PERT Problem." *Operations Research* 11, no. 5: 839–60.

VERMA, H. L., and C. W. GROSS. 1978. *Introduction to Quantitative Methods: A Managerial Approach.* New York: John Wiley.

VILLEDA, R., and B. V. DEAN. 1990. "On the Optimal Safe Allocation and Scheduling of a Work Force in a Toxic Substance Environment." *IEEE Transactions on Engineering Management* 37, no. 2: 95–101.

WAGNER, H. M. 1969. *Principles of Operations Research.* Englewood Cliffs, NJ: Prentice Hall.

———, and T. M. WHITIN. 1958. "Dynamic Version of the Economic Lot Size Model." *Management Science* 5, no. 1 (October): 89–96.

WALLER, E. W., and T. J. DWYER. 1981. "Alternative Techniques for Use in Parametric Cost Analysis." *Concepts--Journal of Defense Systems Acquisition Management* 4, no. 2 (Spring): 48–59.

WALPOLE, R. E., and R. H. MYERS. 1978. *Probability and Statistics for Engineers and Scientists.* 2d ed. New York: Macmillan.

WARNER, J. C. 1992a. "Design of Experiments for Project Managers." *PM Network* 6, no. 5 (May): 36–40.

———. 1992b. "Design of Experiments for Project Managers: Part III." *PM Network* 6, no. 8 (August): 69–74.

WEBSTER, R. 1985. "Expert Systems On Microcomputer." *Computer and Electronics* 23, no. 3: 69–73, 94–95, 104.

WEGLARZ, J. 1981. "Project Scheduling with Continuously-Divisible, Doubly Constrained Resources." *Management Science* 27, no. 9 (September): 1040–53.

———, J. BLAZEWICZ, W. CELLARY, and R. SLOWINSKI. 1977. "An Automatic Revised Simplex Method for Constrained Resource Network Scheduling." *ACM Transactions on Mathematical Software* 3, no. 3: 295–300.

WEIGHT, A. C. 1986. "Computer Integrated Manufacturing." *Manufacturing Issues.* New York: Booz-Allen & Hamilton, Inc.

WEISS, S., and C. A. KULIKOWSKI. 1984. *A Practical Guide to Designing Expert Systems.* Totowa, NJ: Rowman & Allanheld.

WEITZ, L. 1989. "How to Implement Projects Correctly." *Software Magazine* 9, no. 12 (November): 60–69.

———. 1993. "New Project Tools Offer Host-Class Function for LAN Platform." *Software Magazine* 13, no. 14 (May): 61–68.

WELSH, D. J. 1965. "Errors Introduced by a PERT Assumption." *Operational Research* 13: 141–43.

WESEMAN, E. H. 1981. *Contracting for City Services.* Harrisburg, PA: Innovations Press.

WHEELER, T. F. 1990. *Computers and Engineering Management.* New York: McGraw-Hill.

WHEELWRIGHT, J. C. 1986. "How to Choose the Project Management Microcomputer Software That's Right For You." *Industrial Engineering* 18, no. 1 (January): 46–50.

WHITE, J. A., M. H. AGEE, and K. E. CASE. 1989. *Principles of Engineering Economic Analysis.* 3d ed. New York: John Wiley.

WHITEHOUSE, G. E. 1973. *Systems Analysis and Design Using Network Techniques.* Englewood Cliffs, NJ: Prentice Hall.

———. 1983. "A Comparison of Computer Search Heuristics to Analyze Activity/Networks with Limited Resources." *Project Management Journal* 14, no. 2 (June): 35–39.

———, and J. R. BROWN. 1979. "GENRES: An Extension of Brooks' Algorithm for Project Scheduling With Resource Constraints." *Computers and Industrial Engineering* 3, no. 4 (December): 261–68.

———, and B. L. WECHLER. 1976. *Applied Operations Research: A Survey.* New York: John Wiley.

WIEST, J. D. 1964. "Some Properties of Schedules for Large Projects with Limited Resources." *Operations Research* 3: 395–418.

———. 1967. "A Heuristic Model for Scheduling Large Projects with Limited Resources." *Management Science* 13 (February): B-359–B377.

———. 1981. "Precedence Diagramming Methods: Some Unusual Characteristics and Their Implications for Project Managers." *Journal of Operations Management* 1, no. 3 (February): 121–30.

———, and F. K. LEVY. 1977. *A Management Guide to PERT/CPM with GERT/PDM/DCPM and Other Networks.* 2d ed. Englewood Cliffs, NJ: Prentice Hall.

WILEMON, D. L., and J. P. CICERO. 1970. "The Project Manager: Anomalies and Ambiguities." *Academy of Management Journal* 13, no. 3 (September): 269–82.

WILLBORN, W. 1989. *Quality Management System: A Planning and Auditing Guide.* New York: Industrial Press.

WILSON, M. 1993. "New Paradigms for Project Plans." *PM Network* 7, no. 6 (June): 39–40.

WINSTON, W. L. 1987. *Operations Research: Applications and Algorithms.* Boston, MA: Duxbury Press.

WIT, J., and W. HERROELEN. 1990. "An Evaluation of Microcomputer-Based Software Packages for Project Management." *European Journal of Operational Research* 49: 102–39.

WITTRY, E. J. 1987. *Managing Information Systems: An Integrated Approach.* Dearborn, MI: Society of Manufacturing Engineers.

WOMER, N. K. 1979. "Learning Curves, Production Rate, and Program Costs." *Management Science* 25, no. 4: 312–19.

———. 1984. "Estimating Learning Curves from Aggregate Monthly Data." *Management Science* 30, no. 8: 982–92.

WOOD, O. L., and E. P. EERNISSE. 1992. "Technology Transfer to the Private Sector from a Federal Laboratory." *IEEE Engineering Management Review* 20, no. 1: 23–28.

WOODWORTH, B. M., and C. T. WILLIE. 1975. "A Heuristic Algorithm for Resource Leveling in Multi-Project, Multi-Resource Scheduling." *Decision Sciences* 6: 525–40.

WORLD BANK REPORT. 1993. "Project Failures Increasing." *Norman Transcript* (April 12): 8.

WORTMAN, L. A. 1981. *Effective Management for Engineers.* New York: John Wiley.

WRIGHT, T. P. 1936. "Factors Affecting the Cost of Airplanes." *Journal of Aeronautical Science* 3, no. 2 (February): 122–28.

WU, N., and R. COPPINS. 1981. *Linear Programming and Extensions.* New York: McGraw-Hill.

YAU, C., and E. RITCHIE. 1990. "Project Compression: A Method for Speeding Up Resource Constrained Projects Which Preserve the Activity Schedule." *European Journal of Operational Research* 49: 140–52.

YELLE, L. E. 1976. "Estimating Learning Curves for Potential Products." *Industrial Marketing Management* 5, no. 2/3 (June): 147–54.

————. 1979. "The Learning Curve: Historical Review and Comprehensive Survey." *Decision Sciences* 10, no. 2 (April): 302–28.

————. 1983. "Adding Life Cycles to Learning Curves." *Long Range Planning* 16, no. 6 (December): 82–87.

YUNUS, N. B., D. L. BABCOCK, and C. O. BENJAMIN. 1990. "Development of a Knowledge-Based Schedule Planning System." *Project Management Journal* 21, no. 4: 39–45.

ZAHEDI, F. 1987. "Artificial Intelligence and the Management Science Practitioner: The Economics of Expert Systems and the Contribution of MS/OR." *Interfaces* 17, no. 5: 72–81.

ZALOOM, V. 1971. "On the Resource Constrained Project Scheduling Problem." *AIIE Transactions* 3, no. 4 (December): 302–5.

ZEYHER, L. R. 1969. *Production Manager's Desk Book*. Englewood Cliffs, NJ: Prentice Hall.

Index